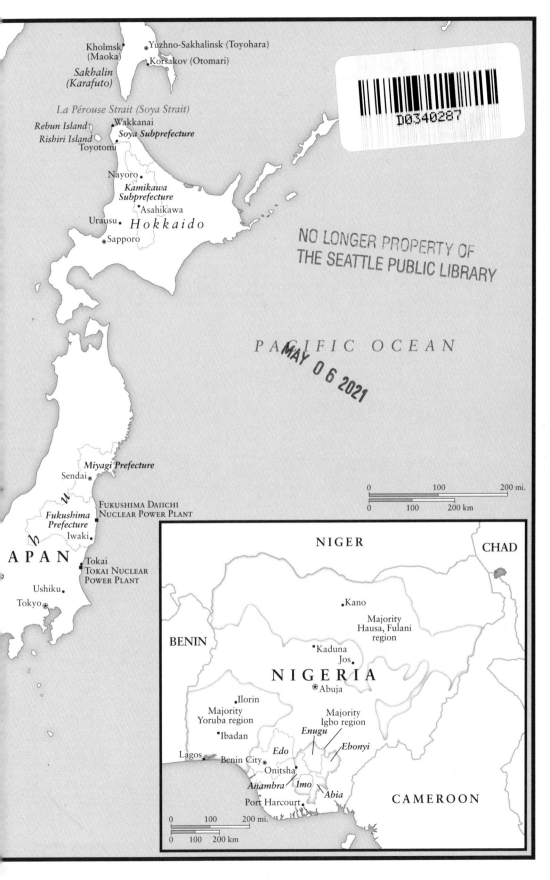

Kholmsk
(Maoka)

Yuzhno-Sakhalinsk (Toyohara)

Korsakov (Otomari)

Sakhalin
(Karafuto)

La Pérouse Strait (Soya Strait)

Rebun Island
Rishiri Island
Toyotomi

Wakkanai
Soya Subprefecture

Nayoro

Kamikawa
Subprefecture

Asahikawa

Urausu

Hokkaido

Sapporo

NO LONGER PROPERTY OF
THE SEATTLE PUBLIC LIBRARY

D0340287

MAY 0 6 2021

PACIFIC OCEAN

Miyagi Prefecture

Sendai

FUKUSHIMA DAIICHI
NUCLEAR POWER PLANT

Fukushima
Prefecture

Iwaki

A P A N

Tokai
TOKAI NUCLEAR
POWER PLANT

Ushiku

Tokyo

| 0 | 100 | 200 mi. |
| 0 | 100 | 200 km |

NIGER

CHAD

Kano

Majority
Hausa, Fulani
region

Kaduna
Jos

N I G E R I A

Abuja

BENIN

Ilorin

Majority
Yoruba region

Enugu

Majority
Igbo region

Ibadan

Edo

Ebonyi

Lagos

Benin City

Onitsha

Anambra

Imo

Abia

Port Harcourt

CAMEROON

| 0 | 100 | 200 mi. |
| 0 | 100 | 200 km |

EVERY HUMAN
INTENTION

EVERY HUMAN INTENTION

||||||||||||

Japan in the New Century

Dreux Richard

Pantheon Books
New York

This is a work of nonfiction, but the names of certain
individuals as well as identifying descriptive details concerning
them have been changed to protect their privacy.

Copyright © 2021 by Dreux Richard

All rights reserved. Published in the United States by Pantheon Books,
a division of Penguin Random House LLC, New York, and distributed
in Canada by Penguin Random House Canada Limited, Toronto.

Pantheon Books and colophon are registered
trademarks of Penguin Random House LLC.

Library of Congress Cataloging-in-Publication Data
Name: Richard, Dreux, author.
Title: Every human intention : Japan in the new century / Dreux Richard.
Description: First edition. New York : Pantheon Books, 2021.
Identifiers: LCCN 2020027141 (print). LCCN 2020027142 (ebook).
ISBN 9781101871119 (hardcover). ISBN 9781101871126 (ebook).
Subjects: LCSH: Nigerians—Japan—Social conditions.
Nuclear industry—Social aspects—Japan. Japan—History—Heisei
period, 1989– . Japan—Social conditions—1989– . Japan—Emigration
and immigration—Government policy. Japan—Population policy.
Classification: LCC DS832.7.N54 R53 2021 (print) |
LCC DS832.7.N54 (ebook) | DDC 305.896/69052—dc23
LC record available at lccn.loc.gov/2020027141
LC ebook record available at lccn.loc.gov/2020027142

www.pantheonbooks.com

Jacket photograph by Warchi/E+/Getty Images
Jacket design by Emily Mahon
Map by Mapping Specialists

Printed in Canada
First Edition
2 4 6 8 9 7 5 3 1

For Laura Macklin,
my mother

for Charles Macklin,
her father

and for William Macklin,
his brother

EVERY HUMAN
INTENTION

Introduction: The Japan I Am

The history of unattached American intellectuals expatriating to Japan is long, particularly crowded after World War II, and many of the resulting books are only as relevant as the era of Japanese economic ascendancy that inspired them. Americans read about East Asia when they are afraid, and now China frightens them more.

Another category of book has endured better: the prose left behind by expatriate literary interlopers of the same era, mostly men (many of them quietly homosexual) and a smaller number of women whose oeuvre is divided between the canon of modern Japanese literature they translated into English and their written reflections on the lives they led in the process. In the 1970s, Donald Richie's autobiographical novels established a global audience for this group's experiences. Since his death in 2013, his literary epitaph praises his lyrical observations of life between cultures, where he was, in his own words, a citizen of Limbo, "this most attractive, intensely democratic republic."

My mentor in Japan was Hideo Levy, the first Westerner (American, culturally Jewish) to abandon the comforts of Limbo by succeeding as a literary fiction writer in the Japanese language. Before his books won prizes, there was no such thing as "Japanese literature." There was only "Japan literature" and its suggestion of an indispensable connection between language, ethnicity, and nation. Some would rather it remain so, and haven't anything nice to say about so-called border-crossing writers like Levy and his Korean contemporaries, the label itself a pejorative nod toward the illicit act of fence hopping.

In his former life as a scholar of Japanese poetics, Levy's first student was a woman three years his senior, born in occupied Japan to an American father and a Japanese mother. Her name was Norma Field. In 1989 she wrote *In the Realm of a Dying Emperor,* a book that made it possible for English-language writing to change Japan's politics and culture; previous generations of expatriate writers had been content to interpret and analyze.

Field wrote about three people, call them dissenters. They had in common their respective decisions to ruffle the ornate curtain obscuring the recent death of Emperor Hirohito, even if that meant speaking candidly about his executive role in prolonging Japan's wartime agonies. Two were private citizens, whose acts of protest attracted media attention and exposed them to violent harassment from right-wing militants. The third, Nagasaki's mayor, was shot through the chest by one of these militants, at city hall.

When Hirohito died in 1989, a growing minority of Japanese economists were nervous about the nation's asset price bubble, but they were yet a minority. To call attention to the psychic and social costs of Japan's postwar achievements—to oppose the "solidity of everyday life in a successful society," as Field put it—was not merely impolite. It was regarded as willfully wrongheaded. In 1991, the year Field's book was published, asset prices collapsed, plunging Japan into an economic purgatory that was labeled the Lost Decade and lasted much longer than ten years. The dissenting voices in Field's book suddenly seemed prescient.

When I arrived in Japan, three months before the Fukushima disaster, the chronologically acute anxiety of the Lost Decade had been replaced by a more diffuse and enduring economic malaise, this time as a function of demographics, which point to a dramatic decline in Japan's population: from 128 million in 2010 to 97 million in 2050. In Wakkanai (Japan's northernmost city, where much of this book was written), closures of schools, medical facilities, and local businesses tell an old story, distinct in its details but not its theme: Throughout Japan, youth is scarce.

Increased immigration presents a remedy, but it would need to occur on a large scale. Economists estimate the Japanese workforce will require an infusion of roughly 33 million workers. Immigration remains unpopular, eliminating political incentives for comprehensive reform. In 2014, Japan's chief cabinet secretary, Suga Yoshihide, declined to comment on reports that the prime minister was considering plans to admit an additional 200,000 immigrants on an annual basis. The administration's intentions eluded the press until 2018, when a reform bill was rushed through parliament, circumventing the normal vetting process. Elements of the law were borrowed from Japan's existing guest worker program, faulted for its facilitation of human trafficking by the U.S. Department of State.

In public discourse about the nation's economy, hope—or despair, depending on one's politics—attaches to the achievements of Japan's longest-serving prime minister, Abe Shinzo, whose aggressive reforms generated a recurring cycle of skepticism and praise from ambivalent economists. Japanese voters have developed the accurate impression that they are inhabiting an economic experiment, founded on a controversial hypothesis. Abe left office in September 2020, but his policies will predominate until late 2021, when the "continuity cabinet" installed by Suga, Abe's handpicked successor, will be forced to consolidate power or relinquish it.

For the past decade, the question that has been posed to Japan's body politic most directly is the question of domestic nuclear energy production. Before the multiple meltdown disaster at Fukushima, Japan ran fifty-four commercial nuclear reactors, accounting for one-third of the nation's energy. In 2020, eight were operating.

Even in prefectures where the nuclear industry has been responsible for a significant share of economic activity, Japanese voters want to divest from nuclear power. Japan is a nation where antinuclear partisans can build a compelling narrative that begins with the bombing of Hiroshima and Nagasaki, includes the irradiation of a Japanese fishing vessel during nuclear bomb testing at Bikini Atoll, and leads inevitably to the Fukushima disaster on March 11, 2011, which occurred in one of Japan's poorest regions because these were the regions where nuclear utilities had been able to create the political possibility of siting several reactors.

The opposing narrative favors economic exigency: Japan can't afford to withdraw from commercial nuclear energy production, not at a moment of grave economic uncertainty. Instead, the Abe administration and its successors insist, the nation's oldest nuclear plants should be decommissioned and the rest should be subjected to stringent restart guidelines, then allowed to operate until they've aged beyond the intended longevity of their designs.

Immigration and nuclear energy formed the poles of my professional life in Japan. I covered Tokyo's African community for *The Japan Times,* and Japan's nuclear regulators adopted me as a confidant because I responded to the events of March 11 by investigating American nuclear facilities, contradicting the notion that Japanese bureaucrats were exclusively responsible for a disaster that implicated a global industry. These were not literary subjects—not until later, when I gained distance and perspective—and one distinction made my life different from the lives of my literary predecessors: They weren't working in Japan illegally. I was.

A pause must be inflected to explain my understanding of the word "illegal," which appears rarely in this book, and the word "extralegal," which I use. I do this not to insist, as advocates for immigrants often do, that a human being should never be considered illegal. This argument already has its share of eloquent partisans. I do it because, strictly speaking, I was not an illegal immigrant to Japan. Very few people are. In contemporary political discussions, "illegal" is synonymous with illegal entry or illegal border crossings. But Japan is an island nation, and immigrants reach it almost exclusively by plane. In the proverbial 99.9 percent of cases, if you meet an immigrant in Japan, they entered the nation lawfully and they can prove it.

As a reporter covering the African community, I dealt with immigration officials. Over the course of my six years on the job, people read my stories and remembered my face, so it was often the case that I would enter a detention center and submit a fraudulent request to visit one of the detainees (fraudulent in the opinion of the detention center's administrators because there's a separate procedure for journalists, which I declined to follow), then wait to be screened for entry by people who knew I was working without an appropriate visa.

My sources in the Immigration Bureau told me I shouldn't worry; I was protected by a peculiar idea bureau officials had developed about my profession: They believed if they arrested me, my career would benefit. I remember the only time I used this immunity. I was photographing one of my interview subjects for a story, and a group of police officers patrolling nearby made him open his van so they could

search it. He earned his living out of the van, buying and selling secondhand clothes, so it took a long time to complete the search, and
by the end they knew who I was because he'd explained, hoping it
would make them reconsider. It had the opposite effect. They worked
slowly, apologizing for how long it might take "to inspect this entire
warehouse—sorry, *vehicle*." When they finished, I followed them to
the station, where I expressed my displeasure for several hours, until
it was clear that every officer in the room regretted my involvement
in their evening.

The next day I called my interview subject, and he thought I
was as brave as I had convinced myself to feel. Good trick, he said,
making myself so insufferable that the police forgot to ask his name
and put it in the system. But there wasn't any trick, and if I hadn't
been distracted by my indignation, I might have paused to wonder
whether he could afford to have his immigration status checked.
Were he detained on my account, it wouldn't have helped my career,
and I wouldn't have felt brave.

Without meaning to, I'd taken a few pictures after he noticed the
police approaching. The photos were dim and the focus was soft, but
my editor insisted on using one. There were other photos to choose,
none of which captured the subject looking directly into the camera
and expressing the idea the people quoted in the story seemed to be
trying to express. That was how my editor explained it. I asked: What
idea? and my editor pointed at the photo on his computer, at the face
that showed the final moment of a cheerful expression, the beginning
of disappointment, the strange congruity of this overlap. He said,
"Looks like for him, the idea is: Of course."

As an American who went abroad intending to write about what I
learned, I share with Donald Richie what one professor of literature
has called "the rage to explain." By following this instinct, I have
invited the question I least welcome: What is it about Japan I would
like you to know? When Norma Field finished *In the Realm of a
Dying Emperor,* it was published with a postscript, written to thwart
a notion many readers adopted anyway—that Field was a "Japan
basher." Japan's soft power attracts partisans on both sides, who refuse
to believe that the object of their attention isn't always on trial.

Some people described in this book feel that Japanese society's collective shortcomings are responsible for the hardships they've endured. I've tried to show them in the moments their discouragement formed, or when they resisted it. I tried not to entertain the question of how bitterly a person has to struggle before he earns his embitterment, or to put in moral terms the question of whether their lives would have been better if they lived somewhere else.

Other people I've written about would never consider the possibility that their private difficulties suggest a defect in Japan's national character. I was attracted to people whose experiences led them to develop a sense of possibility about life in Japan and who guarded that sense at considerable cost: immigrant workers who chose to remain in Japan when the bubble burst and the demand for cheap labor vanished, elderly residents of Wakkanai for whom the city is not solely a product of its past, nuclear industry professionals capable of holding a mirror to their work without resorting to politics.

What I learned from them I have tried to explain in the way a writer explains, the way an ethnographer explains, and the way a journalist explains. Foremost, I have tried to explain in the way a translator explains. I have assumed that the lives I observed were real before I discovered them and I would comprehend them imperfectly. Between the thoughts and experiences that were shared with me and the words I used to describe them, I could only hope to limit the inevitable introduction of infelicities. I feel that my relationship with the world of Nigerian immigrants, of Hokkaido's northern frontier, of Japan's nuclear industry, is intimate, but these places are not native to me. In the role of eavesdropper, I misapprehended more than I understood.

The title of this book is lifted from a quote once spoken to me by a nuclear reactor operator in Illinois, and adopted as a personal emblem by a retired reactor operator in Fukui Prefecture after I told him the story behind it. He felt it described his first shift, which was the inaugural shift at Japan's first modern commercial reactor. When the reactor reached criticality, it occurred to him that airplanes would be guided by lights he was powering, and mothers would bathe their children in water heated by electricity he sent, and homes would be built from materials fabricated in factories that ran on it, and the country would become more itself, limited in the pursuit of better living only by its collective will, not the shared trauma of the war or the absence of domestic energy resources.

He might not have been so eager to inhabit that memory again if it hadn't grown distant in the years since the Fukushima disaster. The reactor where he'd worked had been the first of its kind in Japan, and by the time he retired from operating it, it was one of the country's oldest. But it ran uneventfully while newer reactors—including two more constructed later at the same plant—experienced headline-worthy accidents. It made him wonder: Had the operators who'd brought the other reactors online felt what he had felt during his first shift? Or did those feelings arise only for the handful of people whose sense of their work's importance was formed before the achievements of their era began to appear impermanent? He was aware that his reactor would be decommissioned now, for partly political reasons, so Japan's largest nuclear utility could say they hadn't tried to restart their entire fleet. And the lesson he felt his career had taught him, *Look how good it can be, done right,* seemed like a vulgar thing to tell people who didn't know anything about the reactor he'd run, but who were intimately aware, if only from watching it on television, of how bad it can get, done wrong.

Japan is a changeful nation, where all the phases of modern history have arrived and passed suddenly. In Japanese literature and the country's historical consciousness, individuals who outlive the era that formed them often appear in a specific role, as the lonely bearers of a secret sadness, whose vestigial lives reveal the folly of our faith in the present. Students in my literature courses complain about it: Who wants to read so much writing about anomie? My students are American and believe so deeply in the individual nature of their values that the notion of having to inhabit them alone doesn't arouse anxiety; they are certain they would live as they always had and would feel no differently doing it. Perhaps because this is one of the differences in culture and perception that deserve better explanations than I can provide, I tried to attach myself, while researching this book, to people who longed to encounter Japan as they once imagined it, or knew it before.

When my partner and I moved to Tsuruga, the hub city of Japan's Atomic Broadway region, we stayed with a local family while we looked for an apartment. Our host father was a nuclear fuel technician. Shortly before Fukushima, he'd started his own firm. Contracts were scarce after the disaster, with the exception of cleanup at the damaged reactors, on the opposite side of the country. My host father was lucky: He and his business partner found work at Takahama, an hour away by car. They came home every weekend.

There was a week when I heard from his wife that he'd been injured at work. It turned out he'd gotten drunk and fallen down a staircase. When he got back, we barely discussed it. Just enough for him to make clear that he hadn't intended to conclude his career this way, boozing with a bunch of other sullen contractors while they waited to see if the fog would ever lift. "We've been told it's important for the utilities to preserve their skilled workforce during the shutdown," he said. "Dutifully, I have pickled myself." The notion that the electric utilities were committed to the welfare of their contractors was among the least convincing distortions I encountered on a regular basis in Tsuruga. I could only imagine how often my host father heard it.

He doesn't take it personally. He once told me that Tsuruga didn't choose him. He chose Tsuruga. It was odd for him to say. He'd grown up there. I believe he meant to make clear that he wasn't complaining about the place he lived or the lives people led there, just proclaiming—in a weary, restive voice—his difficult and complicated love.

Diaspora

||||||||||

1.

When the earthquake struck, Prosper Anyalechi was seated on the floor of his room in a love hotel in Senzoku, the Tokyo neighborhood where courtesan culture had permanently interwoven sexuality and the arts in premodern Japan. Africans weren't supposed to know enough Japanese history to understand what the neighborhood had once been. But the history of Senzoku was the history of Japan's red-light districts, and that was Prosper's history now. It had been since he started working nightlife jobs in Tokyo eight years earlier.

He might have been sitting on the floor for twenty minutes by the time of the earthquake; when I asked later, he couldn't say precisely. He'd been in a reverie, he explained, typical for him since he arrived in Japan. The column of murky light drifting through the hotel room's privacy window had traveled past the open door to its unit bathroom and struck the plastic fixtures inside, in a way that reminded him how far he'd come from the country of his birth and upbringing, where windows admitted light without changing it.

A young Japanese woman slept in the bed he sat beside. Young enough, he'd learned, inspecting the contents of her purse after she drifted off, that she would still be living with her parents. And would confess what had happened if she came to regret it, but in the version her parents would hear, the role of her consent might diminish. So he'd moved out of bed and dressed quietly, planning to leave before she woke. He wasn't sneaking away; he would return to work on the same street in the same red-light district the next night, where they'd met a few hours earlier. Probably, she wouldn't try to find him, and he assumed this wasn't the first time she'd lied about her age outside the entrance to a love hotel.

When the room began to shake, Prosper's mind did not register that he had already dressed, and he thought to himself, "I'm naked." The woman woke and they made eye contact, Prosper now standing, looking down. She wrapped herself tighter in the blanket, a gesture that also reminded Prosper how far from home he'd come, to a nation where people trusted that buildings would not collapse. Prosper's instinct was different, and he followed it into the hallway, where a Japanese couple—one wrapped in a sheet, the other a towel—had opened their door halfway. They had been deciding how to react to the temblor's unusual persistence, Prosper guessed. Now, seeing a foreign man emerge alone from the room across the hall, their concentration had broken, and they turned in unison to observe him. He was not the only Black person in Tokyo, but perhaps the only one who had seen them seminude, in a vulnerable moment.

As the door to his room shut behind him, Prosper thought he heard the woman in bed call, *Omae!* and continue yelling, muffled through the door, as he turned toward the emergency exit and crossed the hallway. *Omae,* he knew, is a way of addressing someone in the second person, reserved for moments of affection or disdain. In Prosper's life, it had been used as a preamble to abuse, from people who thought his ethnicity made him deaf to gentle speech. Offered softly, the word was a caress. He would wonder about it later. *Omae.*

Prosper made his way to the exit and began to descend the staircase outside. He missed a step, slid down a few more on his back, and found himself on a landing, peering past the metal railing at the street, two stories below. He braced himself against the railing and fastened his gaze on the featureless concrete wall of the adjacent building, which appeared to bulge and contract as it swayed.

When the shaking subsided, Prosper descended to the street and joined the crowd gathering there, among whom he could already hear news of the earthquake's origins around Sendai, or about a fire that had broken out on the roof of a high-rise in Odaiba, an upscale development on reclaimed land in Tokyo Bay. It no longer seemed to matter—in the way it had to the couple he'd encountered in the hallway—that he was Black and present in a vulnerable moment; he received only as many second glances as he would on any other day, in any part of the city.

Phone circuits were jammed, so he couldn't call his cousins in Nigeria to tell them he hadn't been hurt in the earthquake, which they would hear about. There was one severe aftershock, then everyone around him turned their attention to the task of getting home, examining maps displayed on their phones to orient themselves for a long walk. Prosper hadn't slept since he woke to work the night shift twenty-one hours earlier. A nearby alley ended in a cul-de-sac of ventilation units where he sheltered from the gaze of passersby and noticed, as he lay down, the abnormal quiet and stillness of the machines.

Prosper woke to a tap, which he was not surprised to discover had been delivered by a police officer, though he was surprised the officer hadn't used his baton, choosing instead to lay his hand on Prosper's shoulder.

The officer brought Prosper to an interview room in the back of a nearby *koban* and asked him if he wanted a can of coffee, which Prosper accepted. "We got a call from some guests at a love hotel," the officer said, "about an African man." He was sitting opposite Prosper, at a small table. He was young. His utility vest would have fit a thickset man better; it gave his gestures a thwarted, impatient quality.

"I was scared," Prosper said. "In my country, we don't have earthquakes." He thought of the unhappy possibility this discussion presented: The officer had gone to the hotel and met the young woman, whose distress Prosper had ignored when he fled.

The officer waved off Prosper's explanation. "People act foolishly in this kind of circumstance."

"I'm very sorry," Prosper said, and bowed in his seat.

"Oh," said the officer. "Not you. The guests who called. I hope you don't think all Japanese are like that—racial discriminators."

"Of course not," Prosper said, and paused to detach his thoughts from the difficult conversation he had anticipated. "I love Japan. I love Japanese people."

"You don't have to love them, but I hope you don't blame them."

A long silence passed, and it seemed to Prosper, who still thought he might be arrested, that the officer wanted him to ask the next question.

"Japanese people are accustomed to disasters," the officer said. "They behave in a particular way. Did you live in Japan during the Great Hanshin earthquake?"

"I was newly here."

"You remember the media reports?"

"I lived in a dorm for laborers. We didn't have TV."

"It's possible to live in Japan without really living here," the officer said, in the manner of making an admission. "To live that kind of life."

"My first four years in Japan."

"The world was impressed with Japan. Some people said the government was slow. They would criticize the government. People came from all over Japan to volunteer. You could never criticize the Japanese *people*. I wonder if everyone considers themselves part of the 'Japanese people' in a situation like that."

The term the officer used for "Japanese people" had always struck Prosper as strange, even when spoken unselfconsciously. The officer had emphasized each skeptical, staccato syllable: *ko–ku–min*. People belonging to the country. There was a corresponding word in Prosper's language: *Ndigbo*. But the words had little in common. *Kokumin* described a people who dwelled within the *koku*—the land—it referred to. *Ndigbo* expressed a bond defined by the significant proportion of Igbo people who lived in diaspora, particularly the generations born after Biafra's defeat in the Nigerian Civil War, when their ethnicity was dispossessed of its economic and political prospects.

"A Japanese is a Japanese," Prosper said.

"It wasn't the same for everyone after the Hanshin earthquake. Some people lost more than others. Some people never got it back."

"Japanese people?"

"Ask one," the officer said. "Ask a Japanese person."

There was a knock on the door and the officer was summoned away. When he returned, the *koban* was quiet; his colleagues had left. Now the officer told Prosper that he'd brought him in for his own good, to keep him off the street, where a certain segment of Japanese society might be inclined toward paranoia about foreigners in the wake of a disaster. The officer wheeled a TV into the room and they watched the news. The officer had already seen helicopter footage of the tsunami. Prosper hadn't. "Look at all that—houses, cars—all inside the water," the officer said, tracing a circle on the screen with his finger.

"And human beings," Prosper guessed.

"What about in Africa? In a disaster, everybody gets helped the same?"

"Rich people decide."

"Who's rich in Africa?"

"Politicians," Prosper said. "Oil."

"Someone told me Nigeria has two hundred languages. Can you tell someone is from a certain group just by looking at them—before you hear their voice?"

"Usually."

"So you have discrimination, too."

"We call it tribalism."

"What makes a tribe?"

"Time. My people have been living together a thousand years. It could be three thousand."

"What did your people do—for three thousand years?"

"We had our own civilization."

"Was it known for something?"

"We had multiple kingdoms. It's like saying Japan is a place where people make sushi."

"What was your biggest kingdom? The original."

"Nri. In my home state. They were making artwork in metal before anybody else. They also made masks."

"They're still around, or—"

"There's a town, not a city."

"Your people still work on metal? I can go to your part of Africa and see the methods?"

"It's not like Japan—like kimono or tea ceremony. The methods are lost. We teach the masquerade performances, the cultural part."

"What do your people do now? For money?"

"What you see me doing. We travel. We have a saying: If you go to a country and there's no Igbo man living there, you should leave. If it's a good country, an Igbo man would have found it."

The officer slouched, his hands resting between his legs on the chair. He was morose, but wasn't unfriendly. His demeanor reminded Prosper of a policeman he'd encountered at a group home for asylum applicants where one of his friends was living. He let his thoughts wander, remembering that: the policeman in the doorway, surveying the futons laid end to end along the floor of the house, the piles of personal effects and work clothes in every corner. The policeman had turned to the Buddhist clergywoman who managed the property and said, "If you're upset about the phone bill, disconnect the phone." The clergywoman pretended she hadn't heard. She gathered everyone in the hallway and told them to leave. Some residents went upstairs to search for boxes. The sound of an argument filtered through the ceiling. It was winter, and almost night.

The policeman walked onto the house's porch and lit a cigarette, Prosper remembered. When Prosper followed him to ask about calling a taxi, he was gone. Prosper looked both ways down the street and saw him walking toward the local police box, silhouetted by the late-afternoon sun, his shadow reaching nearly back to the house. A gust knocked the policeman's hat off his head. He picked it up, turned to face the lamppost next to him, and hit the lamppost with the hat, gripping it by the brim. He hit the lamppost again, and the hat's insignia detached. Prosper is certain he remembers the metal-on-asphalt sound it made when it landed, though that would have been impossible over the traffic noise.

Prosper let the memory recede. He leaned across the table. "Why are you asking these questions?" he said.

"I'm interested in people who go where they aren't supposed to," the officer said, and glanced, in a way that struck Prosper as involuntary, at the door to the *koban*'s other room. "I might be one."

Prosper felt the sympathetic effect of this disclosure, as if he'd learned the officer's name. He wanted to ask what it meant, but his phone vibrated on the table. "Text message," he said.

"You want me to read it?"

"If it's in Japanese, it's a wrong number." Had he given his number to the woman he'd taken to the hotel? She could have picked up the phone and called herself while he was in the bathroom.

The officer took the phone off the table. "You have children?"

Prosper said he didn't.

"Someone trying to make sure their father is safe." The officer put the phone back. They watched TV. More helicopter images, cycling now, repeating themselves.

Eventually Prosper said, "I hope their father's all right."

"Huh?"

"The person sending the messages."

The officer picked up the phone and tapped a few buttons. "I told them it's the wrong number."

There was a noise in the adjacent room. Another officer had entered the *koban,* talking to someone. "We can call your family from the landline," Prosper heard the new voice say.

The officer Prosper had been speaking with excused himself. When he opened the door, Prosper saw the woman he'd taken to the love hotel reflected in the windows of the vestibule, accompanied by an older officer. "She fell asleep riding the Meguro line," the older officer said. "Who's in the interview room?"

The younger officer closed the door, leaving Prosper alone, listening to their voices. The moment the younger officer mentioned an African, the woman would have known it was Prosper. Her call went through. She told her father the same story, that she'd fallen asleep riding the train and disembarked before the earthquake. The officers were speaking to each other now, talking over her. The older officer said a nuclear plant in Tohoku had been crippled by the tsunami. Prosper looked at the television. An aerial shot of the tsunami panned around the water as it flowed inland, permitting the viewer to watch the wave approach the traffic on a highway. The helicopter carrying the camera entered a cloud of smoke from the fires in the tsunami debris. For a moment, nothing was distinctly visible. Prosper got up and turned off the television. The voices in the other room stopped. He stood in place, not knowing why, but trying to be absolutely silent.

2.

I met Prosper Anyalechi in an elevator, on my way to visit an after-hours nightclub, the place where nightlife workers wait for the morning train. He was employed by a club on one of the lower floors of the same building; he had pressed both elevator call buttons in order to board as soon as possible. When the door opened and I saw him, he had a drunk Japanese man slung over his shoulder, and he was fending off his own boss, who was trying to drag the man back into his club, ostensibly to persuade him, because he was in no condition to resist, to buy a few more bottles for the club's hostesses. Prosper was able to enter the elevator with the Japanese man and close the doors. The three of us rode in polite silence, Prosper waiting for the elevator to reach my floor and change directions. I was crossing the corridor that led to the after-hours club when Prosper called out, "Stop the presses," maybe for fun—to startle me—or maybe he wanted a witness. I waited for the elevator to return and rode back down, then walked onto the street, where I saw Prosper put the Japanese man into a cab. Prosper's boss fired him on the sidewalk, while their colleagues and compatriots watched. The scene earned my curiosity. If he was unwilling to cooperate with dishonest employers, Prosper wasn't financially distressed or impatient to acquire wealth. At least one of these descriptions applied to every nightlife worker I knew.

For my first two years covering the African community, I woke in the afternoon, arrived in Roppongi or Kabuki-cho (Tokyo's largest red-light districts) by 8:00 p.m., then stayed the night. Not because nightlife interested me; these were the places I could spend twelve hours acquiring sources without exhausting the available supply.

Nightlife is not an easy business, nor a particularly honest one (not in those neighborhoods), and Africans working in it had already received more unflattering media attention than they wanted, so I spent most of my nights hearing what I could shove and where.

Prosper was the first person who invited me into his life and described the history of the community I'd been asked to cover. He had arrived in 1991, in the first wave of Nigerian migration to Japan, whose participants shared distinguishing characteristics: a middle-class upbringing, membership in an ethnicity that lived in the southeast or south-central part of the country (former Biafra), some tertiary study or vocational training, and the belief that they could earn enough money in a year or two to capitalize a small business in Nigeria, where they assumed they would return.

Prosper's life in Japan had included several experiences that are also common in the Nigerian community: "eight or nine" jobs at factories and scrapyards in the countryside; the eventual purchase—from a fellow Nigerian, for a considerable sum—of a steadier job at a recycling plant closer to Tokyo, where he worked for several years without a visa but with the open acknowledgment of the local police; a series of romantic relationships with women whose motives he was never able to discern, because the language barrier obscured them and the lingering possibility of deportation made these relationships urgent and fraught.

He was married, eventually, to a woman who said she wanted to see Africa and agreed to visit home with him. She abandoned the marriage after they returned from Nigeria, offering no explanation. By the time they divorced, Prosper had obtained permanent residency in Japan. "If not for that, I would have gone."

Freed from immigration concerns, Prosper tried to assemble the type of life that many of my Nigerian acquaintances idealize: his own business, and the esteemed position in Igbo society enjoyed by entrepreneurs. He suffered a series of misfortunes common to the business of choice among his compatriots—filling shipping containers with secondhand goods and sending them home for resale. In Japan, it's typical to meet Nigerian immigrants who have attempted this scheme several times, with uniformly unsatisfactory results. As Prosper discovered, it is difficult to participate without passing your containers through the shipping yards of money launderers.

Two decades had passed since Prosper arrived, and his failures in business became an occasion to reflect on his community's trajectory. Over time, his compatriots' intention to return to Nigeria had been replaced by the optimism of marriage to Japanese romantic partners, of fatherhood and legal residency, an optimism eventually dissipated by the longitudinal challenges of integration. They wanted to live in a safe place where they could find work, and Nigeria didn't fit that description, but they'd exhausted their tolerance for living away from their culture, in a society where discrimination limited their prospects. Pulled in two directions, they made mistakes about how they spent their money or what they promised their loved ones. Many started leading double lives, married in Nigeria and Japan, with children in both places.

Prosper resolved to follow the example of the self-consciously middle-class uncle who raised him, a logistics agent on the Niger River who equated financial autonomy with its moral equivalent. In nightlife, Prosper earned about thirty-five hundred dollars a month, assuming he worked diligently for an honest boss who paid on time. He rented a cheap apartment and slept when he wasn't working. Until he had a careful plan or a strong feeling about what to do next, he would save his salary and detach himself from the Nigerian community's cycle of speculation and disappointment.

Prosper had maintained this holding pattern for more than a year when I met him, and you could read it in his appearance. He was the only Nigerian I knew in Japan who never wore an ill-fitting suit, because he only bought suits in Nigeria, where experienced tailors worked for ten dollars a day. He bought casual attire at charity shops in Japan, where the quality was better but the secondhand clothes that fit him looked like a gangster's vacation wardrobe—billowing beach shirts with dragons printed across the chest. In a community and a profession characterized by manic, nocturnal notions, Prosper was a sober face, a person who didn't sneer or scheme to distract himself from the indignities of working on the street. I imagined he had been one of those children who knew how to scale a rope without being taught; the way he took potential customers aside from the foot traffic on a busy street, his hand hovering above their backs without touching them (his colleagues were overbearing—they lost customers because of it), was a skill owed to the same attribute of knowing how a single gesture, executed properly and repeatedly, turns effort into effect.

The pattern of my reporting was a product of Prosper's guidance, beginning with my first story for the newspaper. The Nigerian community's civic organizations had launched a fundraising campaign for tsunami victims, and on one particular afternoon raised several thousand dollars. Their generosity convinced me that the Nigerians I met had accepted the compromises their lives in Japan required. Prosper introduced me to colleagues of his who had donated, and they provided my story with nuance, by sharing their ambivalence toward their host country. Later, the organizers of the fundraising canceled the media appearance arranged by the charity that received their donations, a decision prompted by the paranoid suggestion that the media would doctor the resulting footage and allege the fundraising was a scam. Generosity was not the sum of the Nigerian experience in Japan; it was the bright part, surrounded by difficulty.

When I tell people about the work I did for *The Japan Times,* most respond with a question: "There are Africans in Japan?"

Maybe twenty-two thousand, according to Ministry of Justice statistics, which I don't trust. Consular officials at African embassies will tell you, in confidence, that the real number is higher. It remains true, however, that among Japan's modest minority population Africans constitute a small group.

Within that group, Nigerians are the most numerous. Perhaps twelve thousand lived in Japan at the peak of Nigerian immigration; the official figure presently hovers between three thousand and four thousand. Japan's network of Nigerian civic organizations is not the only African polity in the country, but it is the largest and most active. This is often true of Igbo Nigerians living in diaspora, who form town unions based on their natal villages.

The African story in Japan—specifically the West African story—includes well-publicized chapters, when the economic activities of African expatriates attracted media attention. In the 1990s, West African immigrants began to open hip-hop fashion boutiques in Tokyo and its suburbs. Though profitable, the hip-hop apparel business became an incubator of ethnic tension as Japanese proprietors with discriminatory access to affordable commercial leases undercut their African competitors. It was also true that some African business owners had built their credibility by claiming African American identities. Japanese media outlets recognized the deception, which set the stage for a minor media frenzy when a significant number of African boutique owners were caught entering Japan with counterfeit clothing and several were smuggling narcotics in their merchandise. It was the first time the African community had been characterized according to its most opportunistic elements, a phenomenon that would culminate in a statement by the governor of Tokyo blaming Africans for the city's rising crime rate.

The alarmist tone of this rhetoric resurfaced in the late 2000s, after the Lehman shock, when margins were shrinking and opportunism was increasing in the nightlife business. Reports began to circulate that customers had been drugged at nightclubs in the city's red-light districts. When victims woke up they'd been divested of their cash and credit cards, and later learned that outrageous charges had been made to their accounts. The U.S. embassy issued a list of clubs to avoid. Several were Nigerian owned. Some were raided, their owners arrested. The press attended the raids, tipped by police.

Taken together, these periods of media scrutiny represent a disagreement between cultures: West African diaspora culture, which prides itself on establishing quick prosperity anywhere, and Japanese culture, which isn't always comfortable with immodest displays of wealth and is rarely comfortable with immigrants. The amount of media attention dedicated to Africans in Japan, particularly Nigerians, is remarkable in light of the size of Japan's African community. One reason to accept that community's experiences as a barometer for what might occur (or continue occurring) between Japan and its foreign residents is the possibility that, on a per capita basis, Africans living in Japan claim a larger share of the public consciousness than any other immigrant community. Their experiences demonstrate the cycle of fascination and discrimination that other immigrants will encounter as Japan's foreign population grows.

Since 2007, Japan has experienced a surge of asylum applications. The result has been a dramatic repositioning of refugee narratives, from the periphery of Japan's immigration debate to the absolute center. This shift leaves Japan open to international criticism—sometimes condemnation—as immigration authorities react to the increase in applications by rejecting almost all of them. (In 2017, Japan accepted 20 out of 19,628 asylum applications.) During the years when application numbers rose most rapidly, West Africans accounted for the fastest-growing segment. Although this didn't rank them near the top of the list of applicants by nationality, they constituted the largest demographic of clients served by Japanese nonprofits established to assist refugees. In several respects, they have become the face of Japan's struggle to cope with asylum claims. They confront high barriers to integration and pursue available assistance tenaciously, bringing them into contact with neighbors and bureaucrats in working-class communities, people unaccustomed to the presence of foreigners, astounded by the presence of Africans.

In the final year of my reporting, I spent more time with these new arrivals. The stories of how they emigrated were becoming more uniform, the mechanisms more monetized: Buying a visa to Japan from a broker wasn't one way among many anymore; half the people I met had done it. When they arrived, they were connected with a parallel network of labor agents who were paid by employers once an immigrant arrived at a job site. Among those who were fortunate enough to avoid scrutiny and detention when they landed at the airport, many worked in a succession of physically demanding jobs that barely paid enough to reduce their debts to their visa brokers.

A migrant can't work in the factories and scrapyards forever. Eventually he (and it was uniformly men) would be loading a container with stolen goods when the police visited. Some would manage to save up and move on, perhaps to nightlife work or their own export business, and in both cases the temptation to commingle legal activities with their illegal equivalents would ensure that immigrants were available to commit the high-visibility offenses that Japanese criminals preferred to avoid.

I met these people at the detention center on the day of their release, or on the street in the red-light districts, or through their friends at events held by African civic associations. When the story on my desk called for it, I learned about their lives. But they rarely chose to contact me, because I had nothing to do with the path they were on or—more significantly—the path they wished to be on. Although my work had occasionally truncated an immigrant's incarceration or turned a civil suit in a source's favor, that was mostly by coincidence. The odds that I could help "a desperate someone" (as Prosper called these recent arrivals) to fix their residency status and find a livelihood were slim, and most immigrants who arrived on broker visas made a habit of figuring the odds.

Instead, when my phone rang, it was usually to remind me of the exceptions to the generalizations I've made. It was the asylum applicant who insisted on promptly declaring his intention to reside in Japan and following the prescribed legal process. I stood in municipal offices all over Tokyo, showing West Africans who had recently arrived how to register their addresses and sign up for health insurance. I accompanied them to banks and cell phone stores—critical, language-heavy errands too time-consuming for immigration nonprofits to organize. Most vividly, I remember escorting these new arrivals to their language classes, where the retired office workers who volunteered to teach often regarded their African students as learning-stunted victims of circumstance. Said one of the teachers to me, knowing I was a reporter, "I pretend to believe in them, but they'll never speak Japanese."

The African student who'd prompted this comment was the only one in a class of twelve who was not educated at a secondary level, who was not married to a Japanese national, who was not from Asia and hadn't been exposed to Japanese before arriving, and so on. As a result, his classmates learned at a reasonable pace while he did not, permitting the instructor to ignore the possibility that an experienced teacher might have made a difference, or a classroom not shared by two other classes, or the use of a textbook that did not insist on forgoing English as a bridge language. There is at least one excellent language school in Tokyo where enrollment is open and tuition is free. Many African immigrants who speak Japanese first studied there. This school is staffed by qualified teachers, each with his or her own classroom. Its curriculum is designed for students who may need to use a bridging language or improve their basic literacy before commencing language study.

One can't record an immigrant's entire early life in Japan so their Japanese teacher can observe the difficulties this student overcame to secure housing, to travel ninety minutes on four trains to the

English-speaking food bank in Tokyo (then make the return trip laden with groceries), to assemble enough simple stability to permit the luxury of attending language classes. One can't force a new immigrant to understand how the deck has been stacked against him, to raise his self-discipline to an extraordinary level, adequate to the unfamiliar challenges he will confront. It is only human to wish the relevant information were available to all parties, the failures of compassion and fortitude equally visible, the consequences of these failures imposed on the policy makers and vested interests who invited them.

"It is a miracle I am here," an asylum applicant once told me, while the Ebola outbreak engulfed his home country. We were sitting in the empty, sunless apartment the Japanese government had allocated him, in a neighborhood that was a suburb of a suburb—no public transit, and the only business within walking distance was a cigarette kiosk. The rent would continue to be paid so long as he did not seek work. If a friend gave him a gift of clothes or food, he was required to report it so his benefits could be reduced correspondingly. Later that afternoon, I accompanied him to the town office, where he asked if there were activities he could join; the loneliness of his apartment was beginning to overwhelm him. "Nothing I can think of," a municipal employee demurred, ignoring a nearby bulletin board that advertised community events. Then, in consolation, "Everyone is lonely here."

A miracle of many intentions, I concluded, and envied my acquaintance's attention to the few that were good.

The first time I visited Prosper's apartment, he pointed to a poster on the wall printed by one of the Nigerian civic organizations in Japan. It displayed the names and faces of prominent Nigerian expatriates. He'd crossed out the people who'd been incarcerated or deported or had chosen to go back to Nigeria. He said, "Maybe 60 percent of Nigerians in Japan, they were doing stolen car business, or touch money from a wrong place." I imagine I looked confused, or looked as if I wanted to take notes, because he added, "It's not news." Next to the town union poster were others, advertising memorial services for elderly relatives in Nigeria, none of which he'd been able to attend.

"In Nigeria, the same person will never steal cars, steal nothing," he continued. "They come here, in the beginning their hope may be high. Now they see they can never achieve, so it becomes this get-rich-quick mentality."

These two notions—that a significant percentage of Nigerian immigrants have participated in illicit activities, and the overwhelming majority of them wouldn't have become criminals if they'd never left Nigeria—would become familiar to me during the six years I spent writing for the newspaper. I remember listening to Slip, the chairman of one of the larger town unions, talk about car thieves, money launderers, and drug traffickers he knew in Japan who belonged to upstanding families in Igboland. When he was young, Slip had been one of the more successful scrapyard owners and shipping agents in the Nigerian community. Export enough containers and the police will want to know you. Slip developed relationships with several detectives in the organized crime unit and made the mistake of asking why they allowed many Nigerians to avoid incarceration. The detectives explained how investigative units dismantle criminal organizations, including the cultivation of surveillance targets and informants. It dawned on Slip that the detectives were not referring to a criminal organization that had taken shape in the Nigerian community. In the minds of the detectives, the criminal organization *was* the Nigerian community.

How does one distinguish criminal-mindedness from opportunism? Among developed nations, Japan is a comparatively easy country to visit (to enter legally, that is) from Nigeria. It is also a much harder country to earn a living in, especially for Nigerians. It happens that cars are cheap in Japan, presenting entrepreneurial immigrants with the tantalizing prospect of exploiting price discrepancies. In Nigeria, importing certain used vehicles is illegal; others are impractical to import, due to tariffs. Exporters targeting the Nigerian market must break these cars into parts. In Japan, a typical day for the yakuza involves insurance scams that require the disappearance of used cars. The disappearing is accomplished at scrapyards in the countryside, owned or fronted by immigrants, whose limited language ability ensures they never learn enough about their suppliers to identify them. Immigrant owners of legitimate scrapyards inevitably discover that it is a difficult business unless they run it crooked, and their employees learn that breaking down cars—if nobody asks where the cars came from—pays a living wage, rarely available elsewhere.

Slip lives in Lagos now and rarely visits Japan. The last time I saw him before he left, we were in a bar in Chiba, where he keeps the office the detectives know about. Slip is an inveterate talker and a curator of his community's gossip, so we arrived at a question I always asked: Anyone in jail I should know about? The smile went off his face and he said the name of one of his friends, who was an undiagnosed diabetic. Slip had noticed his friend driving a new car, then another new car, and a few weeks ago told him, You'll get sick in jail. His friend said, Let me touch it. (The money, he meant.) Now he was in jail, and Slip was in the bar, imagining his friend sweating, shaking, getting faint, and prison staff ignoring it because they're under pressure to minimize the prison system's medical bills.

The waitresses at the bar wore tight cotton tube tops and miniskirts, bare midriff. Slip was looking at them; they were picking up ashtrays and leaving coasters on the tables. Slip always had Japanese girlfriends. Women who wanted to be pop singers, models. The look on his face, watching the waitresses, was the look of a person unnerved by the smallness of his destination in the distance. He said, "I guess that's it," and except for our pleasantries saying goodbye, he wasn't an inveterate talker anymore. Then I heard he hadn't shown up when the union voted on whether to reelect him, so he wasn't the chairman.

Discussing how many Nigerians had been involved in criminal activity since coming to Japan, and how many of them might have made better choices in other places, Prosper would call the Nigerian community in Japan "a failed culture." He'd point out you could go to China, and there was auto theft, there was money laundering, it wasn't paradise, and the Nigerians living there would go home if they weren't making money, but at least they didn't keep their addresses and businesses hidden from each other, because a smaller percentage of them were involved in illicit activity, and they hadn't been targeted the same way by the police, and people didn't settle their differences with fellow Nigerians by calling the immigration authorities so livelihoods ended and families broke apart over words exchanged in a nightclub.

Only once did Prosper involve himself in the Nigerian community's social turmoil, and the experience disabused him of the urge. The chairman of the largest town union had pledged his personal wealth to build a community center, where Nigerians in need and recent immigrants could access the benefits of union membership, and Japanese citizens could encounter the public-facing, civic-minded elements of Igbo diaspora culture. Rather than endorse the project, his rank and file insisted on diverting the union's resources to construct a shopping mall in Nigeria, a scheme that failed when government officials in the union's home state laughed off the union's request for preferential access to undeveloped land. Prosper had fought to pass the community center proposal. When he wasn't working, he canvassed customers at the oldest Nigerian-owned business in the Tokyo area, an unlicensed gambling parlor. Many of these customers had lived in Japan as long as him—a quarter century—and said they couldn't countenance the expenditure of union funds on Japanese real estate, because Japan, notwithstanding their Japanese jobs, spouses, and children, belonged to strangers.

In the history of the parlor, Prosper had known several members of the Nigerian community who gambled away their savings. Some were deported for crimes they committed to pay their gambling debts. A few, including his former neighbor, died in the commission of these crimes. Still, he never met a person who disliked the parlor or its proprietor, and the reason, when he inquired, was always the same: If we must lose our wealth and our dignity, if we must lose our sanity and ultimately our lives, at least these losses will begin in a place that belongs to us.

3.

Prosper's apartment, where I spent my afternoons and Sundays, was a small room, nine square meters, with an attached kitchen of another meter in length, and a unit bathroom, made from a single piece of injection molded plastic: toilet, sink, tub. There are millions of these apartments in Greater Tokyo. My Nigerian friends who are not married or raising children live in them. Prosper's neighborhood was in the far west of the city, in an area that had been undesirable when he moved in, but developed into one of the trendier shopping districts for the young and affluent. His building was spared the wrecking ball because a corporate landlord bought several of its apartments and advertised them to tourists, at triple the market rent.

On the day in 2013 when I asked Prosper to recount his experience of the earthquake and he told me about the text messages he'd received while he waited in the *koban,* he suggested I accompany him to an appointment in Ueno. We rode the breadth of the metropolis on aboveground trains and I recorded his comments about the lives Nigerians led in Japan, particularly the increasing number who intended to repatriate.

"I believed I can make my life here to matter. In the beginning, I would never believe it. Eventually, I'm thinking, Ah, I can make money. I can marry. You don't forget who you are, but you are more than the place you are from. This is a lie. Because the marriage can never succeed. And most of us never want to keep these factory jobs. We start to believe we can make it when we have nightlife jobs, hip-hop jobs selling clothes. We mistake one good moment for the permanent thing. The second lie is the one we have now. When we are dreaming of Nigeria, we are thinking of the Nigeria we are free to escape. To really go home, you can't leave. I can say I'm an Igbo man. I can say I'm Nigerian, because my citizen is there. Can I say it's my country? Everything that made me to leave, it's true. I've lived a different life here."

I remember thinking, while he spoke, about photographs Prosper had taken during his early years in Japan. When I saw them, I was struck by the simple fact of his youth. He was lighter, by at least fifty pounds, and showed optimism in every expression. Many photos were taken on ferries that departed from Tokyo and Yokohama, where Prosper and his wife spent his days off, to feel that the hardships of his factory job and their interracial marriage were briefly distant.

When Prosper and I switched trains to ride the last few stops to Ueno, a mother and her talkative son sat beside me. The boy, who might have been seven, didn't think we could understand him. He asked his mother about our appearance—my tattoos, Prosper's skin. Before the train arrived at Ueno, Prosper leaned across me and said to the boy, "When I was Japanese, I got cold, so I decided to be African." The boy exhibited a moment of wide-eyed incredulity, then drew back against his seat, pulled his head to his chest, and lifted his arms to hide himself. Soon he peered at Prosper through the space between his arms and the brim of his baseball cap. His mother observed in silent astonishment, apparently uncertain whether Prosper had insulted her son or tried to be friendly.

We disembarked at the next stop. On my recording of our conversation, the sounds from the interior of the train are replaced by the song that plays at Ueno Station when the doors of a departing train are about to close, then the outdoor noises on the platform: synthetic birdsong piped through speakers, a campaign speech from a sound truck, an escalator announcing to the blind which way it runs.

"In Nigeria, living with my mother and uncle, I'm too young to have children," Prosper says. "The time I came here, I didn't have anyone to get advice."

Leaving the station, we watched a well-dressed man argue with a uniformed employee of the railway; he had tried to pass through the automated gate using a child's fare. The man had been poor once, Prosper concluded. The sound a gate makes when it captures a full ticket would have bothered him for hours. Instead, he chose embarrassment, an emotion affecting only the senses already too dull to use.

In 2011, after three hours in the *koban*, Prosper found the streets empty. He'd never thought Senzoku was a busy neighborhood, but like anywhere else in Tokyo it had its rhythms—the sound of aboveground trains, the whir of delivery scooters, revelers walking home from Asakusa. It took him forty minutes to reach Ueno. Halfway there, he joined a growing current of office workers proceeding on foot. Ueno Station includes multiple substations, run by several train operators. All of them, Prosper observed when he arrived, were full to overflowing. People were lingering at the information desks, apparently in an attempt to determine how they might get home, and at least as many seemed to have no intention of leaving the area without shopping for perishables first. The impractical necessity of acquiring as many goods as one can before beginning a long walk home reminded Prosper of Nigeria, and the familiarity of the sight made him realize he would need to walk to his apartment, twenty kilometers away.

He navigated by following the train lines southwest to Iidabashi, where the flow of workers walking home swelled to improbable proportions, many seeming to follow—as Prosper did—the river or the tracks. It was for him, as it was for everyone, a peaceful and solemn act to walk for hours surrounded by people doing the same, aware of the anxiety and gratitude of their pilgrimage: Some would be worrying about loved ones in the disaster area; all would be grateful to return to loved ones closer by. The longer he walked, the less he was bothered by the solitude of the apartment he was returning to.

The red-light districts were deserted during the days that followed. He had a few customers who lived nearby, and he would see them in the neighborhood when they left for work in the morning. "You're still in Japan?" they would say. "It's not safe." He would reply, "This is my home." They might bow, tell him how grateful they were for the support of Japan's foreign community. At first, it agitated him. Hadn't he spoken the same words every time a volunteer member of the neighborhood patrol in Roppongi told him to go back to Africa? Or every time a detective—reading from the little gray phrasebook of investigative interrogatories and banal small talk in English—asked him what he was doing in Japan?

But it didn't agitate him every time. By the end of the week it felt good hearing nice things about his decision to stay and, by extension, his decision to migrate in the first place.

Workers in the red-light districts don't take nights off. A handful find a way to give time to their families, but most are nocturnal; they're asleep while their kids are awake. After the earthquake, the clubs were full of their own employees, who played a nervous game with their bosses, drinking the bar stock, wondering if they'd get paid, knowing the free drinks would be invoked in the event a paycheck didn't materialize.

Prosper kept going to work but didn't drink. He sat in the back with his boss and watched football. None of his colleagues felt capable of predicting how long the slowdown might last. Like Prosper, they weren't sure they minded it. For the first time since the club opened, a week went by without any visits from the police, then two weeks. The club's two bartenders, a Nigerian and a Ugandan, attended the same Catholic church, where the foreign and Japanese congregations held separate services and often argued over petty issues that seemed, to the foreign congregants, to bespeak a deeper prejudice. At the church's bilingual vigil for disaster victims, they were sought out by members of the Japanese congregation, befriended, and invited home for dinner.

Owing to the circumstances of Prosper's birth, he wasn't connected by kinship to many of his Igbo compatriots in Japan. A given Igbo nightlife worker in the red-light districts might see three or four people from his village of origin just walking to work. Prosper was vaguely aware of his extended family's relationship to the family of one other person working in Roppongi—a man Prosper happened to dislike on the basis of a failed business venture they'd undertaken together. We'll call this man Sunday, in honor of his tendency to describe himself as a pastor.

Sunday showed up at the club two weeks after the earthquake. Someone had called his phone looking for Prosper. He would have hung up and never mentioned it, except they asked for Prosper by his Igbo name, and Sunday was under the accurate impression that nobody knew it who didn't know Prosper's family. When Sunday related this, one of Prosper's co-workers volunteered that he'd received a call from a Japanese speaker asking for the same person a few days earlier but hadn't known to connect the name with Prosper. The number matched the one that texted Prosper while he was in the *koban* on March 11.

Prosper liked his name private. If a Japanese man was looking for him and used his name, a Nigerian had put him up to it. He stepped into the elevator and dialed. Someone answered around the time he walked onto the street. The voice was young. Nervous. Not what he'd expected. His first thought had been yakuza. There was a time, only a few years earlier, when it wasn't difficult to invent a false debt and hire a gangster to collect it. When the voice confirmed his name, he said, "I'm someone else."

"Do you know Ms. Oda Yumiko?"

Prosper's ex-wife. He knew Africans who'd been contacted by Japanese men who were dating their former partners. They called without explaining why, out of insecurity.

Tohoku. The word appeared in Prosper's mind. *Tohoku tsunami.* Yumiko was from Tohoku. But she lived in Tokyo. Didn't she? After the divorce?

"This phone number belonged to my father," the voice said. "I think he's a foreigner, because my skin is dark."

Prosper cut the line. The air was cold enough to see his breath. Something seemed to hover above or behind him, buzzing. (That's how he describes it.) He looked down the street, at the people working for nightclubs, at the glowing signs, passing cars, the Tokyo Tower in the distance. Just like that, he thought. One phone call, then the world looks different.

"In Nigeria, family is everything," he later reflected. "My feeling is like I didn't have a family. When I learn I have a son, I'm thinking I am not meant to be a father. Even if I become, it should never happen in Tokyo."

Prosper was born in 1968, in Benin City. A year earlier, the city had been occupied by the Biafran army during its invasion of Nigeria. Situated between the majority Yoruba southwest and the Igbo southeast, Benin City has long been home to a large Igbo community, but its indigenous residents are members of minority ethnicities. Until it was recaptured by the Nigerian military, the city's allegiance remained an open question.

Prosper was the only child of his single mother, a circumstance he learned not to explore because the subject seemed forbidden. Benin City was home, but he traveled to his family's natal village for weddings, festivals, and funerals. Once, visiting for Christmas, he encountered one of his cousins at the market, a woman he'd seen at family events but never spoken to. When he approached her, she fled.

Prosper told his mother. She explained that he had been conceived when she was raped by a Nigerian soldier. This conversation is not a vivid memory for Prosper, or so he claims.

His mother died two years later, of an embolism. He went to stay with an uncle in Onitsha, a trade hub on the Igbo side of the Niger River. During secondary school he worked for his uncle's shipper-carrier agency, assisting the agency's most important customer, a Lagos-based auto parts reseller who convinced Prosper to continue his education.

After he began studying electrical engineering at Anambra State University, Prosper became reacquainted with a young woman he'd known in Onitsha. By the end of his freshman year, he wanted to marry her. He visited his uncle and disclosed the relationship, which—in the innocent way of the churchgoing class—had been kept secret, then asked for his uncle's help arranging marriage. His uncle refused; her family would never accept it. Prosper volunteered the possibility that he would never be able to marry, because his mother had been raped by a Nigerian soldier. His uncle replied, without hesitation or guile, "When our compound was requisitioned, the soldiers were Biafran."

To date, Prosper regards his uncle's remarks, however elliptical, as the best explanation he has received for the possible identity of his father, offered by a decent man in a moment of surprise and candor. Prosper can recall, with as much mirth as regret, many hours spent in front of a mirror diagnosing the non-Igbo features of his physical appearance, and just as many hours un-diagnosing them after his uncle's explanation. Some memories are solemn, including a visit to a rehabilitation settlement for wounded Biafran soldiers, to see his mother's father.

By the time Prosper visited, residents lacked food assistance and medical care. If traffic backed up nearby, they emerged from their dwellings and helped each other (many could not walk) to the roadside, where they begged from motorists. Some dwellings were robbed a few days before Prosper's visit. The culprits, upon learning whom they'd stolen from, returned the items and piled them in the center of the camp. The pile remained for several weeks, left by the veterans to express their bitterness.

Prosper rode to the settlement on the back of his uncle's motorcycle. He found his grandfather occupying a cot in the corner of a temporary building (a shed, you could call it), along a path that led to the neighboring leper colony. He stood beside the cot, watching the sun through the doorway illuminate the corrugations of the building's interior walls. The leaves of a breadfruit tree collided in the harmattan outside, isolating the room from the noise of the village.

His grandfather asked who was there, remaining in the posture of sleep, eyes shut.

Your grandson, Prosper said.

The eyes opened.

Uncle told you I'm coming.

I don't talk to anyone. I'm just living here, like this.

Sit up. See my eyes.

He sat up. I don't want you to stay, he said.

It's better I ask. Otherwise I'll come back.

There was softness in his grandfather's eyes as they adjusted to the light. They were not yellow from fatigue or disease, not bloodshot from a nervous disorder. They were clear white, the lucid eyes of a man whose mind still obeyed.

You look well, Prosper said.

I'm fortunate, his grandfather said, touching his hip where the shrapnel had entered. It's pain. But there's nothing I can't do. I could stand up. I could go.

Will you leave?

Who will walk at night? To find the men in wheelchairs—when it rains, if they're stuck. Should they sleep in the rain? Should they soil themselves, waiting for their absence to be noticed?

Mother told me she was raped by a Nigerian soldier.

Prosper's grandfather waited, his face indicating nothing except the changes in the light as passersby cast their shadows across the building's doorway, his eyes moving minutely as they registered the difference.

I look at myself, Prosper said. Everything about my face is Igbo. If a soldier raped her, it must have been a Biafran soldier.

His grandfather lay down and turned on his side, his back to Prosper.

Grandfather?

I'm remembering.

Prosper placed a hand on his grandfather's side, touching—but not realizing until he'd done it—the old wound. He lifted his hand and let it hover, wanting, for reasons he couldn't understand, to touch the spot again.

There was no rape, his grandfather said, turning on his back, keeping his eyes closed.

Who is my father?

Some boy from Benin City. He's not Igbo. Your mother had boyfriends. One boyfriend, they were together all the time.

His name?

I don't remember.

His village?

I don't remember.

What did he look like?

You could have starved during the war, his grandfather said. Three of your cousins starved. One was a baby, but her mother couldn't give milk because she herself was never eating.

I'm going, Prosper said.

There are children here. The men went to their villages, pleading for brides. The children do not know they were born to care for the lame, they do not know their fathers asked for money and settled for women who would bear children. They do not know they are livestock.

Prosper crossed the room and stood in the door. *Igbo Kwenu*, he said.

Ka udo di.

Prosper left. From behind him came the clatter of items across a floor, as if someone had upset a table in his grandfather's room. He reached the road and told his uncle he was ready to leave. If his uncle was surprised the meeting had been brief, he didn't show it.

Igbo Kwenu is a greeting offered by Igbo orators, which Prosper's grandfather would have heard at every speech given to Biafran troops during the war. The first half is a reference to Igbo culture and ethnicity. *Kwenu* invokes consensus, the type shared by entire communities, produced by the harmony of living together in an ancestral homeland over the course of centuries. "All Igbo, past, present, and future," seems grandiloquent, but this much is intended.

Prosper's grandfather's reply is one of a few phrases an audience might offer in response. It means, "So peace will come."

Riding home on the nearly impassable roads connecting Igboland's two great cities, Prosper considered what it meant that his uncle had maintained his modestly middle-class life, running a business that honored the permanent attributes of Onitsha—a river city and a trade hub, a place where cultures met before political boundaries attempted to constrain their relationships. Already, Nigeria was becoming a difficult place to live, and the farther one traveled from Lagos, the greater those difficulties became. As conditions in Onitsha deteriorated, families like his, anchored by patriarchs like his uncle, retained their positions as celebrants of the city's old symbols. Their resilience, Prosper realized, resulted from an abiding belief that they belonged in Onitsha.

When they reached Onitsha they lingered in the courtyard of his uncle's compound, awaiting the return of their equilibrium after the long ride with the engine buzzing underneath them.

"It's time I'm leaving," Prosper said.

"There are ways to send you," his uncle replied.

A practical discussion occupied their evening, concluding with advice that Prosper was too young to understand: You can become what hope makes you become, or remain impartial to hope until circumstances inspire it. Prosper, his uncle observed, had chosen the early hope, and must remember that his life depended on meeting its demands after its promises were broken.

4.

Prosper doesn't believe his grandfather's story and will tell you he rarely recalls it. As for my suggestion that he might have stayed in Nigeria if his visit with his grandfather had provided a reconciliation or better knowledge of his origins, this is not a connection he is willing to make.

He admits he thought of his grandfather as he rode north on the bullet train to Sendai in 2011, several weeks after he'd spoken with the thirteen-year-old who claimed to be his son. Prosper had decided this was probably true. A pregnancy would explain his former wife's return to Sendai, to parents who deplored her marriage to a foreigner.

Prosper was surprised not to see any tsunami damage at Sendai Station. He couldn't comprehend how a tsunami would choose some neighborhoods but not others, as it had in Sendai, where more than a thousand residents died. He wandered the city's *shotengai,* doing brisk business shortly after the disaster, and for a moment entertained the notion that the most embittered among his compatriots were right: God was at work in the tsunami, and it had struck where iniquity was greatest. By the time the thought had occurred to him, he found himself dismissing it; God had missed the parliament building (and all the worst people he personally knew) by almost two hundred kilometers. The agony of the people whose loved ones had drowned and the agony of the immigrants whose lives in Japan had led to bitterness seemed somehow to intensify each other.

Prosper had agreed to meet at a chain restaurant overlooking the *shotengai* and sat in a booth to wait. Shortly, a tan-skinned teenager with curly hair entered the restaurant. Prosper stood up to bow, but the boy extended his hand and said, in Japanese, "That's all the English I know." Then in English, tentative, and he blushed as he said it: "It's a joke."

They talked about nothing in particular. Was the boy the only half-Japanese student at his school? Did he live downtown? Together with his grandparents, or with his mother in their own apartment? The boy didn't look like Prosper, except he was clearly bi-ethnic.

"What did she tell you?" Prosper said.

"I was young and she told me, 'You're Japanese.' I got older and she told me I was half-Malaysian. I saw half-African people on TV, so I knew."

"How did you find me?"

"When the earthquake happened, I realized I could have died without knowing who my father was, so I searched her closet. She wrote your name on the pictures from the maternity clinic."

The boy handed Prosper a photograph of his ex-wife holding a baby in the snowy parking lot of a squat concrete and glass building. She was looking away from the photographer, over her shoulder, and holding the baby in the nervous way one holds a puppy, as if she expected him to kick loose and flee.

"She looks happy," Prosper offered.

The boy's expression went rigid, as if reacting to impoliteness. "She always looks like that," he said.

I met Keiji two years later, at a café, on the day Prosper asked me to accompany him to Ueno. Waiting for us, Keiji had emptied sugar packets onto the table and arranged the crystals in a spiral pattern, and now observed us while we sat down, to see if we noticed it. I was reminded of the adolescent male children of my Nigerian friends and their Japanese wives, boys whose intelligence is quiet and prickly and bespeaks a childhood spent beyond the reach of elementary education's condescending essentialism. One of my friends, a father of two boys like that, has coined a phrase to describe them: "See pass say." (Why speak, just watch.)

Keiji, too, who sat quietly while Prosper explained—to my surprise—that I had developed a panoramic comprehension of Japan's Nigerian community, spanning two generations of immigrants and their children. I was apparently expected to tell Keiji that half-Nigerian children who chose to live with their fathers in Nigeria were happier for it.

Some were happy, and embarking on lives that benefited from obtaining secondary education in English, in a country where bi-ethnic adults can choose their vocation (in Japan, many jobs are closed to them). Others wished they could return to Japan. In the years I'd covered the African community, I'd learned that the voluntary breaking up of families often left an indelible impression on children, in a way that other traumatic events (untimely deaths, deportations) didn't. I had promised myself I wouldn't counsel anyone—immigrants, spouses, their children—to choose one family member's desires over another's.

"Your mother, she—" I glanced at Prosper, whose pained expression stopped me from mentioning her trip to Nigeria. "She doesn't talk about what it would be like if you lived somewhere else."

"My father showed me pictures from his village. It looks poor, but I can tell the people take care of each other. Do you like Nigeria?"

"I like bargaining. People from your father's part of the country think they're the best."

"If you don't know the price?"

"You overpay."

"Stressful."

"Or fun."

Nothing in our subsequent conversation was remarkable. Keiji trusted his father's description of Nigeria as a difficult place with redeeming qualities. My involvement wasn't necessary, and Prosper's desire to overcome Keiji's reticence affirmed an impression I'd developed about my Nigerian acquaintances: As the hour grows late in the extended experiment of their migration to Japan, many of these men need more from their children—in the form of validation—than their children need from them.

There was only so much that Prosper and Keiji had learned to say to each other by that time, and less they were going to say in my presence. I received a phone call while we were in the café and went outside to answer. I left my recording device on the table.

"Your grandparents—" Prosper says. "How do you feel about them?" (His tone suggests this is a subject he previously avoided.)

"They're old-fashioned. Grandmother asked me where I go when I leave Sendai. I told her I visit my father. She asked if I'm sure it's my father. Then we talked about other things."

"She told me herself she doesn't mind if I die tomorrow."

"You didn't come to Japan to make friends with my grandmother."

"It's since I met you I'm thinking why I came."

"I made this," Keiji says. (Prosper is looking at a flyer Keiji designed for a school event.)

"Can you learn this kind of thing in university?"

"I prefer to teach myself."

"Early days in Japan, I would think what kind of person I would have become if I'd been a Japanese boy."

"I don't get to choose."

"You choose that you like computers, choose to study, get a job."

"Children don't think that way. In Nigeria maybe it's different. Maybe for people who live in villages."

"I don't know," Prosper says.

"You don't know?"

"I don't understand. My Japanese isn't fluent."

For a while they don't talk. Nearby, two young women express surprise and happiness, old friends running into each other; their children have grown in the meantime.

"I wasn't who I thought I was," Prosper says. A pause, and the women talking. He says, "Young."

"If you go back to Nigeria now, you're a failure," Keiji says. "You're teaching me it's better I give up from the beginning."

"A different young, I meant." Someone sighs—Prosper—and his next words are in the exhalation. "Japan can really kill me," he says.

There is some talk about what Prosper means, and Keiji describes a time his mother checked herself into a hospital for psychological evaluation. She returned as a less animated version of herself, Keiji remembers, her troubles intact and her feelings about them unchanged.

Prosper says, "There is nothing that can happen to a Nigerian in Japan that makes him suffer more than what happened to him in Nigeria. If you want to punish a Nigerian, his brothers and sisters must suffer."

"I don't have brothers or sisters," Keiji says, in a trailing-off way I now recognize from other occasions when he deferred to the depth of adult experience. "Suffer how?"

Despite reforms adopted in 2019 and 2020, observers of Japan's immigration system tend to agree that it's broken. Conservative critics argue it presents too many opportunities for unwanted immigrants to enter Japan and lacks an effective means for removing overstayers. The liberal opinion laments Japan's inconsistent consular practices and argues that immigrants who take advantage of the resulting gaps are faced with the inhumane conditions of Japan's extralegal labor market, then detained for unreasonable lengths of time. The substance of each argument is identical: Japan attracts immigrants of economic opportunity, then those immigrants get stuck.

The notion that this indicates a policy failure is a curious one, requiring the application of public morality, whether liberal or conservative, to the habits of the institutions involved. In a neutral analysis, the Japanese system operates close to perfection—perfection being not what the public believes the system *should* do, but what the objective circumstances that influence its operation demand.

National economies in advanced stages of development, Japan included, sort potential immigrants socioeconomically: affluent, middle-class, poor. There are wealthy immigrants, whose assets and relationships with global enterprise are universally desirable and who, in the interests of the nation, are welcomed. This is why most developed nations offer investor visas, which often result in naturalization, and it's why Japan has overhauled its mechanisms for obtaining residency to include a points system that expedites the process for high-value migrants, a gesture meant to include skilled and educated immigrants but that, so far, has been used primarily to attract immigrants with access to capital.

There are middle-class immigrants, people who don't possess enough individual wealth to make any difference to Japan's economy but whose aggregate spending is a core component of global commerce and whose purchasing power can be courted, because their collective habits are mediated by broad categories of consumer activity—tourism, for instance. Developed nations prefer not to encourage the large-scale migration of middle-class people, because the middle class is where national identities are formed and argued, and because middle-class people pursue livelihoods that entitle them to government-funded benefits. In the economically ideal iteration of Japan's relationship with middle-class foreigners, these people would visit, but wouldn't be able to settle, and would be forced to spend their money at an exaggerated rate—a touristic rate—to experience Japan in a manner consistent with their accustomed standard of living.

In every year from 2013 to 2019, Japan saw a record number of foreign visitors (31.9 million in 2019). For those of us who were living in Japan in 2011, this was a stunning reversal of expectations. Following the nuclear disaster, resident foreigners and tourists alike fled to the airports and camped in departure lobbies waiting for the first empty seat on an international flight. Newspapers forecast an irreversible decline in tourism revenues. The following years proved that Japan's tourist appeal is immune to domestic catastrophe. It has taken a pandemic (and Japan closing its ports of entry to foreigners) to curtail growth in the tourism sector.

Finally, Japan receives immigrants who are economically disadvantaged. Their presence fuels debates about the efficacy and fairness of the nation's immigration regime. Since the end of the bubble era, Japan's relationship with migrants from the developing world corresponds to the Japanese economy's need for cheap labor. Many of these immigrants work for sub-legal wages in facilities with poor safety records. If the difficult conditions they encounter don't lead to burnout and voluntary repatriation, their lack of immigration status provides a fail-safe: detention and deportation. The Japanese government engaged this fail-safe in 2003, when the Ministry of Justice was directed to reduce the number of immigrants without legal residency from roughly 220,000 to 110,000. Plainclothes "detection squads" fanned out across Japan's cities, surveilling ethnic restaurants, community meetings, guesthouses, and places of worship. At a Tokyo church favored by African immigrants for weddings and baptisms, priests had to physically remove detectives from Mass so many times they lost count, and thereafter shelter their congregants—sometimes overnight—until the detectives gave up waiting outside.

According to the Ministry of Justice, the campaign was successful, a determination that reportedly depends on counting anyone who was subjected to a deportation order among the number removed from Japan, even if the order was never carried out. For my sources in the Immigration Bureau and the immigrant community, the campaign succeeded chiefly in shattering morale. Being a dragnet, it resulted in the arrest of immigrants who had no choice but to live in the open. Parents of school-age children, for instance. Gainfully employed, with no criminal records. For the mid-level officials responsible for the bureau's casework, many of whom had persistently lodged internal criticisms of the loophole-ridden system that lured migrants of opportunity, the campaign confirmed their worst suspicions about their profession: that they were engaged in the overt regulation of the extralegal labor supply as a kind of commodity—like crude oil or metal ore—in a glutted, politically volatile market.

For immigrants, including many of my African friends, the campaign's profligate, racially charged visa checks and detentions erased their hard-won sense of Japan as a society that rewarded diligence and good behavior. They could remember their interactions with police during the bubble era, when immigrants tended to work in the manufacturing sector. The police knew their employers needed them at a below-market wage, so the deal was straightforward: If you're working and you stay out of trouble, your life in Japan is a private affair. As one civic leader in the Nigerian community put it, "Before all of those arrests, I had more in common with law-abiding Japanese people than Nigerian criminals. After those arrests, I would never speak to a Japanese police officer again, no matter what I knew."

In 2007, Japan instituted universal fingerprinting for arriving foreigners. Within two years, the number of immigrants seeking asylum in Japan had increased at an unusual rate. If you've made the connection I've implied between fingerprinting and asylum requests, you're doing better than immigration officials were doing in 2009, when many admitted they weren't sure why the numbers had gone up, and repeated a careworn mantra: because Japan is a nice place to live—never mind the nation's high language barrier, its low political tolerance of immigration, or the futile nature of an asylum application in Japan (one-tenth of 1 percent were approved during the period when application numbers rose precipitously). The fastest-growing category of applicants came from nations where false passports were cheaply available. For decades, immigrants from these countries had avoided the consequences of overstaying their Japanese visitor visas by returning under assumed identities.

History has not recorded the name of the first immigrant who thought to address the inevitable discovery of his false identity by claiming refugee status, nor, to my knowledge, has it recorded what nation it happened in, or whether the immigration authorities there understood what was taking shape. If dates and statistics can be trusted as circumstantial evidence, it happened in Japan in the months following the establishment of universal fingerprinting. Awareness of the temporary immigration relief provided by the asylum process spread among immigrant communities, eventually reaching the immigrants' nations of origin, where visa brokers learned that an immigrant who entered Japan legally on a short-term visa could apply for asylum and would be able to obtain housing, welfare payments, and a work permit. Leaving out the less rosy elements (a high risk of immediate detention at the airport, long wait times for approval of government support during which applicants may not work, and the near certitude of detention after the initial asylum application is denied), visa brokers were able to sell prospective customers on a deal too good to pass up. Combine the salesmanship of these brokers with the dwindling immigration solutions available to long-term extralegal immigrants already in Japan—many of whom had nothing to lose by filing an asylum application because they'd already been detained—and a refugee crisis with no apparent origin engulfed Japan's immigration system, prompting blanket dismissals of asylum applications, supported by Immigration Bureau investigations that deliberately stripped applicants of their rights under international law.

As political moderates in the Immigration Bureau often explained to me, fingerprinting is valuable, and offering work permits to asylum seekers is principled, but only if the diverse effects of these policies have been considered beforehand. "Japan's refugee crisis," one of my bureau sources lamented, "is the sole crisis of its type that is self-inflicted." Other nations have implemented similar policies without catalyzing the rapid increase in asylum claims that resulted in Japan, a country where almost every arrival has been approved in advance by immigration authorities (there are no land borders, no maritime borders easily accessible to refugees, only commercial flights).

If you visited a particular street near the Embassy of Japan in Nigeria during the years when application numbers were highest, you could meet an entertaining assemblage of aspiring migrants and the reprobates who worked in the cottage industries that served them, swapping last-minute advice about letters of invitation and phony bank statements. All of this involved some jollity, because it was a gesture toward a better life and included the thrill of putting one over on a reputedly racist nation. Some of the advice had its own humor. For a while brokers were telling their clients to imbibe before entering the embassy, because it would make them prone to providing answers the consular officers would attribute to limited cognitive ability, and dumb immigrants don't integrate; they work in factories.

The few Japanese consular workers I've known have been capable bureaucrats who learn, after their arrival in Nigeria, that the difference between the number of disingenuous visa applications the embassy receives and the number it rejects is considerable. Once, during a reception at the embassy, a consular officer asked me what I thought he could do to curtail the submission of suspect applications. I said, "The fakes should be easy to spot. Reject them." He became visibly annoyed—annoyed to the point of anguish, perhaps—and said, "That won't work."

I've met other people with stories to tell about the embassy, depicting a type of dysfunction you can't make up: annual spending sprees to justify the embassy's budget for the next fiscal year, resulting in piles of unused plasma TVs and other big-ticket items collecting dust in the embassy's spare rooms; frequent grocery trips to London, provided to insulate embassy workers from the fear-inducing prospect of eating a local diet; wives of embassy officials weeping over the loss of their affordable domestic workers when their husbands' terms in Nigeria expired. That's the colorful stuff. Never mind the chronic pay parity issues and unequal treatment of local employees.

Not that most embassy employees particularly care. But the few who do—the same few who would like to stop funneling extralegal laborers into Japan—find the experience more comprehensively Orwellian than they otherwise might.

I describe all of this as a means of following the trails that link Japan's immigration apparatus to potential immigrants before they arrive in Japan. Ostensibly, Japan's immigration system is broken. Yet it procures for Japan precisely what the nation needs: from the wealthy, a long-term commitment; from the middle class, a globally competitive share of their disposable income; from the poor, a steady supply of cheap labor, adjustable according to political and economic exigencies.

Keiji wanted to know what happened to people who got stuck in the gears. So we told him: detention for a year if you're lucky, or until you get sick enough that the Ministry of Justice doesn't want to pay your medical bills. The detention center is difficult to reach from Tokyo, even if your loved ones can spare the three hours each way by public transit. Visits are half an hour maximum. When you're out, it's on provisional release (a form of parole), which means no work and no traveling anywhere not specified on release documents. Get caught crossing any lines, and you go back. If you don't get caught or don't cross lines, the resulting period of relative freedom and looming uncertainty lasts as long as it takes for the authorities to deny your next appeal, then it's back to detention until you obtain provisional release all over again. Leave Japan at any point during this process and it's unlikely you'll be permitted to return, no matter how long you've been here, no matter how many loved ones you leave behind.

In short, as Prosper reiterated, "They kill you." He meant: They might as well.

The next time I saw Prosper, he stopped at my apartment before leaving the city to survey the scrapyards in Ibaraki, Tochigi, and Gunma Prefectures, with the intention of reacquainting himself with the container business. It had been two days since I'd met his son. We walked to the *shotengai* near my apartment and ate dinner. We talked about Keiji. Teenagers could be cruel, I pointed out. Prosper shrugged. It wasn't cruel of Keiji to call him a failure. Nigerians don't care so much about what their children say, he explained. What matters is you spend time with them.

When we came out of the restaurant, the sky was passing twilight. Surrounding the *shotengai* were several fifteen-story apartment buildings; their size created the impression of ships floating on a tide of smaller structures. We were walking beside one when its exterior lights (one over each apartment's front doorway) flickered on, hundreds at once.

"I love that," I said. "When the lights come on."

Prosper laughed. "Why do you love it?"

"It reminds me of the scale of the city. Every light is a life."

"Americans are romantic people. I never knew."

"You find that affected, huh?"

"When you see these lights, these are the same people who don't like to sit next to you on the train. The ones who think their children should not learn about what their country has done in the wars they started. I don't care how many are alive. It's no different seeing people in the street. Because it's only the light you see, you can pretend you are in love with them."

"Not in love."

Prosper stopped walking and looked at the apartment building, then the sky. "This is also my favorite time, when the sunset is ending."

"The sky is different in Japan. Different in Nigeria. It reminds you where you are."

This made him laugh again. "I can't forget. This is Japan. Winter, I'm never seeing the sun because of this night work. Summer, I can see one hour of light. I get to understand Tokyo is never dark. Even the far neighborhoods, there are lights. Nigeria, there is this power shortage, it's the kind of dark—you can see the dark, you believe you can touch it."

I didn't say anything. Prosper stood in a way that made him look as if he'd paused mid-stride. His weight was over his front foot, his shoulder thrown forward; I thought of a visual cliché in Japanese movies, where an unwelcome guest wedges his foot in a sliding door before inviting himself in.

"Days I'm waking up," he continued, "I'm thinking to myself, 'God, give me a life where I have more light. One hour is not enough.' But the sun is going down, and the way you see is different because the light is not the same. If you wish for sun and somebody is bringing a lamp, you will wonder if God has replied."

5.

Prosper had lifted the shipping container's latch and could have swung the door open, but stepped back, far enough to see around the container into the scrapyard. He waited, staring at the half-stripped diesel generators rusting near the trailer we'd come from. The hand he'd used on the latch hadn't dropped; there was still a trace of the gesture.

"It's funny to remember a certain thing," he said. Whatever it was, it hadn't occurred to him during our visits to other yards, where everything seemed automatic: arrive, exit car, look around, get back in the car before anyone notices you (and you have to pretend out of politeness they've got anything worth selling). Ostensibly, we were pursuing our best lead today, suggested to us by the only exporter Prosper knew who never shipped stolen goods. Hesitating in front of the container, maybe Prosper was remembering that the experienced entrepreneur brings his sense of humor—not expectation—to a procurement safari in Tochigi.

We'd come out of Tokyo in a rental car, first into the consecutive billboards and bunker-shaped chain stores that characterize the Japanese equivalent of America's suburban miracle miles, then into the equivocal prettiness of Tokyo's far exurbs, where rice fields snake between tract housing developments, punctuated by the occasional convenience store, looking like it could have been built yesterday, and more frequently by small factories and warehouses, many of which, showing their coats of rust through roof-high blankets of kudzu and creepers, provide the sole evidence that an era has passed in these places.

It had been five years since Prosper was a regular visitor to the scrapyards operated by his compatriots, in the places where land is cheapest within driving distance of Tokyo. In his lifetime, he'd made the trip too many times to count. Many of his colleagues in the factories and nightlife districts had done the same. There will never be a statistic that captures the cumulative, collective effort all this pilgrimage has amounted to, the bulk of it expended in self-destructive credulity, not knowing (or wishing to ignore) that the viability of the container business in Japan's Nigerian community depends on money laundering and destroys more honest wealth than it generates.

"Entering the center of the universe," Prosper said now, and opened the container.

In the moment it took my eyes to adjust to the container's interior, the smell of baked sweat and urine wafted out. Prosper coughed and startled backward. "Hey—" he said, speaking toward the floor, to three people lying there, wrapped in garbage bags, stirring awake. Two rolled over, away from the light. The third wriggled out of his bag and stood, another bag covering the top of his body, his head emerging from a hole he'd cut in it.

"Here are the rats," he said, and dragged the toe of his shoe along the floor. "Stop chewing our clothes." We learned his name was Horatio, and he meant to say it was thanks to the garbage bags the rats weren't chewing him. He needed to wear a few to be certain, and he was always thirsty the next morning, and covered in sweat. He didn't notice his own smell anymore, but the odor of his colleagues still bothered him, and he seemed to regard this inconsistency as a source of amusement, saying, "Every person special." He told us he'd come to Japan from Côte d'Ivoire two months earlier. His Francophone accent made him sound inappropriately cheerful, but in fact that was his mood. To free his arms, he worked the trash bags up above his shoulders, and they bunched there like epaulets. When he finished introducing himself, he pulled them over his head, wadded them into several pockets in his coveralls, and said, "Maybe you look Sam?"

Sam was the owner of the scrapyard, who'd told us that the computer accessories Prosper came to see were in this container. He operated through a front company, owned on the paperwork by an elderly Japanese man who now appeared at the door to the shipping container, carrying a beer in one hand and a cigarette in the other, transplanted, perhaps, from a cartoonist's rendering. His cheeks were flushed and he swayed on his feet. The sun was on him directly, and I wondered if he might faint from being drunk in the hot weather. He had stubble on his face in gray patches, and long wisps of white hair, matted by sweat to his pate. His coveralls didn't fit, the tight chest and short sleeves exaggerated the asymmetry of his shoulders, one higher, evidently the result of an old injury. "Wrong container," he announced.

There wasn't any other container in the yard.

While Prosper dialed Sam, the Japanese man stood outside the container looking at us, but I at least did not feel watched. Horatio was also quiet, though he nodded slightly when he met my gaze or Prosper's. Later we asked him why he does that, and he told us it was a gesture he'd devised to accumulate goodwill with Japanese people, a way of bowing to them all the time, from the neck instead of the waist so he wouldn't have to break eye contact.

Until Prosper reached Sam on the second try, the only noise in the container came from the breathing of the two men on the floor, trying to sleep, and the sound of the air trapped between the parallel walls. Prosper and I have a mutual friend, long since returned to Nigeria, who slept in a shipping container for a year when he launched his export business in Yokohama. Fourteen years later, he still has the same dream that first occurred during that time, and there are variations (where it's set, who's in it), but it's always about the trapped-air sound of the shipping container; the dream never includes a container, but the sound plays like it's being piped in. He searches for the origin of the sound until he walks off a bridge, or until a car-sized boulder of red clay soil falls on him—the kind of soil underfoot in his family's village—or until his pulmonary system shuts down spontaneously in the dream and he wakes up.

Horatio followed us to the office and waited outside while Sam tried to sell Prosper the phantom computer accessories, which he claimed he could obtain this week. On our way back to the car, I felt Horatio's hand nudge mine near my hip. Without looking, I clasped his hand, a gesture I'd learned passing tips to workers under the greedy eyes of their supervisors in Nigeria. He gave me a slip of paper with a phone number on it.

We parked the car down the road and called. Horatio walked to meet us, and we bought him lunch at a *konbini* nearby, parking at the edge of the large, empty lot and eating off the hood of the car. We could see across the lot into the store, where the employees were doing a poor job of pretending not to stare at us. This appeared to make Horatio nervous. Maybe Sam had warned him against attracting attention. Maybe in the period since he'd come to Japan, he'd learned that lesson himself. He was on a clean six-month visa, issued to asylum applicants who enter the country legally and request refugee status before their entry permission expires.

His story sounded as if he'd told it before. He'd fled Côte d'Ivoire to escape practitioners of black magic who wanted to kill him and use his body in a ritual sacrifice. The longer he talked, the more the real story—about his straitened life, supporting a young wife and four children—crept in, making the chronology of events he'd written into his asylum application impossible. Here, too, he didn't know how little the distinctions mattered: His application would be rejected, regardless.

"You forgot Boko Haram," Prosper said. "Next time you go for refugee interview, you can ask them please add this one."

"Of course," Horatio said, wanting to be agreeable, not having understood what Prosper said. He needed a shower and another set of clothes so he wouldn't have to live in his coveralls. He needed a better place to sleep. In Tokyo he could get some of these things. Church ministries and charity groups provided clothes. Volunteers hosted people in their apartments and helped them find jobs where they'd be housed in a dorm. Horatio had been through that when he arrived, then struck out for the scrapyards, believing he'd come to a wealthy country, refusing to settle until he'd paddled in a stream that flowed toward wealth. Chalet du Sam, corrugated steel, was as close as he'd come, and he turned down our offer to take him back to Tokyo.

Instead, he asked to work for us. Prosper, he assumed, owned a business. And I was one of Prosper's investors, my presence—my American presence—proving that Prosper was important. Prosper had this new car, had enough money to drive around looking for goods to containerize. Prosper attempted to explain his present circumstance: One unprofitable container could ruin him; the car was a rental; he lived in an apartment not much bigger than the car, in a building where the only Japanese tenants were the elderly poor because nobody except foreigners and dementia sufferers would put up with the water-stained walls and insect infestations. None of this had the effect Prosper intended. Horatio would have liked to live in his own apartment, to rent a car, to pay the table stakes of the containerization business.

"You say how you small," Horatio put it. "But I see the big man." This comment was intended as an obeisance, spoken with an intensity—leaning over the hood of the car toward Prosper, close enough we heard his breath in his nostrils—that proved there's no word for experience in the language of hope.

Prosper watched Horatio eat the egg sandwich and rice ball he'd chosen for lunch. He had a bag of similar items to take with him, had piled his basket after a nod of permission from Prosper. He ate and looked at the bag while he did it, with morose determination, as if steeling himself for the inevitable repetition of these offensively novel flavors. Prosper's eyes followed Horatio's to the bag, watched the lines of Horatio's cheek muscles while he chewed, taking large bites, putting the food to one side of his mouth, where he wouldn't taste it so much.

"That building," Prosper said, pointing to a moldering warehouse down the road, "is what?"

Horatio squinted, swallowed his food.

"Diapers," Prosper said. "The Chinese, they come to the countryside, they find these old warehouse, sometimes they rent, sometimes, if nobody there, they don't take any permission. They buy disposable diaper, which you can't get them in China like you get here, they fill these warehouses so they can wait when the price in China is too high. The price is like that, they send everything. You can even read in the news, sometimes the warehouse owner will say, 'Someone has been using this my warehouse.' Police now come, thinking maybe it's drugs or something illegal, but they break the door and it's many thousand of diapers, reaching to ceiling."

"Huggies," Horatio said. "Best hot Africa." He was looking at me and made a you-tell-him gesture, as if he'd shared something I would know about. "Huggies from America," he said.

"Money coming and going from Japan all the time," Prosper said. "Only because I know about this money doesn't mean it will come to me."

How much of this Horatio understood, we weren't sure. He promised if he heard about anybody selling diapers, he would call us.

When we dropped Horatio off a few meters down the road from the scrapyard, Prosper stiffed him, or that's how Prosper would usually describe refusing to donate money to someone desperate. Horatio lingered in the car after we pulled over, waiting for alms to materialize. Eventually he unclipped his seat belt, smiling in spite of the insult implied by our silence. He couldn't help—once he was out of the car—putting out his hand, and he was beginning to ask for an indulgence when Prosper pulled away.

In pictures of Prosper with his co-workers from the recycling plant, Coup is one of the other Nigerians, several of whom came into their jobs after Prosper convinced the owner of the plant that his compatriots could learn quickly and work like beasts of burden. Prosper remembered the disappointment of arriving in Tokyo to discover that no one in the Nigerian community was willing to help him, and thought he could mark himself for a life separate from their unkindness by giving jobs away instead of selling them to other immigrants. He used his days off to visit the guesthouses where he'd stayed when he first arrived. He brought toiletries, and Nigerian candies his cousins sent, and in this way became acquainted with the Igbo tenants. He gave one of the first jobs to Coup, who told Prosper he had also been raised by his extended family after his mother died.

Coup took two paychecks from the job at the recycling plant and quit. Combined with the money he'd obtained by subletting his room at the guesthouse for the final month he was there (and not paying his rent), it was enough to purchase space in a shipping container bound for Nigeria. Around the same time, several copy machines at the refurbishing center next to the recycling plant disappeared. There was never any proof Coup took them, but Prosper knew Coup hadn't earned enough money to put anything in the container. Coup's container business was quickly profitable, and invitations to his wedding were circulated, with his parents' names on the list of honorable guests because he could afford to fly them from Nigeria. The story about his mother's death had been invented to stir Prosper's sympathies. In the meantime, Prosper's boss reduced his hours to punish him for recommending a worker who quit without giving notice.

He saw Coup several times afterward, at the handful of Nigerian restaurants in Tokyo, at fundraisers for members of the Nigerian community who had suffered medical emergencies. Coup always showed Prosper great affection and spoke about the factory days nostalgically, as if he had worked there for years, his hands on his bootstraps, waiting for the accumulating moments of exertion to reveal their great reward. This was the story he told when it was his turn with the microphone during the fundraisers, at the high table where he always sat. *Like you, my brothers, I began in the factory.* There was a list of injuries he'd mention, how he'd worked until his hands cracked and bled, his hair fell out, and everything he ate tasted like the fumes he'd inhaled; it went on like that.

For a while Prosper thought he'd overestimated Coup, that he wasn't the shrewd Lagos-bred operator he appeared to be. He was just another person who wanted eyes on him. Then Coup stopped coming to the fundraisers, stopped sitting at the high tables and cultivating his reputation. He'd done it long enough to know the people he needed to know, then disappeared. The day Prosper and I went to Sam's scrapyard in Tochigi, it had been six years since Prosper had seen Coup. He would have let another six pass if his encounter with Horatio hadn't made him impatient to know whether he was wasting most of his time, or all of it.

Prosper thought Coup would have made a habit of changing phones, but he answered on the first ring and knew it was Prosper calling. Coup named a time and place to meet before Prosper had a chance to ask, then wasted five minutes talking about family and friends before cutting the connection without saying goodbye. Prosper reasoned that Coup didn't like short phone calls because they'd look suspicious if detectives were reading his phone bills. The call was enough to remember how Coup conducted himself. He wouldn't change his phone number. He'd keep the same nickname, and if anyone wanted to find him, they could. He'd dress like a poor man and borrow from his Japanese in-laws to pay rent. Anyone bothering to look would see a guy down on his luck, someone who had fallen from a dull and modest height.

On the day they agreed to meet, it didn't strike Prosper until he spotted Coup approaching from the other end of the main drag in Roppongi that his former acquaintance was now, in the words the organized crime detectives would have used, the African auto theft king of Tokyo. The only containerization entrepreneur with a bigger business had been arrested a few months earlier. Coup was already in the final stages of extricating himself from anything the police would care about, and had enough money saved to stop sending auto parts if something spooked him.

Roppongi was Coup's preferred meeting place, because the neighborhood thrilled him with his own anonymity. The Nigerians on the street knew his reputation but couldn't connect it to this diminutive man, who avoided eye contact and responded to strangers' greetings with a nervous cough (both behaviors were performances). To meet Prosper, he wore a grease-stained sweatshirt and a pair of corduroys with cuffs that had come unstitched so they dragged on the ground. It was hard to imagine him coordinating a crew at a chop shop, and he didn't anymore, he had people for that, but the talent involved would always be the one he relied on.

He wasn't much different from the younger version of himself in the pictures Prosper had shown me. His face had the alert features hunger develops in a person—when he looked at you, you felt he'd lived an ascetic's life. It wasn't true, Prosper was the one who'd lived that way, but seeing them next to each other, Prosper better dressed and heavier, anyone would have guessed wrong about their hardships.

They said nothing at first except "good evening," then stood together and watched the street. They both knew the bouncer at the club next door, who came to Japan from Nigeria with his fiancée and left her for the first Japanese woman who showed interest, then spent the next eight years looking penitent the way you do if you back your car over the family pet. They knew the owner of the hostess club in the adjacent building, who invited customers to stay at his apartment and fed them drugs—they stayed a week, a month, chose addiction over jobs and families—to keep them visiting the club every night, until they'd emptied their bank accounts, now arguing with one of his hostesses at the entrance to the jumbo sushi restaurant (a matter of owed wages; "I don't owe you what goes up your nose"). In a way, this was all the catching up that Prosper and Coup needed, because they could intuit what it had been like for each of them in the meantime, living in the world of their shared familiarities.

"I'm a landlord," Coup said, and showed Prosper a few photos on his phone, new construction in the Festac neighborhood of Lagos. "Flats."

"Is it everything?" Prosper asked—everything Coup did for income.

"A few containers," Coup admitted, and made a piano-playing motion with one hand to show he wasn't sure how many.

Coup's remark caused a young nightlife worker to cock his head in their direction. Prosper and Coup guessed his thoughts simultaneously and caught each other doing it, which was enough for a laugh, and the young worker stormed off, making of their mirth some judgment against him. But the funny part was the smallness of their world and how clear the rules were.

"Long time to see you again, but I'm happy," Coup said, ducking into the staircase that led to the hostess club owned by their payroll-challenged acquaintance. "I'm sure it's not apartment in Festac you are looking for."

"You know this Sam, in Kobana? Yesterday I went. The goods were not there."

Coup shrugged, and in this small gesture there was self-assurance, the emphasis a person gives who knew all along he'd reach a point where as much or as little truth as he wanted to tell, he could tell it.

"Newly arrived peoples, just refugees," Prosper said. "Like he's getting direct from the airport, not paying these boys. Even they're sleeping in container."

"He's just like—what do they call it in U.K.?—temp agency. If you don't have your own boys, you bring your goods, use his boys."

Prosper absorbed this, and the remainder of the narrative Coup spun to contextualize it: Jobs were thin on the ground. The way the cars got chopped and shipped, law enforcement figured it out a decade ago. The smart operators who could afford it were mostly out of the business, leaving the work to less experienced players who didn't care if their employees were capable, as long as they were cheap.

"The only surprise I'm hearing is you tried to go for container," he concluded. "Many years you don't send anything for Nigeria."

"I had a son."

"In Nigeria? You don't invite me for wedding?"

"You attended."

"The Japanese?"

"She never tell me the pregnancy, then the divorce finish," Prosper said.

"You prefer container? Roppongi is easy for you. Choose some girls; open your club. Take few rich customer for ATM. The money is enough."

"It's a stranger's garden, *nwanne m,*" Prosper said, using the Igbo word for sibling. "It doesn't continue forever."

"Neither what I do."

"So."

"Nigeria—" Coup said, then started again in a way that made it sound as if he'd changed his intended subject. "Georgette is late now."

Prosper nodded. Georgette (a man's nickname, after the cloth he exported) had been a friend to them both in Japan, had died suddenly, violently, senselessly during an extended visit to Nigeria, and left behind a mess of family and business considerations in both countries, imponderable because neither of his two wives—in Japan and Nigeria—knew about the other, or knew where he'd kept his money.

"I saw him in Lagos. He was happy, one week before he died," Coup said. "He thought he was home."

"Soon he knew," Prosper said. Knew it when he was removed from his car at gunpoint, Prosper meant. There wasn't any judgment in the words, any sense the death contained a lesson. Georgette was like anybody else; if he could afford to be Nigerian by visiting home, if he could afford to lead a safe, peaceful life by coming to Japan, he would do both until something made him choose.

"I travel anyway next month," Coup said. "You can come. See Festac side."

"Maybe," Prosper said. He put his hand out. Coup flinched; he hadn't expected the conversation to conclude. Prosper was walking past him before the handshake ended.

Prosper turned the corner. When I caught up with him, he was scanning the street, squinting against the light from the electric signs. He started to run but stopped after a few strides. "I saw her," he said. For weeks, he'd been claiming he had seen the woman he was with three years ago when the earthquake struck. She had been pushing a little girl in a stroller, the girl's skin had been dark, and she looked about two years old.

6.

Coup offered to buy Prosper's plane ticket if he would take responsibility for clearing two of Coup's containers through customs in Lagos, where Prosper could use his late uncle's connections. Not car parts, Coup promised. Factory discards, Chinese air conditioners that prospective buyers might mistake for the Japanese items they resembled, and what a shame if they did. Prosper tried reaching Keiji to make sure there wasn't anything he wanted to do in Tokyo on the proposed dates, but couldn't get him on the phone. They hadn't seen each other in a month. Their conversations had become terse, Keiji's reasons for avoiding Tokyo more perfunctory.

As if in response, Yumiko called and insisted on visiting.

They met at the station nearest his apartment. Prosper waited outside the gate, in the sunlight, where it was too bright to make out the faces of the passengers riding the escalator, so he scrutinized their strides when they disembarked, wondering if he would recognize her. She appeared out of his peripheral vision, a few feet away, having chosen the wrong exit. She said, "I never see Black people in Sendai!" and removed the straw hat she was wearing, but didn't bother to bow. Seeing her face again, he remembered the fastened-on quality he'd ascribed to it before the divorce. He had been raised in a culture that relied on reading faces. In Igbo, the phrase that means "I love you" goes, "I see you at the eyes."

She was taller than Prosper, taller than most Japanese men. He had wondered whether she would have considered marrying a foreigner if all the men she knew hadn't been intimidated by her height. Both of her parents were short, maybe not genetically; they hadn't gotten enough to eat during their postwar childhoods, in an era that Prosper often thought he'd like to learn about because it produced the first marriages between Japanese women and dark-skinned foreigners. Some couples never had a language in common.

Prosper and Yumiko had met at a church attended by her Filipina co-workers, shortly after she arrived in Tokyo and took a job cleaning hotels. This, he later decided, was the moment when her needs and his—despite differences of language and culture—were most compatible. As his Japanese improved, he learned about the experiences that led her to leave Tohoku. She had been bullied, often violently, by classmates whose fathers blamed her father, the local representative of an agricultural supply company, for the company's decision to cut off their credit. An equivalent situation was impossible to imagine in Igboland, where title and lineage ensured that influential men, not the clerks in their businesses, answered for injuries to the community. The better he understood Yumiko, the less he understood Japan, and the two could not be separated.

These were old thoughts, and they occupied Prosper while he walked Yumiko to his apartment, where he heated a pot of leftover bitter leaf stew, which he knew would offend her palate. "It's like chewing headache pills," she said.

"The *real* bitter leaf, the fresh one, it's good."

"I've had it," she said. "It's worse! For a Japanese, it's worse."

Amid the familiar bustle of occupying a shared space and the good-natured kidding about the bachelor life that Prosper had resumed since the divorce, the conversation halted on the difficult matter between them—the time they had spent in Nigeria. Yumiko filled a bowl with stew and ate to fill the silence. Prosper checked the pockets of his work suits; it was something to do that let him face the wall.

When she had finished half of the soup, Yumiko pushed her bowl to the center of the table and leaned over it, inspecting the remainder. "I'm surprised I could eat that much," she said. Then, straightening her posture and folding her hands formally on her lap, "What has Keiji told you?"

"I can't think of anything he told me you wouldn't want me to hear."

"No—what *you* wouldn't want to hear," she said. "Any parent wouldn't want." Her hands rested on her legs, where she worried the fabric of her dress by gathering it between her knees. "Keiji has problems in school and life."

"He told me he doesn't go sometimes. He stays home to finish his homework."

"Keiji hasn't done homework in months."

"Just say," Prosper offered, "whatever made you come here."

She did, in a fragmented way. Keiji didn't come home for days at a time, then lied about where he'd gone. He locked himself in his room and wouldn't talk to her—or anyone else, including his grandparents—if he came out to eat or use the bathroom. The hints about what he'd been doing outside the house, or even in his room, were bizarre: Elderly neighbors approached her and said how much they enjoyed Keiji's visits; girls and boys Keiji's age who didn't go to his school showed up at the apartment and went into his room with him; the police visited last week because they'd received a complaint from an apartment nearby that someone had been ringing their doorbell late at night, and security footage showed Keiji doing it.

Prosper couldn't understand why Keiji hadn't been forced to give an explanation. This swirl of incidents and anxieties conformed to his broader misgivings about the emotional lives of Japanese families: Yumiko and her parents hadn't loved Keiji warmly enough to make him feel comfortable expressing himself, nor had they been firm enough to shape his behavior.

"For several weeks he asked me to call him by a different name," Yumiko said. "Not Keiji. Some other name. I don't remember it. I remember, actually, but why say it?"

"You've told him lies. He learned it well."

"You're talking about something specific."

"You told him he's Malaysian. Of course he's unhappy. He doesn't know himself." He resolved to take his son's side and was surprised by how noble it made him feel.

Yumiko shaded her face with her hand so Prosper couldn't see her. "Nothing was kept from him," she said. "He knew he was Igbo. I told him if he wanted to meet you, he could." She dropped her hand. Her face explained: She felt responsible. She knew Keiji's deceptions would wound Prosper, in light of their apparent shape, in light of his family history.

The shock did feel familiar, the way his ethnicity had been used against him, to make him believe the most convenient explanation for a family member's absence. He never would have questioned it. His friends' wives had done it to their children, taken them away and deceived them about their fathers' identities.

He imagined himself in the places he'd seen his married friends take their children: an elementary school, a church, a city office. It came to him that the life he led in the one-room apartment, with nothing inside it he wouldn't mind leaving behind, was not the worst outcome, not the best, not an outcome distinct to him, though he could sleep and it would wait, though he knew the eggshell color of the apartment's textured wallpaper intimately enough he had recognized it in a dream, illuminating the windows of a passing train. The places he'd lived in Japan, the places he'd worked, his marriage to Yumiko—he had allowed himself to believe that fatherhood would occur apart from them, would restore a gentleness he was owed.

He crossed the apartment and opened the door. His neighborhood was in a valley. The afternoon sun was behind the apartment building, casting shadows that reached up the incline toward the skyscrapers in Shinjuku. At the far end of the building, someone climbed the metal staircase and the frame continued to resonate when they stepped off. The staccato, treble chirp of a bird reached Prosper several times, nearby, going past, farther away, the birdsong he always heard in Tokyo, but he'd never learned the bird's name and every Japanese friend he asked said the same thing: They always heard it but didn't know what it was called.

He realized Yumiko was talking to him and turned to listen. She talked about Nigeria. She talked about praying, what God had shown her, what a privilege ("a spiritual experience") it had been to see Prosper at home, where every path in the village led to the compound of a family whose children had been his co-initiates in a secret language of symbols and sounds, distinct to their generation and the place their mothers were born. She noticed her pregnancy during the trip, but she hadn't taken a pregnancy test, could attribute the symptoms to travel, could convince herself that nothing she needed from Prosper would prevent what she could see—what God told her—he wanted.

To go home.

7.

A rotating cast of journalists, the same activists and church volunteers every week, thousands of anxious family members clinging to the irresponsibly sanguine assessments peddled by immigration lawyers: They've all felt what Prosper said he felt, riding the train, the bus next, then walking the rest of the way to the East Japan Immigration Center in Ushiku, Ibaraki Prefecture. Most begin to feel it on the bus, when the first half of the long ride has elapsed and Ushiku's 120-meter-tall Buddha statue appears in the distance past the rice paddies, on a stretch of the drive that is neither bucolic farmland nor semi-abandoned bubble-era tract housing, but one edging into the other in an irregular pattern that makes the landscape appear incongruous and stitched. They feel: Where am I? The answer suggests itself: too far away. The remoteness is part of the punishment, taking fathers and mothers away from children and spouses, enacting this burden of time, train schedules, bus schedules, round-trip fare, setting aside a day to make a half-hour visit through reinforced glass.

The center—"Ushiku" among immigrants, there's no other reason for them to know the town—is one of Japan's largest detention facilities. Carceral institutions (unless you call them immigration centers) are not popular neighbors; that's why, if you tell a taxi driver at Ushiku Station you need to renew your visa, he'll drive you to the detention center, where asking about a visa renewal makes as much sense as ordering a pizza. I once bet one of my Immigration Bureau sources he couldn't find a taxi in the station queue to take him to the prison, if "prison" was what he called it. He pissed off every driver in the queue, all of whom knew exactly where to take the guy wearing the uniform of the Immigration Bureau's parent ministry, none of whom knew the uniform was worn, in Ushiku, primarily by jailers.

A week after Yumiko's visit, Prosper and I went to the center to pick up his friend Uche, who had obtained parole. Uche was one of the other Nigerians in the pictures of Prosper and Coup when they worked at the recycling plant, where Uche continued working for nine years before taking a nightlife job. His timid mannerisms, of a variety rarely seen in the red-light districts, worked to his advantage until the nightlife business entered its lean years. Customers trusted his reticence. Long enough ago that nobody remembered who, someone compared him to a duck floating quietly in the reeds, and the comparison stuck. His colleagues called him by the animal's Igbo name. When Prosper's early attempts to join the container business failed, Uche visited his apartment every afternoon, to be certain he was awake and preparing for work. "The crop is lost but you are not," Uche would remind him.

Uche hadn't been able to renew his spouse visa after his wife left him two years ago. When he was detained, his colleagues in Kabuki-cho called it inevitable: Wasn't his wife paying for their apartment, and wasn't he hiding income from her? Hadn't he invited discovery by remitting to his other wife, in Nigeria, from the joint account? The person whose words were least kind—this shouldn't have surprised me—had barely managed to elude the consequences of a similar predicament. Shortly before his visa expired, his estranged wife stopped by to complain about a loan he'd taken out in her name. While she used his bathroom, he rifled through her bag for her personal seal, then stamped it on several visa renewal forms for later use.

A hat went around on Uche's behalf, and after a week it had three thousand dollars in it. The money didn't help, and maybe it hurt: Uche hired an administrative solicitor—not an attorney, but empowered to practice immigration law in Japan—who advertised in Tokyo's English-language periodicals. Like many solicitors I've met in Ushiku, he wasn't equal to the task of representing detainees and had somehow persisted in this ineptitude longer than experience should have permitted. Solicitors often tell detainees' family members that their loved ones will be out in a few weeks, or ballpark the likelihood of a swift resolution above 90 percent. After nine months, or a year—or two years—the same family members, transformed by desperation, find themselves paying up front for a ticket to send their spouse, child, or sibling back to their country of origin.

Three thousand dollars later, Uche was still in detention and his solicitor was still advertising in the magazines. Uche got out the same way nearly everyone gets out. An activist who visits detainees walked him through the paperwork to claim refugee status, apply for provisional release, and promptly appeal every subsequent denial of his requests until the authorities understood he wasn't going anywhere.

In the waiting room at Ushiku, while Uche was completing his discharge, Prosper and I watched two detainees who'd been paroled a few moments before we arrived. Their release documents were in English, but they had never learned to read, so the church volunteers read the forms aloud and showed them where to mark with the pen. "I am meant to call this," one of the detainees said, and pointed to a phone number penciled in a notebook. Next to it someone had written, "gives work." The detainee borrowed a church volunteer's phone and placed a call, the contents of which were transparent: He'd come to Japan through a visa broker and wanted to follow up with the labor agent he would have called if he hadn't been detained at the airport. In the meantime, the agent seemed to be saying, the cost of the job had gone up, but they could put it on credit and garnish his wages. Prosper stared at the Immigration Bureau employees working within earshot of the conversation. They were inspecting a pile of items donated to the inmates by the church group: toiletries, instant cocoa, Bibles, and rubber sandals. In a lighthearted moment, one of the staff put a sandal on his hand and pretended to stamp his paperwork with it.

I had been visiting Ushiku a long time and it didn't dishearten me anymore. Solidarity forms between the people in the waiting room who aren't irretrievably on the wrong side of the system, and occasionally this includes the detention center's employees, who are generous when they can afford to be. Some visitors leave behind symbols of defiance, which restore the sense of a shared human presence to the bureaucratic process of visiting. In one of the visiting rooms, KURDISTAN has been scratched into the seam of the steel counter. For several years, until detention center employees thought to cover the seam with gauze, anyone who used the room was reminded that

Japan approved visa-free entry for Turkish nationals without accounting for ethnic strife in Turkey, resulting in the mass incarceration of asylum applicants who soon arrived.

When Uche was released, Prosper gave his address as Uche's new residence. The apartment was barely bigger than the two of them lying down. They would share it until Uche found something better. We walked to the bus stop. Uche was upbeat, overstimulated, wanted to talk. He said in Nigeria he married at eighteen and became a father at nineteen. His village was one of the few in Igboland that relied on agriculture, not remittances, and no one born there could afford to attend university, so he gave his savings to a visa broker. The broker came up empty at the American embassy but managed a Japanese visa. Uche described driving to the airport with his wife and his one-year-old son to board his flight to Tokyo, how his son had refused to sleep the night before and cried to keep them awake, then fell asleep in the car the next morning and they couldn't rouse him at the departure gate.

After Uche arrived, he made a few attempts to obtain a job that provided immigration sponsorship but found himself working in factories, on an expired visa, looking for a wife so he wouldn't have to bet everything he'd built in Japan on the continued inattention of the authorities. His first marriage in Japan was childless. After he divorced and remarried, his second wife discovered he was skimming from their joint account, the same amount on the same day every month. He didn't fault her for leaving, he said, and in retrospect he was only surprised it had taken her so long to learn where the money was going—maybe she had needed the illusion of his fidelity as much as he needed her to believe it. But it had never occurred to him that she could take their children and refuse to let him see them. Their daughter was seven. Their son was four.

Prosper had told me that Uche persuaded his second Japanese wife to have children when she was in her mid-thirties and wasn't sure she wanted any. Most of the Nigerian immigrants I knew who were married in Nigeria wished for women to marry in Japan who didn't want children, to make the double life easier. While we walked, Uche described his plans for raising his Japanese children together with their Nigerian siblings, in a Lagos neighborhood where diplomats and billionaires lived, in a life I knew he couldn't afford. "Children are wealth," he said, and his voice contained none of the irony or bitterness I often heard when my Igbo friends used the same aphorism, or pointed out the first half of it: "To the poor man."

A plot of farmland planted with cabbage paralleled the road where we waited for the bus. Prosper uprooted one of the plants and threw it across the road. He kicked another and it broke apart, the petals scattering on the sidewalk. "This place," he said, ripping another out of the ground and throwing it at a nearby greenhouse.

Uche talked about the waiting room at the center, about something that was playing on the TV when he came out. It's nice, he said, that they installed a TV so visitors have something to watch.

Prosper stopped tearing out the plants and glowered at Uche. Only briefly, but before it turned to sympathy, to resignation, before Prosper had said, "Nice of them—yeah," Uche saw it and shook his head. "It's not the worst for me," he said. He talked about people with valid visas, detained despite legal appeals from their Japanese spouses. A mother of five, her children all Japanese citizens, facing deportation because one of the employees working for her family's business let his visa expire and she didn't fire him.

Prosper looked down the road. The rest of the visitors were making their way from the detention center to the bus stop. A few stopped to buy drinks at a lone vending machine, positioned at the front of a small scrapyard that marked the turnoff for the center. "Every money they can make they are making," he said. "See now this vending machine in middle of nowhere."

"You want to drink?" Uche offered, and went into his pockets for a coin.

"No," Prosper said, after a moment of incredulous silence.

Uche picked a cabbage petal off the sidewalk and put it in his mouth. "Bitter," he said. "Maybe sweet later." He pointed at two contrails in the flight path that passes over Ushiku. "Coming or going?" he said, then answered himself: "It's going."

Uche wanted to get off the train in Shinjuku and eat, specifically at a Nigerian restaurant in the red-light district. It wasn't open, but the owner was asleep in the adjacent apartment. She woke and put a container of leftover soup in the microwave. Uche said how much he'd missed Nigerian flavors, and she asked if he'd just moved to Japan; he explained he came into the restaurant a few times a week until a year ago. She couldn't remember him.

Uche ate the soup with a handful of pounded yam, in the West African style, while we walked back to the station. People stared. He put his ticket in the station gate with his left hand, his right sticky from the soup, and it was only on the platform, with the lid of the soup open again, that he noticed the attention he attracted. He cleaned his hand with a towelette, the buoyancy of his gestures diminishing as he registered the sensory overload of Shinjuku Station. He had been detained for more than a year.

On the third night Uche spent at Prosper's apartment, Coup brought Prosper's plane ticket. "Open-ended," he explained. "Go when you feel you are ready. Do something for me while you're there."

They sat down to eat. "I knew you," Coup said, after they'd prayed over the food, then waited to see how Uche would reply, which would show how Uche remembered him: as the manipulator and thief who'd stolen the goods he needed to launch his business, or the successful entrepreneur his initial transgressions allowed him to become. Uche gave Coup what he wanted, a few words about the talents of self-made men, about how rarely Coup's name was on anyone's tongue because he kept his accomplishments quiet.

Coup offered Uche a job and a place to stay near the work site (the apartment where Coup slept when he stayed overnight). Prosper objected: Nothing could be worse than Uche going to work in a yard where cars were chopped and exported. Coup promised he'd send Uche on errands, make sure he was never in the same place anything stolen was being containerized.

"One good deed per year," Coup said. "This one for 1996."

That got a laugh from Uche, but not Prosper, who looked at Coup over his bowl, through the steam rising off the soup, and said, "If he goes inside again, you will go with him." Prosper's intensity startled Coup out of his good humor; before any tension could form, Uche began telling Coup about his children. The three of them discussed the lives they'd led in Japan, which were—from the perspective they felt they shared—a map of fate's erratic intentions.

Uche packed after dinner. By the end of the week he'd left Prosper's apartment, as vulnerable to being detained for working without a visa as he had been when his first entry permit expired twenty-one years earlier, intending to go precisely where he and Prosper had both started: into the extralegal labor supply.

Instead of packing it, Uche insisted I take his immigration file. "If you read, you know," he said. The paperwork was roughly a thousand pages, including notes and correspondence. He wasn't being specific or coherent; mostly he dwelled on the improbable notion that he was the sole target of Immigration Bureau machinations. He felt he had been persecuted. I would discover he wasn't wrong.

8.

Prosper worked overtime in the red-light districts so he wouldn't have to spend down his savings in Nigeria. When business was slow he worked his phone, getting in touch with his second cousin and a cousin-in-law at Apapa Customs, finding out what had changed since he last cleared a container—who was in charge, how much money they would want, what words they liked to hear.

He called Keiji, too. A few times every night, but Keiji never answered. He avoided calling Yumiko, whose version of her reason for leaving him would perhaps help him to heal, but had also left him wondering about her in the same way he always had, trying to determine how much of the Yumiko he knew was made up of the neuroses she couldn't express in her own culture. She would never tell a Japanese person that the Buddha or a Shinto deity had visited her with instructions about their relationship.

He reached Keiji three days before leaving for Nigeria, after coming home from work at 11:00 in the morning, showering, and getting into bed with the lights off. He was falling asleep when Keiji answered. "You saw Mom," Keiji said.

"Tell me if she's lying or if it's you," Prosper said, and tried to sound like a father being strict. To himself, he only sounded impatient.

"Of course it's me," Keiji said, and his tone wasn't aloof the way it had been the other times they'd spoken or been together. It was frustrated, and there was sarcasm in it, in the form of false cheer.

"I wish I could understand your reasons."

. . .

"Do you believe me?"

"No."

"You live in a safe country. Probably, you don't feel safe. I grew up in a dangerous country. Most of the time, I felt safe." Prosper was again aware he was treading at the edge of his relationship with the Japanese language. He had meant to explain how you learn, growing up in Nigeria, that there isn't time or reason to be angry at someone whom you love but don't understand.

. . .

"Your mother is worried. You probably think it's important how you behave. It's not so important. Nobody's going to notice except her. Family is important. So if you want attention, you can just ask your family."

"I'm not trying to get anybody's attention."

"You bring strangers home?"

"We play on the computer. I don't like anybody watching me."

"Family isn't anybody."

. . .

"You want to live with your mother, or come with me to Nigeria?"

. . .

"I'm going next week. I can arrange schooling for you."

. . .

"You tell me."

"I don't mind life how it is."

"How?"

"Time passes. I have friends."

"And?"

. . .

"Did you consider your grandparents can die anytime and your mother can even get old and she can't help you anymore?"

"There's a woman I know who's ninety-seven. She says, 'If the ants come, we'll make it rain. Because the ants are really mechanical; they have electricity inside of them.' Sometimes I help her shop. She knows the price before the clerk adds it up. She says, 'The shoes I had when I was your age looked like this and that.' I looked it up. It was a shoe people wore."

"Do you want to come to Tokyo?"

"It was easy before because I told you the things that made it simple for you to understand."

"Something happened to you, either you never told me or you never told your mother or maybe nobody knows."

"For example, I've been beaten up at school."

"Everywhere is crazy," Prosper said, and tried not to listen to himself. The sound of his voice in Japanese intensified the prickling sensation of not having slept.

. . .

"I'm going to pray."

. . .

"Dear Father, If we move in the right direction, send a sign so we know."

The phone was quiet. Prosper held it above his head and pressed a button. Nothing—the battery was faulty, sometimes it quit. He felt around the floor near the bed in the dark, looking at the blurry outlines of the mess in his apartment. It had been thirty hours since he'd slept. He was surprised at himself for failing to speak any words of discipline during the conversation. But the voice that answered the phone wasn't the voice he'd imagined. He'd imagined a voice that spoke only English, a voice without inflection or idiosyncrasy, a silhouette of a son.

Falling asleep, he remembered that he hadn't known the traditional taboos from his mother's village until he'd been a university student and learned them from the house girl who worked for his uncle: Don't take the bridge across the stream if your wife is pregnant; if you harvest on the night of a full moon, carry everything home under a thick cloth; any snake that dies, bury it as you would your mother. In his vivid thoughts, drifting off, he is a young man, hearing these stories in the restless predawn hours when he would wake to the house girl humming a melody that entered his sleep before interrupting it. "Who taught you?" he asks her. The house girl says, "A teacher was never needed."

Prosper called me out of an appointment and asked to meet in Kabuki-cho. We waited outside a restaurant where the woman he thought he'd been with on the day of the earthquake was eating dinner, then I followed her onto the train. It was a long ride, to the Tokyo-Saitama border. When I examined the mailbox she checked on her way into her apartment building, there wasn't any name on it or any mail to peer at through the slot.

I told Prosper he'd need a name and a birth date to ask the city office for information, and he'd need a reason, which he couldn't get, because Japanese courts rarely order paternity tests. He described his fear that he wouldn't feel capable of boarding his flight to Nigeria if his questions about her remained unanswered. It would be enough, he said, if I could learn whether she had a child, what the child looked like, and how old the child might be.

I followed her to an appointment at an insurance office, then waited outside her building the morning after that, Saturday. It was a long surveillance, from sunrise until the afternoon, but eventually she left the building pushing a little girl in a stroller. The girl had dark skin and curly hair; if the woman was the girl's mother, the father wasn't Japanese.

After a train ride across the city, they visited an apartment building where they stopped to chat with a woman in the lobby. Both women went up to the fourth floor, and I rode the elevator with them.

I gave them my back, which put me next to the stroller. The little girl was half-asleep. She had pulled her mother's purse over her face to keep the light out. Judging by the woman's voice, the conversation she carried on, and the brief, close glimpse of her I caught as she moved past me out of the elevator, she was young enough to be the person Prosper remembered. I watched to see which apartment they entered, then went downstairs and checked the name on the mailbox, the Japanese equivalent of Dave Smith.

I thought about what I would say if Prosper asked for the woman's address and decided I'd visit her first, to see what could be learned by asking. I pictured myself knocking on her door, and the image demoralized me. It had been the worst kind of work, every time I'd done it as a reporter, getting to know both sides of a custody dispute between a Nigerian man and a Japanese woman. If the mother had lied to her children about their father's identity, the father usually considered this a heinous moral wrong, as it would have been regarded in his culture. The mother was often bewildered and bitter about her husband's behavior during their marriage. If he'd been leading a double life, or if he'd misused money loaned by her parents, it was typical she would describe the process of finding out by saying it had been like discovering she'd married a criminal or a mental patient.

Though I tried a few times early on, it never worked to say to the father, "If you apply a certain variety of self-preserving duplicity, learned in Nigeria, to your family life in Japan, society here may judge you unfit for fatherhood." It never worked to say to the mother, "It will cause your children pain—and you won't be able to relieve it—if you wish on them a life of not knowing who their father is." By the time the coincidence of my involvement occurred, they had wounded each other in ways neither of them had imagined possible.

I performed one other surveillance in the six years I covered the African community, to help a source file a lawsuit (Emanuel was his pseudonym in the article I published). Emanuel's wife had abducted his youngest daughter and remarried a Japanese man while Emanuel was in Nigeria with their three older children, caring for his mother, who had suffered a stroke. She had also taken the money in their savings account. Stories like Emanuel's arose often enough in my reporting to form a theme, and I was waiting for one of them to become litigious so I could follow the case. Litigation was rare because the likely relief was paltry (once-per-year visitation, unenforceable), and many of the potential plaintiffs I knew had not been model husbands, which made them reluctant to appear in family court.

Emanuel hadn't taken money from his in-laws or tried to lead a double life. Legal experts said if he won, he would be the first African plaintiff to obtain restitution in a case like his. I was certain he would win. In order to secure a divorce without his consent, his ex-wife had claimed that he'd vanished three years earlier. But their youngest daughter was only two when the divorce was filed, and Emanuel's ex had taken her to Nigeria for a visit shortly after her birth. Emanuel kept home videos of their visit.

The court ultimately ruled in Emanuel's favor, but the legal process required him to spend a year away from his children, who remained in Nigeria to continue their schooling. When his mother recovered from her stroke, she began misappropriating the money he remitted for their care. He returned to Benin City to find his children leery and reticent. Finally, they confessed they hadn't been fed regularly. His oldest daughter had been sick with typhoid, and his family doctor, upon examining her, said malnutrition had nearly made her infection chronic. I once asked Emanuel whether he would have stayed in Nigeria if he had known his absence would provoke his mother's opportunism, and he would be forced to evict the woman whose illness had summoned him home to begin with. He was a devout Christian and replied that every decision he made, he took to God in prayer. "Recently," he said, "God is familiar with your question."

I remember riding the train with Emanuel's ex-wife and daughter when I followed them home. The girl threw a tantrum. When her mother leaned down to soothe her, she pulled her mother's hair, hard enough that her mother nearly fell. When the girl let go, the effort and concentration of the tantrum stayed on her face; her mother looked merely patient, in the pure and blank way patience has of making a face seem uninhabited. I pictured myself waving a hand between them, to interrupt their discomfort. The thought had a compulsive quality—I nearly did it.

I was in my mid-twenties, still shedding the misapprehensions I'd carried with me into reporting. I'd thought I was going to see directly into the lives I intended to describe, the way a telescope surveys a landscape, one coordinate—one life—after another. This was rarely true. Usually, I talked to people. When I was able to observe them, much of their behavior was for my benefit. In the beginning, my coverage was cloying and breathless. Then it was diligent, and for a time it was artful. Near the end, I'd learned too much. Everything came out overreported, wreathed in caveats, compared with some other case. I drooled anecdotes. I'd developed the habit of wanting the people I was writing about to illustrate or disprove some broader phenomenon taking shape in the African community, and of feeling—when this failed to occur—that I'd never understand anyone I reported on because I hadn't been around when money was still flowing from the apparel and nightlife businesses, leading so many Nigerians to entertain the improbable notion that they'd soon be an integrated part of Japan's middle class.

I wanted to develop a sense of occasion about my work, and discovered one when it became clear how many people thought about returning to Nigeria, talked about it with each other, or had started living there part-time. Some had already repatriated, especially among the first wave of Nigerian immigrants to Japan, mostly from

Edo State. If you added the minority of people who made an earnest effort to live in Japan as if it were Nigeria—by finding ways for their Nigerian wives to join them—that left very few who shaped their lives as if they intended to stay, and fewer still whose reasons were not purely pragmatic.

The following two memories are adapted from notes I took when I was interviewing long-term Igbo residents of Japan about their intention to repatriate:

1. A successful Igbo businessman I know immigrated to Tokyo when digital photography was replacing film in Japan but had barely appeared in the developing world. He bought color film processing equipment, sold it in Nigeria, then left the business before it declined. He was oblivious to his good timing until he'd already made his money. If he started an export business today, he admits, he wouldn't be able to repeat his success: "Many have worked much harder, more carefully, only to receive much less." He was among the first members of the Igbo community to return to Nigeria, and his compatriots praise him whenever the virtues of repatriation are discussed. His Japanese wife accompanied him. So did their children. The oldest recently graduated from a prestigious Nigerian university, and all seem happily adjusted. But his wife spends more and more time in Japan, and he recognizes the likely outcome of her absences: She'll want to go home.

His younger brother still lives in Japan and earns his money in nightlife. Lately, this younger brother is trying to establish a livelihood that will allow him to live in Nigeria, with his Nigerian wife and their three young children. If he finds a way, it will be easier for him than it was for his older brother. His marriage in Japan is purely transactional, a visa marriage, no kids.

He's not home yet, and who knows when he will be. He says if he could trade places with his older brother, he would. His brother says the same about him.

2. In the Ajao suburb of Lagos, near the airport, many roads aren't paved. Local wisdom says the deprivation is tribal in nature: Ajao is an Igbo enclave, its residents vote for Igbo politicians,

and the Yoruba-dominated Lagos government excludes it from the infrastructure budget. The wives and children of my Igbo acquaintances in Japan live there. On my first trip to Nigeria, Ajao was the first place I went. An acquaintance's wife took me to visit their children in school.

Classes had ended a few minutes before we arrived, and the principal led me into the schoolyard, where children were waiting to be picked up. "Raise your hand if you have a relative in Japan!" the principal said. Eleven hands went up. We gave a few of them a ride so I could ask questions. My friend's wife drives a sports car he bought when the nightlife business was lucrative and he thought he'd never have money troubles. It bottomed out every few meters on the unpaved roads and the kids hit the ceiling; they were having a wonderful time. *What are your relatives' jobs in Japan?* I asked. Everyone's father or uncle was a CEO, a businessperson, owned a company; that's what the kids thought. Later, when my friend's wife drove me across town, she asked me what her husband does. He's an entrepreneur, I said, but you know how it is—it's a bear until the right opportunity comes along.

It had been a year since she'd seen him. He planned to come home in a few months, for Christmas. The month of pay he intended to set aside for his plane ticket disappeared when the hostess club he worked for closed. He was disconsolate, and it made me remember that his children had played with my field recorder when I'd visited them, singing pop songs, inventing the speech they'd give if they were elected president. I took the recordings to his apartment, copied them to his computer— Merry Christmas—and left. He called me that night and left a message without meaning to, a few minutes of his children's voices playing in the background. He lay in bed and listened to the recordings, I guess, and his phone was on the bed or near it, because I could hear the creaking of his mattress springs and the rustling of his sheets. Briefly, I forgot what I was listening to and expected to hear him address his children, but the room he lived in, I remembered, was too small for a family, if one had been near enough to join him.

—

When the woman Prosper asked me to follow came out of the apartment building where she'd taken her daughter, I followed them to the train station. The girl was awake now and looked at me as she passed in her stroller, wearing a perplexed expression, a young child contemplating the riddle of her environment. For an instant, her face reminded me of the face of my closest friend growing up, a musically gifted girl whose Yoruba father and American mother divorced acrimoniously during her childhood. I remembered sitting in her father's kitchen while he carried on phone conversations in Yoruba, which was how he spent his time during her visits. She sat next to him, waiting for the person on the other end of the call to talk, then played her guitar as loudly as she could, angling the soundboard toward his uncovered ear. He ignored it or didn't notice, and the sound, intended as an interruption, became an accompaniment. For a confused moment, I thought I was watching them perform together.

Entering the train station, the woman I was surveilling had a problem with her fare card. The gate wouldn't accept it. While she was talking to the stationmaster, her daughter began climbing out of the stroller and was nearly out when her mother pushed the stroller forward. The woman caught the girl as she began to tumble, but the girl's face struck the gate and she began to cry. Her mother stroked the hurt place on her cheek, and the stationmaster showed her an LED light on his key chain. She calmed, but put her hands on the stationmaster's wrist and when her mother turned to leave, they had to be pried away. Later, I told Prosper how the girl had behaved, and he asked what the crying had sounded like. "You never know what it tells you," he said.

Prosper and I visited Uche before leaving for Nigeria. He was working the phones in the apartment near Coup's scrapyard, contacting the owners of the Lagos warehouses where Coup sold his merchandise. Uche told Prosper he hadn't worked on anything risky, anything related to cars, but the bills of lading on the desk listed auto parts.

Coup was on the phone with someone in Nnewi, the spare parts capital of Nigeria. The window to the apartment was open, and anyone in the neighborhood could have heard him yelling, but the area wasn't a neighborhood, just a cluster of industrial buildings. Most seemed vacant.

The apartment's sliding glass doors faced the gate of Coup's scrapyard, 450 square meters of untaxed enterprise. The possibility that Coup would choose a licit livelihood had been eliminated decades earlier when he spent his first two years in Japan shuffling from one factory to another, laid off, spit on, screamed at, and physically abused by his Japanese bosses, his Korean co-workers, anybody less marginal than he. When Coup invented a painful childhood to gain Prosper's sympathy and the job that came with it, he had intended to use the resulting wages to leave Japan forever.

Moods in the apartment were bright. Coup was confident that his shipment would arrive during a shortage of similar goods in Nigeria. If Prosper could move his share of the merchandise past customs, the profit would be tidy. For the first time since he was detained, Uche had remitted money to his wife and children in Nigeria; his oldest daughter was overjoyed when she learned she could return to the school she attended last year.

Coup played a highlife record that West African club owners in Tokyo often play after turning on the house lights in the morning. My Nigerian friends would tell you, after a few drinks, why people living in diaspora were the only ones who could understand the song's bittersweet homily on love and privation. The bandleader who'd written and performed the song was half-Nigerian, only partly raised in Nigeria. When he was a teenager, he fled to Cameroon to escape the civil war. The song made him one of Africa's few globally recognized musical acts, but he soured on the music industry and retired to the Nigeria-Cameroon border, where he opened a hotel, telling his friends he could only live in both countries at once.

> When I wan sleep, my mother go pet me,
> she go lie me well well for bed.
> She cover me cloth, say, "Make you sleep,
> sleep sleep my pikin oh."

When the chorus played, Uche stood on his chair, dancing in the masquerade-meets-disco style that was fashionable when he was young. Prosper and Coup danced with him. They froze in place at the parts where the band drops out and the bandleader sings "stop stop," as a mother to her crying child. While he danced, Coup talked about staying at the bandleader's hotel once. It was clear he'd never been there, but this didn't diminish the general gladness of the moment, which Coup had tried to own because that was his language, what he owned and how much of it.

Sweet mother, I no go forget you
for the suffer wey you suffer for me.

When the song ended, the apartment smelled like sweat. We took a moment to catch our breath and let the melody leave us. Uche began to talk, in the half-conscious manner of a person allowing a compulsive thought to become verbal. He told us he had decided not to sign the divorce agreement his Japanese wife had sent. A lawyer had told him how long it would take to secure yearly visitation with his children and how rarely Japanese mothers complied with visitation orders. Instead, he planned to cling to the marriage until the financial pressures of single motherhood led his wife to seek welfare benefits, available only if the marriage had ended. He would leverage his cooperation to see his children. If the immigration authorities detained him before these events occurred, he would endure an additional period of incarceration. Nothing could persuade him to leave Japan while his children were lost to him.

He described his childhood, his father managing a soap factory, the urban poor who ate the fish killed by the factory's runoff, and the truism that Igbo people never starve because every village is food rich, every village welcomes its lineal descendants; when he was young in Nigeria, without a job or prospects, he learned why people chose starvation before the village. At Ushiku he learned to starve in a wealthy country, with your stomach full and money in the bank. *Can't you make a living in Nigeria?* the detention center employees asked him, every day. Detainees imagined—believed in—a committee of Japanese psychologists who supplied demoralizing questions for the guards to repeat. *Won't your children be happier if they believe they're Japanese?*

He was crying. He put his head on his knees. Prosper stood next to him, laid a hand on his back. Coup walked to the sink, filled a glass with water, placed it near the bed Uche was sitting on, then crossed the apartment and went outside. He left the door open. The trail of his cigarette smoke reached us. Uche was quiet. He had wept, then stopped. We heard Coup answer his phone and tell whoever had called that they were garbage, a waste of life. Now he was speaking in numbers. Low numbers at first. A little higher. And the chatter of the voice on the other side of the line.

Prosper grasped a handful of Uche's shirtsleeve and coaxed him to sit up, then squatted, and they regarded each other at eye level. Uche rested his chin on Prosper's shoulder. Prosper's breathing caught, but his body didn't move, and his next breath was normal, his expression sympathetic; he put his hand on Uche's back again, and his eyes wandered to the clock radio. The hour struck while he watched it. He faced the doorway and nodded, as if someone had signaled him. From a bus shelter at the bottom of the hill, the evening melody began to play.

9.

Uche was right. His file was a mess.

Its contents didn't suggest, as he claimed, a conspiracy against him, only the Immigration Bureau's overriding need to maintain institutional equilibrium: with limited resources, under fluid circumstances, always accounting for political interpretations of immigration statistics (either the welcome mat is out, or it isn't). He had submitted a questionable asylum application to an agency that couldn't afford to examine it in good faith.

By the time the bureau reviewed Uche's submissions, he was already a victim of his administrative solicitor's negligence. The solicitor told Uche he could abandon his spouse visa and apply under a category reserved for people in "exceptional circumstances." Uche had been in Japan for two decades. He would lose contact with his children if he couldn't resolve his dispute with his wife before leaving. What circumstances could be more exceptional? Or so the solicitor reasoned.

The Immigration Bureau denied Uche's application on the grounds that he'd applied under the wrong category. There's no pity visa in Japan: You can't stay in order to avoid exposing yourself or your family to adversity. And yet bureau investigators interviewed him more than once, and their questions were about the adversity he was likely to face. Uche says the interviews convinced him he was eligible for the type of relief his solicitor had recommended. Why interview him about how badly he wanted to stay near his children if it had been irrelevant all along?

At the conclusion of each interview, the bureau's investigator summarized what Uche had told him: You've said you want to mend your marriage. You've said you want to spend time with your children and you fear losing contact with them. You've said going back would be a hardship because you've lived in Japan your entire adult life. Can you confirm that there are *no other reasons* why you wouldn't return to Nigeria? Uche would say, Yes, that's it. The investigator would push him: Say it in your own words. And he would.

The stated purpose of the interviews was to determine whether Uche was eligible for the immigration status his solicitor recommended. But his application was meritless on its face. Each interview required hours of work from a bureau investigator and interpreter. The transcripts fattened Uche's file, slowing a review process that could have been adjudicated instantaneously. This seems ill-considered at an agency where backlogs are a source of political vulnerability.

Not ill-considered, however, in light of what Japan's immigration authorities expect from immigrants in Uche's position after they've been denied their visa and detained: To forestall deportation, a significant number apply for refugee status. Faced with the resulting obligations under international law, immigration investigators can never obtain too many statements from prospective asylum applicants suggesting that their reasons for wanting to stay are unrelated to personal safety or political persecution.

Investigators are confident that no one lodging a truthful asylum application would say something like "I would go home if I didn't have children in Japan." They're aware that applicants have often been tricked into believing they've answered an unrelated question: Do you love your children? Do you consider Japan your home? But the rationale goes: If we pose the real question, the answer won't be candid.

Aside from maneuvering to prejudice Uche's eventual asylum application, bureau investigators performed a "forensic review" of Uche's case. The resulting reports include a Nigerian civics lesson ("The applicant states he is a Nigerian citizen. This is inaccurate. He is a non-Japanese citizen of the Federal Republic of Nigeria"); some advice about the use of figurative language ("Applicant says his life is 'rooted' in Japan after living here for two decades. Trees grow their roots from the very beginning, and he was born in Nigeria, so in fact that is where his roots are"); and a personal finance tip ("Applicant says it would be difficult for him to pay child support from Nigeria. We have conducted research, and we are confident it is possible to remit money from Nigeria to Japan").

Large segments of the reports are speculation: The applicant probably has family in Nigeria who will help him. The applicant can probably find a job if he goes home. The applicant probably hasn't been threatened by Boko Haram, although we "don't really understand the situation in Nigeria."

Over and over again, investigators commit the same notion to paper: "Many applicants from this region make similar claims, and they often turn out to be untrue." This statement appears so frequently because it serves dual purposes: first, to apologize for the insult of treating the applicant as if he were lying, and later to explain why the Immigration Bureau has concluded that the applicant's claims should be treated as falsehoods, after all.

None of this bothered Uche. It wasn't news to him that Japanese people were capable of racism, and he was aware that Nigerians living in Japan lied on their asylum applications. The speculation-heavy passages struck him as fanciful, and he couldn't prove a negative: If the bureau thought family money and a good job awaited him in Nigeria, he would have to take it as a compliment. The official name of his country, his choice of metaphors—he wasn't the one who would have to find a way to take himself seriously after writing that stuff.

Uche identifies the bureau's least defensible failures elsewhere, in the reports that address themselves to the primary evidence in his file, where he could have proven the investigators wrong if he'd been given a chance, and where the bureau's review takes the form of ad hominem attacks on Uche's attitudes toward his family.

The phrasing is direct: The applicant's relationship with his wife does not appear to be legitimate. The applicant's interest in his children does not appear to be legitimate. If the applicant loved his wife, he would have tried to reach her through an attorney X times. He would have taken Y and Z steps to obtain access to his children. His wife would have visited him in detention. His wife would have brought the children to visit him in detention. He would have attempted to contact his in-laws. And so on. This reasoning is accompanied by repeated assertions that the immigration status Uche has applied for cannot be obtained on the basis of family hardship. If his family life was immaterial to his application, Uche points out, what possible reason could there be for the bureau to repeatedly insist that he doesn't love his children?

The longest bureau response in Uche's file concludes with an analysis of the dates when he petitioned for mediation with his wife. It says: The applicant made only two requests to initiate mediation, and nearly two years elapsed between these requests without any follow-up through the courts. The second request was filed on the same day that the applicant also filed a petition to prevent his own deportation, indicating that the applicant's proclaimed desire to communicate with his wife and see his children is disingenuous and opportunistic. Because the applicant made no other attempts to communicate with his wife during this period, it is difficult to reach any other conclusion.

In fact, Uche's solicitor, who initiated the first mediation request, which Uche's wife ignored, discouraged Uche from trying again on multiple occasions, telling him that another request would be "an expensive waste of time" and failing to note that other types of correspondence could be channeled to his wife through the courts. When Uche fired his solicitor two years later and obtained a competent attorney, the attorney immediately advised Uche to file a new request for mediation. The additional petition to prevent his deportation and the coincident timing of all his paperwork were the decisions of his lawyer, who quickly determined that Uche's solicitor had neglected several deadlines and who—like most well-organized attorneys—does all his filing in a single court visit.

Regarding the absence of evidence that Uche tried to contact his wife between mediation requests: Phone records indicate he called her literally hundreds of times. If Uche had known the investigation process would take such an odd turn, he would have provided these records preemptively. But he couldn't have known, and feels it was the bureau's responsibility to demonstrate the minimum investigative initiative—to ask him about these supposed inconsistencies, especially in light of how much time the bureau dedicated to interviewing him.

My sources in Japan's immigration bureaucracy would dismiss any criticism related to Uche's case. His application to remain in Japan was meritless under the category he chose, and his asylum application was premised on a fabricated death threat. I've examined several files like Uche's. The dishonesty of the applicant is always invoked to shield immigration officials from acknowledging the long-term consequences of poor forensics, which proliferate, affecting important investigations.

In September 2015, at the height of the refugee exodus from Syria, Japan's Ministry of Foreign Affairs informed the global media that the country would not accept Syrian refugees. Prime Minister Abe supplied a political rationale: Japan's mono-ethnic history had left the nation ill-prepared to host displaced people; instead, Japan would provide the UN with generous financial support to assist with resettlement elsewhere. Immigration Bureau officials took a different approach, characterizing themselves as rule followers whose options were limited by international law. Any refugee who says he or she left home to escape a conflict zone must, according to Japanese immigration officials, be denied asylum, because their fear of death or injury is unrelated to their ethnicity, religion, or political beliefs. To avoid defending this policy in court, the Immigration Bureau preemptively issued residency permits to Syrian refugees (a handful fled to Japan despite Abe's warnings), hoping they would withdraw their asylum claims.

In a functioning bureaucracy with a robust, politically independent investigative capacity, individual casework puts the lie to abdications of responsibility this broad, by revealing that the facts vary too much and shift too rapidly to discard on the basis of nationality. Consider the bureau's inability (or unwillingness) to distinguish between military conflicts that harm civilians collaterally and conflicts that assume and intend civilian harm. That's how most nations interpreted their legal obligations to refugees fleeing Syria, a war zone populated by dozens of ideologically driven non-state actors. The bureau's version, where other countries flout the law because it's the sympathetic thing to do, is a self-serving fantasy.

Detainees have died in the custody of Japan's immigration authorities, at least fifteen since 2007. In 2010 it was Abubakar Awudu Suraj, a Ghanaian man with no criminal record who had been living with his Japanese wife for twenty-two years. He died aboard a passenger jet during a forced deportation, when bureau personnel bound and gagged him—doubled over, in a stress position—to prevent him from struggling during takeoff. The bureau admitted that restraints had been applied in violation of bureau rules, leading to a landmark district court verdict in favor of Suraj's wife and mother, who had filed a wrongful death suit. The Tokyo High Court reversed that decision, in a ruling that went out of its way to affirm the legality of the same restraining techniques the bureau had voluntarily repudiated (the bureau's appeal had focused instead on the role a minor heart condition might have played in Suraj's death).

If Suraj had claimed refugee status, the Immigration Bureau would have been forced to postpone his deportation until his claim had been reviewed. He could have applied for provisional release in the meantime and appealed for immigration relief after leaving detention. He was deported—and died—because he didn't know to make a disingenuous asylum application.

Uche's asylum application contains a few news clippings about Boko Haram, from major newspapers. That's all the supporting evidence he submitted. The bureau official who interviewed him said it was sufficient and couldn't give Uche a clear answer about how to provide witness testimony or certified documents from Nigeria. Uche thought he'd been brushed off because his story was flimsy. Other detainees told him it had been the same in their interviews. This included the handful of asylum seekers at Ushiku whose claims were considered legitimate by fellow detainees. It shocked Uche's conscience to learn that the game the bureau was playing with willing participants like him—in effect, trading weak asylum applications for provisional release—had been forced on everyone, indiscriminately.

In our interviews about his experiences in detention, Uche dwelled on the malingering phenomenon affecting Ushiku. The bureau's eagerness to move sick detainees off its budget had made illness the fastest route to parole, with predictable results: People began claiming they were sick or attempting to harm their own health. Detainees with acute medical issues, including several who died after bureau staff ignored their symptoms, were the victims of this mutual dishonesty. "You blame yourself for half," Uche said. "Soon you wonder: I am not the doctor who works for Ushiku, I am not the lawyer who works for Immigration. Because of me, their promises are difficult, but they are the ones who break them."

10.

In Lagos, a sign at the gate of Ladipo Market proclaims, "Home of Japanese Engines and Spare Parts." These arrive in shipping containers arranged by independent importers, most of whom earn their living on a handful of containers per year. According to the Japanese government's external trade organization, economic relations between Nigeria and Japan have been in a valley for decades. Traders at Ladipo disagree: Japanese automakers facilitate a high-volume business in Nigeria, wittingly or not. Their products are widely desired and widely available. If they haven't been able to service this demand in a comfortably corporate fashion, as they would in a country with a predictable business climate, so much the better. Money stays in the container business.

Ladipo stirs early every morning, so Prosper arrives at 5:30. He would have left his hotel earlier, but doesn't trust the neighborhood in the dark. Already, trucks and taxis have jammed the entrance, where Prosper stands and chats with the *okada* boys. A delivery van has broken down inside the gate, drawing a crowd: those who want it moved, those who are offering help (for a price), the owner (who prefers not to pay), and several of the owner's colleagues (whose notions of justice are involved). Spectating with Prosper, the *okada* boys discuss the possibility that the van is not in need of mechanical attention; maybe the driver and the people offering to move it are working an angle. Prosper smiles and rubs his thumb against his forefinger. "Human electricity," he says. "From friction."

The names of the warehouses have changed since his last visit. The rest is familiar: ankle-deep mud from yesterday's rain, blackened by motor oil; the opioid gaze of the dolly boys, whose shoulder muscles pulsate like drumheads while they haul transmissions, mufflers, engines, pulling them through the muck on handcarts; the market's cooks, women shuttling between their umbrella-shaded stoves and whoever calls out for a meal of *mamaput,* as in *mama, put the food here;* the girls who work as the market's nurses, dozing on their feet at the entrances to the warehouses, a bottle of antiseptic in one hand and a roll of gauze in the other, neither available until a price has been negotiated; the occasional spectacle of a local official making his daily rounds, resplendent in his freshly laundered *agbada,* disembarking from a luxury car and glad-handing his way into a warehouse, as if he has emerged from the pages of a protest novel and only he doesn't know it.

At a warehouse Prosper visits, an old friend and Upstanding Member of the Nigerian civic community in Japan has converted the front office into a visa brokerage. While Prosper lingers, a nineteen-year-old spends his savings (earned in a slaughterhouse) to acquire the Upstanding Member's sponsorship for a short-term visa to Japan. The Upstanding Member tells the youth about places to buy car parts and admonishes him not to overstay his visa. This is for my benefit, not Prosper's, and the youth doesn't make any effort to conceal his confusion. "The journalist doesn't care if you're selling visas," Prosper says, in Igbo, and after repeated reassurances the Upstanding Member gives his young client some halfhearted, half-accurate advice about applying for asylum.

Prosper visits a far corner of the market, where another friend, Austin, has rented the I Am Saved Warehouse from its owner on a turnkey basis. He runs it with his Nigerian wife (his wife in Japan knows; she doesn't mind—he claims). Really, the Nigerian wife runs it. When Prosper arrives, Austin is browsing a home appliance catalog. He wants to build a house in his village, but business keeps him in Japan half the year, Ladipo the rest. He sends for a housepainter, and a university student shows up with color cards, grateful for the interview, or otherwise very practiced in displays of clumsy, stuttering humility. Austin says, "You have big ladder? Master bedroom ceiling, maybe you can't reach. Let me see red—no, I want green. You will paint six rooms, seven rooms, it's all right? How it's better for the rooms of my children? I should use same colors? My children are many—up to twelve." He doesn't mention that the house he's describing is theoretical. And eleven of the kids.

He is waiting for a promised visit from an aggrieved customer who bought an engine in the warehouse two days earlier and paid a dolly boy to deliver it. The dolly boy disappeared.

Austin is six feet three inches, 240 pounds, and carries his size convincingly, but when the customer arrives with three friends Austin chooses courtesy, asking his other customers to vacate the chairs near his desk so the complainants can sit. He turns the TV-VCR to face them, wagging a finger at his own cashier, whose Nollywood habit he is interrupting. He sends an employee to start the warehouse's larger generator, then activates an air conditioner, directing the cool air at his visitors. The commotion of deliveries and bargaining persists nearby, but the area around the desk and chairs begins to feel serene in comparison.

When the group demands a refund, Austin resorts to realism: He will never recover his money if he cannot find the thief, and he cannot find the thief without enlisting the police, who will treat the crime as an opportunity to identify a petty violation in his warehouse and propose an outrageous fine. If the complainants wish to contact the authorities, they may. He suspects, however, that they have not filed a report because their fears resemble his.

Disagreement follows, in the form of voluble appeals to the unspoken rules of Ladipo Market, on which point Austin is a strict constructionist: He signed a lease in Ladipo, not a charter. He allows the group to form a semicircle of shouted abuse around his desk. He remains seated, responding occasionally: "I hear you," or "I see what you say." Gradually, he enters the moment's aggressive atmosphere, his voice rising in increments, until he is shouting, too.

In the crescendo, Austin grabs the largest of his guests by the shoulder. The man swats Austin's hand away and the possibility of violence settles around the front of the warehouse. The customers detect it. They modulate their movements, gestures, voices. The cashier assesses the moment with a glance, then turns the TV back in her direction.

The rest is premeditated. Austin summons his wife. "The employment records," he says, not in a tone that would suggest they're married or she manages the warehouse.

"They are confidential," she says. "You have said it."

"Bring am!"

He's overselling, manic. She's underselling—mild amusement, half-concealed. None of their guests are watching for authenticity. They notice only a woman talking back, and they're as startled as Austin pretends to be.

"Second drawer from top, left side," she says.

Austin withdraws a stack of papers, each with an employee's particulars and photo on it. He offers the stack to his visitors. They can identify the thief and pay him a visit, he proposes. "Unless," he says, and can't help recovering his slick self-possession, "he is really one of your boys, and you are wanting to cheat me." The suggestion is not far-fetched, that they paid the dolly boy to disappear so they could claim the engine was never delivered.

With the papers in hand, having been led to believe that Austin put himself at risk by handing them over, the group leaves. Austin and his wife share a smile across the desk, and Austin reaches for her hand but catches only her finger before she withdraws it. "You are a rude man," she jokes. He says, "God will punish me." She gives a dramatic shrug, throwing her arms wide and spinning on the ball of her foot so her back is to him, then moves to the warehouse's entrance, where she accosts a wandering orange vendor, demanding free samples if the woman wants to sell inside.

Prosper descends from the pile of fenders where he observed the argument. "Soviet Union of capitalism," he declares, a turn of phrase he invented before I met him. He means that daily life in Nigeria, where everything and everyone can be purchased, requires a nimble wit, a thousand-yard stare, and sufficient clarity to discern fate in its natural form: court jester. "In Japan," he says, sitting now in the chair recently vacated by Austin's dissatisfied customer, addressing the younger warehouse employees, "any regular problem, whether mind or body, they will prefer to take medicine. Any Japanese who find small weakness inside must give it chieftaincy title. It's better you learn Ladipo way."

He clasps hands with Austin. "The real life," he concludes. "Watch this your *oga*. He knows how to play."

And must play. Most people in Ladipo work jobs that will leave them sapped or maimed if they don't find a way out. The dolly boy made his attempt. Austin's position further up the food chain hardly exempts him. He risked violence today, and now turns to the next problem on his list: His landlord hasn't provided rent receipts for the last three months, which puts him at risk of summary eviction if a prospective tenant with deeper pockets appears.

In the evening, Prosper visits the restaurant Austin recently opened in Oshodi. The two of them eat outside, on a patch of weed-covered sand near the intersection of two busy unpaved streets. "My Lagos lawn," Austin calls it. They know each other from the days when town unionism in Japan's Nigerian community reached its early apogee and Austin was the union officer responsible for making joiners out of skeptics like Prosper. The union "overheated" (the word its members use for the period when it grew too fast, into a mess of petty grievances), and Austin dropped out for a while. When Prosper heard of him next, it was their mutual friends talking about a container deal, where the other partners had paid their share and Austin backed out, compromising the livelihoods of two honest men with stable, sincere marriages to Japanese women, and well-adjusted children studying at Japanese universities. What happened to these men, Prosper never learned. They don't live in Japan anymore, and for a time neither did Austin, who was deported after his former partners reported him to the Immigration Bureau.

Of all the Nigerians Prosper knows, Austin had been the last to reenter Japan on a bought passport before fingerprinting started. Afterward, he was overcome by superstition; people who encountered him reported that he was always apologizing for old slights, as if any ill will toward him would reverse his narrow luck. Prosper didn't know Austin during the years he'd contended with his enlarged sense of peril, only in his expansive, politicking, entrepreneurial years, which he eventually resumed. Recently, in Japan, Austin goes to wake-keeping ceremonies for the deceased relations of union members and delivers comic monologues on the theme of mortality, using a stage name that winks at his past immigration difficulties: Last of My Kind.

As Prosper recalls, the only person to precede Austin in coming home to open a restaurant lost his business in a fire. The police suspected arson, but no witnesses came forward, just a cattle herder who said when his animals smelled the smoke, they acted the way they do when they're downwind of a gas station.

Prosper thinks: The pieces never form anything except the question of what they form; that's the trouble with Nigeria. He watches the steam rise from the food and evaporate in the torchlight that illuminates the neighborhood, which rarely gets power from the grid. It's simple food, cooked skillfully. The service is quick and solicitous. In the beginning, Austin was a short-order cook at a chain restaurant in Japan, a fact Prosper remembers him advertising, without irony, as an extraordinary culinary achievement: Who needs a Michelin star when you're working for a franchise owner who takes in tens of thousands of dollars a day at seven locations? The world votes with its wallets. In Oshodi, the logic works. Four months in, the restaurant makes enough money that it wouldn't ruin Austin to lose the warehouse.

"If I'm you, I'm beginning to feel Nigeria is really for me," Prosper says, tilting his beer in the direction of the restaurant's entrance. "How much time you spend on Naija side?"

Half of each year, Austin concedes, if the year is lucrative. The real money is in the spare parts business, which requires him to be in Japan as often as he's in Nigeria. So does his visa marriage, so does his active participation in the town union. "This my Ghana brother in Roppongi one night says I have bring this on myself, to have two lives. I tell him: Nobody have two. I have one-half is here, one-half is there, anywhere I'm living, it's half, not two."

"Some people will rather believe whatever they can touch, they can touch it tomorrow," Prosper says. "But the travel is too much, stolen parts for car business is too much, and the two marriage." It's a home truth, and requires no answer. Prosper takes another pull on his beer, then adds, "Even Ladipo, too much. Same like Japan, just build everything in the place where is too much stress, whether earthquake or Nigerian kind of problem, and when trouble come to finish everything, many people will be surprised—Ah, neither this one is forever?"

"How long some people are working Ladipo? Twenty-five years?" Austin says. "My brother: Japan forever no be same forever like Naija forever."

Prosper squints in the dark to discern the shapes of people crossing the street on the far side of the intersection. He catches himself wondering what happened to the streetlights—There were never any, not here. The tapering, tidal sound of people's voices comes to him reflected off the walls of the buildings nearby, giving him the disoriented sense of being addressed by someone at the far end of a corridor. In Japan, it's morning. He would be sleeping. He hardly gets jet-lagged; going from nocturnal to diurnal cancels most of the time difference. Sunlight makes the days seem longer, makes him sleep more deeply and wake up wondering what day it is. He tells himself that when his life settles and he stops traveling, he'll remember the difference between the way time passes in one place and another, the sense of how it slips around you, awake or asleep, in the shape and the length of a night.

11.

"You said *limited* English?" the principal of the international school asks, initialing a document and moving the pile aside to indicate her full attention; she has overcome her earlier confusion about Prosper's walk-in admissions inquiry. Her office occupies the entirety of a modular building (in America, it would be called a trailer) that has been grafted onto the roof of the school's dining hall. The sole window faces the schoolyard, where the lower grades are in recess. The University of Lagos is nearby. Professors send their children here. Prosper knows the school because Nigerians in Japan mention it when they discuss persuading their wives to move the family to Nigeria. He can guess what those women would make of the principal's office; they'd see something like a prison watchtower, overlooking a garbage-strewn schoolyard where the students have their shirts pulled over their noses to keep the dust out. Their husbands would notice how the children are ready with a broom when the grit and trash blow in from the neighboring market. They'd like that, the discipline of it.

"Not unusual," the principal continues. "Everybody is bringing their children home. We have two right now who came here at twelve years old, speaking Turkish. We won't hold our breath for any literary prizes, but we accept students who show promise in other subjects."

She explains how admissions tests are graded. It has been a long time since it occurred to Prosper that at any given moment students everywhere have been confined to rooms monitored by proctors. He imagines Keiji's arguments with Yumiko about studying, Keiji sullen in his room afterward. But he hasn't seen Keiji's room. The room he's picturing—the adolescent clutter, the metal folding chair tipped against the wall, and the sun coming through a skylight above the bed—he doesn't know where the image comes from.

"I would love to see the school," he says. He has interrupted the principal.

Confusion stiffens her expression but quickly becomes patience, a mood she seems accustomed to adopting. "A tour," she says, and gestures toward the door.

It rained yesterday, and swarms of winged ants emerged in the evening to take their nuptial flight. In the high school annex it's lunchtime, and students are using brooms to clear away the bugs, which were drawn inside by lights the custodian left on. As Prosper and the principal enter from the connecting hallway, two girls approach them to report a classmate's hijinks with a dustpan of dead ants. One is bi-ethnic. Prosper asks where her parents are from. She says, "Sir, I've been living in Nigeria," in a rhythm so formal it seems to mock itself. "I was born in Germany."

"Say something in German for our guest," the principal suggests.

"In der Not frisst der Teufel Ameisen."

"So?" the principal says.

"You won't starve if you can eat ants," the girl says. She shows Prosper her thumb and forefinger, pinched together as if there's an insect between them. Then, noticing the principal's raised eyebrow: "It's not rude! They say such things in Germany."

"It's difficult to school here when you never knew English?" Prosper asks.

"I was always speaking English with my father. In Germany everyone can speak it, in a German voice."

"You have ants, too, like this one, in Germany?" Prosper asks.

"Ants," she says. The question appears to disorient her, and her formal cadence is absent from her answer. "I don't know if it's same."

"Ask Mr. Omoregie. He will know," the principal says, and dismisses the girls. She takes Prosper to meet a math teacher, then an English teacher. The English teacher has a thin gray goatee, meticulously clipped; without getting up, he makes a gesture of welcome, moving the sleeves of his embroidered *agbada* in a casual, grandiloquent way, like a judge in his robe. The principal introduces Prosper, and the teacher incants a few words about the challenges faced by children raised in languages that don't have any linguistic forebears in common with English. "These are the children of Nigerians," he says. "They are multilingual; it's in them. But if you catch a bird too young and later you want it to sing—" He blinks, as if he's surprised the thought came out intact. The principal forces a smile and looks away, to the airbrush paintings of Nigerian novelists hanging on the far wall.

"I believe you," Prosper says. "The same how I feel, going to Japan."

"Exciting," the teacher says, making an effort, glad the moment has passed. He rubs his hands together and makes his eyes wide. "Difficult—but exciting!"

The principal sees Prosper out. The parking space for visitors is empty, and she invites him to wait for his driver in her office. When he says he'll go the same way he came, on an *okada* from the junction, she raises her index finger as if he has said something in need of correction, then seems uncertain how to complete the gesture. Prosper says, "It's a good school." He makes his voice deep and formal, as if concluding a matter of intense mutual interest. He gives her a slight bow, out of habit, and finds himself staring at her hand, which she has offered in the meantime.

Two blocks of market separate the school from the junction. Prosper takes his time walking. He stops to buy grilled corn and lingers long enough to witness a disturbance. Shouting is audible, an argument over a price, then a young woman exits a market stall. She is wrenching a bag from an older man's hand. When she has it, she runs. The man retrieves an item that dropped from the bag and throws it—hard. It hits the girl's legs and she stumbles into a cookstove, collapsing it. Before a crowd begins to close around the girl, Prosper sees her lying facedown, coals from the stove scattered on her back and legs.

When Prosper arrives at the junction, the *okada* boys are already discussing what happened. One claims it was a tuber, thrown in an overhand style, that knocked the girl over. He pantomimes the throw and the fall, sprawling himself over the handlebars of his bike and remaining prostrate while the rest laugh and supply their own punch lines: "One time *yam* pound *you!*"

Prosper listens to this, then walks off, not in the direction of the taxis and motor trikes and minibuses, not in any particular direction. He goes beyond the junction and down a hill, where traffic is backing up to cross the sliver of roadway that isn't flooded from last night's rain. Students walk past him, leaving the university campus. They carry books, backpacks. Their clothes are new, or look it. They walk in groups. Here and there a couple holds hands.

"UniLag?" he asks a young man walking by, who nods and keeps going. "I am happy for him," Prosper says. He looks in the direction he came from. "I need *okada*." He watches while a taxi tries the flooded part of the road, but the water nearly reaches the window and the street is filled with the grinding sound of the driver slamming into reverse. "It's not for laughing about," he says. "If price of yam is high and the money of the girl is small. Now she becomes harmed." He sips from a bottle of water to rinse the city dust off his teeth, and spits. "I saw, but I never feel like laughing."

He goes to Benin City, for no reason known to him except that his time is his own until he returns to Tokyo. He considers visiting the neighborhood where he grew up, but it has changed too much to recognize; that was already true the last time he saw it. He wanders past the Edo State government buildings, to the city's central roundabout, through a few markets nearby. Sapele Road is the traffic artery in this part of the city, and he walks along it. Someone he'd known in Japan reportedly died on Sapele Road, but he isn't sure where. The man—people called him Lotto—was a subject of constant gossip in Japan's Nigerian community. People admired him for his choice of vocation, sleeping with Japanese women so his Japanese wife could extort restitution payments from them, or they despised him for his other vocation, which he never bragged about: claiming he had registered as a police informant, blackmailing anyone whose illicit activities or expired visas became known to him.

In the tin-roofed restaurant on Sapele Road where Prosper sits in the breeze from an oscillating fan while he eats his lunch, an off-duty police officer is drinking alone, so Prosper asks if he remembers the accident. Lotto was in a sports car, Prosper explains, and had four or five women with him. He was driving drunk and hit the back of a lumber truck. The lumber went through the windshield and Lotto died on impact, but the women walked away. Prosper explains to the officer what kind of life Lotto led in Japan, and says, "God reach down with pinkie finger."

The officer says he doesn't remember an accident like that and asks how much someone gets paid for having sex with white women in Japan, or in a regular job if that's all they can find. He adds something in Bini to the owner of the restaurant (who's in the kitchen) and leaves. The owner emerges, looks at Prosper, and says, "You don't speak Bini." Prosper replies that he speaks enough—a long time ago he was from here; then, as now, when a police officer says you're paying for his beer it may be a surprise, but it's not a lie. She brings him a drink of his own and dismisses his attempt to pay. "If you are really coming home," she says, "I will be selling you more."

He is not the first returnee she has served. Before Nigerians emigrated in large numbers, it was Bini people, Esan people, and members of other minority tribes from present-day Edo State who established an immigration pipeline. Benin City hosts the largest civic association for Africans who formerly lived in Japan. Most members lost contact with their children when their Japanese wives left them; any attempt to seek legal assistance would have struck their generation of immigrants as a surrender to fantasy. When their compatriots died in Japan, the remains were never repatriated; there weren't any town unions yet to raise the funds. Many in their cohort died shortly after returning. In Ghana as well as Nigeria, mutual assistance associations for returnees have observed that members who lived in Japan die young, and it's usually their hearts, presumably weakened by stress, that kill them.

Prosper means to leave Benin City on the first bus tomorrow, but learns that a former nightlife colleague has come home to bury his mother. On the way to the funeral the next morning Prosper boards the wrong bus and spends an hour in traffic before realizing his mistake. He disembarks and hires a driver, who calls Prosper's former colleague several times to consult about directions. He uses Prosper's phone, which is a cheap Nokia sold in developing countries. It doesn't have a volume control and it's not loud enough for outdoor use, so people tend to hold it backward, letting the speakerphone play into their ear. In this way, while he rides in the taxi, Prosper hears the reception in the background: Women wail in ritual lamentation, and the invited priest hyperventilates into a microphone while he anoints the surviving members of the family. The last time the driver calls, forty minutes before they reach the house, Prosper's colleague picks up, but says only, "Please one moment," and the cab fills with the sound of the reception's closing hymn until the line cuts.

When Prosper arrives, the man who comes to the door is the secretary of the returnee association. Prosper knew him once, in Japan. The man isn't sure whom he's looking at—his face shows it—then he says, "My God. It's you," and Prosper laughs because there's no reason it would be significant to see him, except that decades have passed. The man looks over his shoulder, then at Prosper. He's trying to decide whether it's all right to send Prosper in like this, when the reception has ended and the family members have resumed their private grief. He says (and you can see he is truly sorry), "It already happened."

Prosper travels east before he returns to Lagos, taking money and gifts to Uche's wife and children, who are living with Uche's parents in the village. Recalling his visit, Prosper says:

"The wife is not knowing Uche's situation, for jail and money. She know her husband is married in Japan. She understand he can keep his visa only if the marriage can maintain. She will easily request from me, with children in same room, if I can advise Uche to keep the Japanese wife. My thinking is how it can affect the mind of the children to hear their parents say any dishonest thing, where other human being is opportunity for money and visa. The smallest boy is two years, by now he is three. He has big eye for world. Like sponge take water, the boy will take.

"For the children and the wife it is only two rooms in that house. The parents of Uche are even sick. The mother—her mind is not good, and father have problem with his eyes and with knee for walking. Uche's wife may need to provide assistance, though she is occupied with the schooling and the needs from her children. You can see the place where she live in that compound have become mess because she don't have time for cleaning. The children have even say they don't like to go inside that dirty room again. They don't know they are the ones who have dirty it. Before, when they have their apartment in Lagos, their cousins will just come to Lagos and clean anything if Uche's wife allow them to sleep in that apartment.

"If you have small money from Japan you can make people in Nigeria to believe you are doing very fine, even if, for true, you can only put hand to mouth. Now she don't go back to her family's compound, where they have different idea, how she is having comfortable lifestyle. She will prefer to stay in the husband's village even if the husband's parents are sick. Maybe there is story she have explained to her own relatives to be jealous of this Japan money, so it will be bad for her mental if they see she have fallen on hard time.

"She just feel like there is nothing she can do. I can tell her, it's better if you sweep the floor or take the rubbish from this room where your children are sleeping, but even if she do it, she will still be feeling helpless. She will permit the rubbish to come again.

"Because she have difficult life, she is type of person who don't know—even if her own husband say, 'See this my friend'—whether she will trust you. If she believe you are sincere, she may flip to other side, she may like to chop money from you. In my own case she can see I am just regular person who have known her husband from back those days. How can I put it?—she lose interest in me, just tell me I should make myself at home. So it was like that, playing with the children and small gift I am giving them. Even walk around village for one hour without anybody accompany me.

"When I am preparing to leave, the younger of the daughters is following me all the time, watching with her eye. When one young cousin take me to road to wait for driver, she is following. Before I'm entering car, I just say to her, 'Is there any other thing?'

"She just say, 'Who are you?'

"I'm surprised. I was thinking the mother have explained. I don't want to mention I'm knowing the father. The little girl may begin to have question, but I don't know how the mother is answering the same one. I just tell my name, say I will look forward to see her again. I depart for long ride, to city where I will meet bus, and inside this car I'm thinking about it, how I don't even know while I am in that house, the way people are seeing me—Who am I? I can be any other stranger."

—

Prosper returns to Lagos the day before his flight and checks into a hotel in Festac. Coup visits in the early evening and buys a few rounds in the hotel bar to celebrate the conclusion of a profitable trip. The hotel's chef hasn't shown up, so they walk to Apple Junction, where the roads linking Festac to the rest of Lagos intersect, and they buy carryout from Mega Chicken, the disarmingly suburban focal point of foot traffic, transit hire, and neighborhood gossip. It's past sundown now, and Festac's bars are open. The area in the direction of the hotel is only halfway developed, and nightspots along the road make their money serving prostitutes and johns. Prosper and Coup are nearly back to the hotel with their food when a candid proposition is put to them by a young woman who is waiting for her *suya* at an outdoor grill.

Prosper has been drinking since the afternoon, and it's the end of his trip; like the last patrol in a combat zone before a tour of duty ends, the last night in Nigeria has a famous way—Nigerians who live abroad would explain—of letting you feel like you're already back in an easier country, then reminding you that you're not. Prosper stops when the prostitute calls out to him. He says he doesn't have any money, but does she want some of his food?

She's in his face by the time he finishes saying it. He's gesturing with the bag of food in order to complete the offer, and she bats it out of his hand; it drops into the open sewer running parallel to the sidewalk. She is not homeless, she informs him—at considerable volume. She has paid for her dinner and she's waiting for it, and she doesn't like being embarrassed in public. Her shouting has caused other women to lean out of the windows of a nearby brothel. They're shouting, too. And men's voices, from the same windows. Someone promises he'll be down shortly, to help Prosper and Coup with their etiquette.

They start to run, but it's a half run because the few blocks back to the hotel are poorly lit, presenting several opportunities to fall into the sewer. The scuff of shoes on the sidewalk bounces off walls, off passing cars, and mingles with the voice-and-traffic noise of Festac at night; the sound suggests they're being pursued, but if they weren't, it would sound the same.

They reach the hotel, and half a minute later the prostitute Prosper offended, with two girlfriends and one gentleman associate, is in the lobby raising complaint. Apparently the prostitutes use the hotel; they're acquainted with the manager. Coup, too diminutive and unkempt to provoke much anger, negotiates a settlement while Prosper waits out of sight, in the hotel bar's bathroom. Coup masters the confrontation with gentle, prodding intensity, and anyone walking in as the woman and her associates walk out would think they're watching old friends part company. Sharing Coup's half of the food—which Coup tucked against his chest like a fullback while they ran—Prosper talks about the girl who stole from the market near the international school, and watching her fall into the hot stove, but trails off partway into the story when he sees Coup can't understand why anyone would bother telling it.

Vitus, a friend from Roppongi who has reinvented himself as a Lagos hotelier (but still spends half the year in Japan with his wife and young daughter), drives them to the airport the next morning. Prosper shares the back seat with an ice machine Vitus has purchased to install at his hotel's restaurant. "Fuck," Vitus says, as they approach the military checkpoint that precedes the airport loop. "Ice machine looks like bomb, or can have big bomb inside." Coup leans forward, trying to make out whether the soldiers, armed with automatic rifles, are indicating that Vitus should stop the car.

"Papers?" Coup says. A receipt might help, something that says ice machine.

"Ladipo," Vitus says. He means: Of course not; could be stolen, could be contraband. Could have the Lindbergh baby inside.

Prosper looks over his shoulder at the machine, which is roughly six feet long, maybe four wide, three deep. He unhooks his seat belt, stretches his arms out, and embraces the machine, covering it like an octopus—covering a third of it, maybe. Coup laughs so hard he has to lean on the dashboard. Vitus, too, loses his composure and rolls up his window so the soldiers won't hear him laughing as he pulls over.

But he doesn't pull over, he reduces speed and keeps going. The soldiers don't wave him on, but they don't flag him down either. Two of the soldiers are asleep; the third is watching a portable television. Inside the car, everyone has stifled their laughter for the moment it took to drive by, and it stays like this until they hit the traffic jam that leads to the terminal. *"Kaboom,"* Prosper says, and the laughter starts again until the car rocks on its tires. Coup explains that

Prosper provoked a foot chase down Festac Link Road the previous night by offering to donate his dinner to a prostitute. More laughter, more stories exchanged in a similar spirit of mirth, then the laughter subsides and leaves behind an interrupted feeling. The pattern of the trip—or so it seems to Prosper—repeats itself as the fog of their amusement lifts and their surroundings reappear: the traffic, the poverty, the institutional dysfunction perpetually signified. And the great uncertainty of their respective futures, the illicit and semi-licit nature of what's near enough to reach, the leap of thought and faith required to get from here to wherever each of them intends to go.

A girl begging for money—maybe seven years old, and not as tall as the hood of Vitus' Land Cruiser—approaches the car next to theirs and takes the passenger's arm through his open window. The passenger waves her off. She doesn't let go, so he signals the driver and the car starts to move but stalls right away, leaping forward, and the sudden movement tosses the girl to the ground. She gets up, wipes the dust from her face, bats it from her hair, spits to get it out of her mouth. The passenger puts his hand out the window now. Without speaking or looking at her, he presses his palm into hers. She walks farther down the line of traffic, smoothing the twenty naira note— five and a half cents—against her thigh.

Prosper, Coup, and Vitus are quiet now.

When they reach the terminal and they're taking their luggage out of the car, Vitus says, "See you next week."

"You come so soon?" Coup says.

Vitus shrugs. "Next week is soon?"

Nobody answers. When Prosper thumps the roof of the Land Cruiser to signal that all the luggage is out, Vitus jumps in his seat, as if he had begun to daydream. Coup recites an Igbo proverb: If time flies and money talks, why would anyone shut up or stand still? "Is that what we are doing?" Prosper asks. Coup says, "Look around." He's right. In the kinetic setting of the airport, someone could mistake them for the only people with nowhere in particular to be.

12.

A week after they returned to Japan, Coup called Prosper at 5:00 a.m. on his day off, waking him. Uche had been arrested. Prosper wasn't fully awake when Coup explained. He'd been sweating under his sheet. His throat was coated and sour, the way it feels when he loses a day to exhaustion.

He made Coup explain again: Uche had been detained by plain-clothes detectives outside his church. They overheard him talking to another congregant, also an asylum applicant, about the risks of working while you're paroled. "He never argue with police to claim they don't hear him properly," Coup said. "Immediately he take bank card from wallet and give to his friend, say this is my secret number, please give to Prosper so he can send remaining money to my family. They just call me to bring bank card for you, this is how I'm learning whole story."

Prosper hung up. He filled a glass from the tap and put a packet of chocolate cookies—the only food in the apartment—on a plate, but left the plate and the glass on the refrigerator and went outside in his slippers, sweatshirt, and boxer shorts. He walked until he reached the nearest major intersection, where a pachinko parlor was still open, leaking fluorescent light through the floor-to-ceiling decals covering its glass facade. Prosper went inside, into the cascade of sense-dulling noise from the machines. A young woman in a pink cleaning service uniform was running a floor buffer up the row of vacant machines in front of him, focusing on her work, on keeping the electrical cord out from under the buffer. The din persisted, and the fluorescent glare, and for a moment the only motion that didn't occur in a dull, repetitive pattern came from the floor buffer, rumbling in place, which occasionally leaped an inch to either side, as if its contents had reached a boil.

The woman flicked a switch on the buffer to shut it off. She left the buffer in the aisle and disappeared into the rear of the parlor. Prosper exited through the front doors. The noise followed him into the street, then the doors slid shut and it was quiet outside except for the distant, canned-air noise of an elevated highway. A stripe of early twilight had appeared below the skyline. He watched a man leave an apartment building across the street and dump a plastic bag full of loose batteries in the garbage bin belonging to a convenience store. Prosper lifted a hand and said, *"Sumimasen!"* but his voice was hoarse and the sound didn't carry.

He walked home and lay in bed. He was certain he would wake up famished and short of breath, and these sensations would persist until he slept again. He was periodically incapacitated by fatigue but could never predict or avoid it. He fell asleep wondering when it had become inevitable he would feel this way, and how he could have known.

13.

Prosper exited Shinagawa Station through the ticket gate, bound for the Immigration Bureau, then paused to observe the station's glass-walled information booth, where a South Asian family had stopped to ask directions. With an eye on the line forming behind them, the young woman working the counter stood on her stool and leaned over the family, making slow, exaggerated gestures. The family left the booth, checked the signage nearby, then chose the exit that leads away from the bureau. Prosper rushed down the stairs to catch them coming off the escalator, but they never appeared. Walking back the other way, he took an indirect route and searched for them on adjacent streets. He talked about the problems it would cause them if they missed their visa appointment, and never mentioned the possibility that the bureau was not their destination.

The bureau's lobby was the same as Prosper remembered it, dominated by a convenience store with an attached dining room where, on a busy day (most days were busy), immigrants would fill the tables, sitting with their lawyers, advisers, and relatives. Prosper had wanted to visit Uche but couldn't make it past the cheerful colors of the convenience store's overhead sign, the sleepwalking expressions on the cashiers' faces while they rang up the purchases, the people in the dining room sitting with their heads bowed—to finish the paperwork and affix the revenue stamps, to pray over the final item before submitting it. The other times he'd been here, the mood seemed different, as if these were people together on a long journey, bearing it. Standing there without the will to cross the lobby and board the elevator, he had an image, and it didn't feel voluntary, of forcing open a cash register and stuffing the money in the cashiers' mouths.

So he left.

Uche called the next day. He wanted Prosper to bargain with Junko, his Japanese wife. If she agreed to petition for his release, he would sign divorce papers so she could claim welfare. Mention the divorce first, Uche advised Prosper, and leave the release arrangements for later.

When Prosper called Junko, she didn't believe him about the divorce, which she said Uche was always promising. Prosper claimed that Uche had given him the papers, already signed. She wanted proof. He made excuses. She struggled with English and he struggled with Japanese.

He started trying to reach her at moments when she might be tired enough to relent, around the time she put her children to sleep. She talked exclusively about Uche's dishonesty when they were living together. Prosper thought he was likely the first person to hear most of it. He could tell it helped that he was Nigerian—made it possible for her to revisit the intensity of her resentments. He was pained by her complaints about money, where Uche had been most neglectful of his children. Junko had paid their rent and expenses, borrowing from her parents to do it. Uche had pleaded poverty, saying the nightclub where he worked had stopped opening on weeknights, or he'd been fired, or his boss cheated him on his paycheck. She knew he was lying, but didn't know how much he earned. Once, she found a heavy gold chain in a pair of Uche's socks and had it appraised. Everything they'd bought for their children that month, her parents had paid for, she said, until she pawned the necklace. Uche knocked over a dresser when he learned it was gone, and the edge of the dresser clipped their son's forehead, requiring stitches.

On another call she talked about learning Uche was married in Nigeria. After coming back from one of his trips home (for a funeral, he'd said), he left his Nigerian phone on the nightstand, and she tried to use it. The phone's wallpaper was a picture of Uche with his other family. This happened at night. She had put the children to bed. He was in the shower, getting ready for work. When the water stopped running, she replaced the phone and pretended to sleep. He got in bed naked and put his hands under her clothes. Telling Prosper about it, she yelled into the phone. She demanded to know what she should have done when Uche undressed her; she thought about hitting him with something heavy and running out of the apartment with the children. Instead, she pulled a pillow against her face and breathed through the corners of her mouth so he wouldn't notice she was crying.

Prosper put his phone down and waited for the sound of her voice to stop. He didn't want to hear the rest of it.

After talking to Junko, Prosper would call Yumiko. They would discuss Keiji, Prosper always intending to set a date to visit Sendai, then hanging up without having done it. If they talked for a minute, the minute dragged. If they talked for an hour, it didn't seem longer than the minute did.

When they weren't talking about Keiji, she would apologize for what happened when they returned from Nigeria, for allowing Prosper to wonder why she left the marriage, an experience that touched the wounds of his childhood. "I feel closer to you than when we were married," she said. "Time helps us understand." He gave part of himself to these gestures; he could remember wanting them once and never stopped wanting them altogether. Another part of him noticed how conclusive her statements had become, in a conversation—about their long past—they'd barely started.

On the nights he went to work, which decreased after Uche was arrested, he rarely caught a customer. If he did, it was an old regular who clapped a hand on his shoulder and startled him out of a standing daze. Every night, he thought he saw the woman he was with during the earthquake. He was working freelance, in both red-light districts. He couldn't have seen her as often as he claimed unless she was following him.

He kept calling Junko, who no longer spoke solely about her grievances. Sometimes she described her day or disclosed a meaningful memory. In the factory town where she grew up, she told him, her father was a foreman. If one of the factory's Korean laborers became involved with a Japanese woman, it was her father's responsibility to inform the laborer that his livelihood and the romance weren't compatible. Her mother would visit the woman at home. Occasionally, Junko accompanied her. The visits were short, the women's families invariably poor. "They haven't the sense to live comfortably," her mother explained. After one visit, Junko repeated these words to her father. He called Junko's mother into the room and slapped her while Junko watched—as far as Junko knew, his only violent moment.

On a call with Yumiko one night, Prosper talked about Uche's oldest sister, who was hit by a car in Lagos, then burned to death in her apartment the day after her release from the hospital. She'd left an electric kettle plugged in long enough to boil down the water and start a fire. Investigators who examined the apartment speculated that the hospital had been too quick to discharge her; she could have been concussed, the hospital didn't catch it, and she fell into a coma after putting the kettle on.

When he finished telling the story, Prosper was struck by a gust of panic, thinking he'd related it to Junko, then remembered he'd ended that call.

Junko no longer said they couldn't meet. She said it would be difficult to find time away from work and the children. She started asking favors: someone to watch the kids occasionally, so Prosper called a former girlfriend; a place to leave her belongings for a few days while she moved, so he introduced her to a friend who owned a scrapyard near her new apartment. She asked if he had time to contact day-care centers about her four-year-old son. Prosper reminded her that an inquiry from a Nigerian would make day-care centers nervous, and he couldn't speak Japanese fluently enough to reassure them. She said, "When you know someone, you forget what language you're speaking."

She told Prosper there had been a time after Uche left when she could look at her children and feel they weren't hers. The former girlfriend he had called to help with childcare told him Junko teased the children, especially her son. She'd pinch him, or slap his wrists, hard enough it plainly hurt. Junko worked fifty hours a week, but only twenty-five were paid. She was tired. Her children were restless. Once, her son asked Prosper's ex for a hammer so he could "fix the toilet." The toilet wasn't broken, she pointed out, and a hammer wasn't the proper tool. In response, his expression was alarmingly adult, as if he'd been asked a question that missed the obvious point, and he said, "It's just something to do."

After three weeks of leaving his apartment only a handful of times to work, Prosper rode the train to Chiba, where Junko was living, to meet her at a fast-food restaurant. The restaurant had been painted recently, and the chemical odor stung his sinuses. It wasn't busy. Except Prosper and Junko, only the employees behind the counter were present, and a salaryman eating his hamburger in a steady, gulping motion without looking at his hands.

"I don't eat this," she said, pointing to the meal on her tray.

"Should I?" he said, and sat.

"If you want. It felt wrong to sit without ordering."

She was older than Prosper had expected. If her daughter was eight, she would have been born when Junko was nearly forty, the son when she was closer to forty-five. She wasn't trying to hide her exhaustion. She leaned on her elbows, blinking slowly to steal a little rest. She was attractive, moderately made up. She wore expensive clothes but didn't seem comfortable in them. Prosper knew she was from Tottori, and recognized in her a studied self-possession he associated with women who came to Tokyo later in life.

He watched her fill out her half of a cooperative divorce agreement; he didn't bother lying again about already having a copy with Uche's signature on it. She asked when he would see Uche next. He demurred by saying Uche was in Ushiku now, and he'd have to take a day off work to visit him. "You could help him," he said. "In order to help yourself."

"Help him?"

"To leave detention."

"I didn't know it was possible."

"If you made a reasonable explanation to the Immigration Bureau about why you never signed his visa renewal, then you sign a new request."

"With a new wife it won't work?"

"It has to be you."

"But if we get divorced and he marries someone new—"

"No. They'll hold him a long time."

"But they would let him go in the end?"

"They can't detain him forever."

"How long, minimum?"

"If you help him, he could get out soon."

"If I *don't* help him—if he doesn't have his legal wife helping him," she said, impatient. "What's the minimum?" She held up a finger, then two. Years, Prosper supposed. Her attitude toward Uche hadn't changed. Only toward him. She thought he was neutral, or closer to it. She trusted him to provide information she wouldn't know how to get.

"No minimum," he said.

"No minimum," she repeated, and looked at her hand, where the years were still counted.

"Maybe you can keep him inside, if that's what you want. He can prevent you from getting government benefits, or remarrying. If you help, it might be two months. Then you get divorced and leave your job, take care of the children."

"The job is okay."

"He had a difficult life. He's a human being. Do what he wants."

She squinted, and her expression matched the aggrieved voice from their phone calls. He hadn't meant to say it like that—*Do what he wants.* Until he said it, she was trying to assess how much leverage she had. Now her thoughts would be less pragmatic.

"What about the next woman he marries?" she asked.

"He hasn't met anyone."

"He won't tell her he has a family in Nigeria. Does another woman deserve that?"

"I'll make sure he tells."

"Will you?"

"I won't," Prosper said. He could at least get the other answer Uche wanted—to know what she said when the terms were explained to her. It disoriented Prosper, trying to throw the switch that would let him speak to her as himself, without the inflated sympathy their phone calls had required.

"Don't you think it's disgusting that he would lie to another woman like he lied to me?"

"No."

"If it's not disgusting, what is it?"

"Fine."

"Fine?"

"If we could go back in time, and you could live Uche's life, you wouldn't be a better person than he is." Prosper didn't believe what he said next, but wanted her to feel the pressure of what he believed, the temperature of it. "You would be worse," he said. "If your places were exchanged, he would never leave you in prison."

She stood up. She stepped around the table and stared down at him. She said, "When Uche's son is old enough, I'll tell him he was born after I was raped."

"My mother told me the same story."

She had barely finished speaking. His reply caused her to fumble her purse. "It's ridiculous if you think I believe that," she said, and continued toward the door.

When she reached it, one of the restaurant employees jogged from behind the counter and called to her. Prosper watched the employee—a college student, probably—offer to put her food in a bag. She apologized, bowed, apologized again. She gestured in Prosper's direction and explained she bought it for her friend. She was polite but animated. She was perfectly natural in these familiar gestures, a woman returning to her sense of the world, accustomed to the art of evading embarrassment.

From his meeting with Junko, Prosper went to Ueno Station and boarded a bullet train to Sendai, arriving shortly before midnight. He called Yumiko from a cab and woke her. She told him to hand the phone to the driver, and Prosper was only half-surprised when the cab arrived at a business hotel. Yumiko had phoned ahead, and Prosper found himself seated on the bed in his room, looking at his reflection in the mirror mounted on the opposite wall, startled by how little was apparent in his own unguarded expression. He waited for a knock, but there wasn't any, and after a while began to fall asleep.

Yumiko called in the morning and said she would stop by after dinner. Prosper slept until noon, then explored the surrounding neighborhood, a working-class area south of the Nanakita River. He walked past a string of auto shops, where men in coveralls sat under the car lifts, sharing lunch. He wandered through cul-de-sacs of tract housing and the courtyards of bubble-era apartment complexes. Everywhere he walked, the foot traffic was septuagenarian or older. One elderly man flagged Prosper down but told him not to come too close because he'd just received a radiation treatment. He asked

Prosper if he liked this area. Prosper said he did, and said the trees looked especially green here. The man showed Prosper a wallet-sized wedding portrait of his daughter. Prosper could hardly see it from the distance the man insisted on. She lives in Italy, the man said, and his eyes were damp. He made an arc from the river to the hillside with his finger and said, When I'm gone, take good care of this place. Prosper said he would, and when the man asked where he lived, he named the area Yumiko lived in. The man absorbed this with a gradual turn of his head, perhaps in the direction of the place Prosper had named, and Prosper noticed a tremor in his neck and shoulders, as if he was laboring to hold his head up. The man said, Sorry for saying strange things to you. And—in stammered English, which made the tremor worse, and it spread to his jaw and his lips and his cheeks while he spoke—*You are nice to acquaintance.*

Prosper told Yumiko about the man when she came to the hotel. She brought dinner and they ate in the room, out of Tupperware, sitting next to each other on the bed.

She said, Everyone his age will die soon, so he wants to tell a foreigner to take good care of Japan. It was kind of him, the way he spoke to you.

He said, Yes it was.

She said, Everyone should be friendly to foreigners. We should allow them to live here in peace.

He said, Thank you.

She said, All the Nigerians I met when we were married were very good people.

He said nothing. Many of them were not; he'd known it then, he especially knew it now. She'd been looking at herself in the wall mirror while she talked. Again, he felt she'd too much intended to say it, had brought it to the conversation from someplace else. On the phone, she'd once mentioned seeking psychotherapy for Keiji, but Keiji refused to go, so she went herself, which the therapist said was better than nothing. It disconcerted him to think of the possibility that someone had planned, as a kind of clinical measure, comments that stirred the affections and resentments he'd experienced during his recent phone calls with Yumiko, in the vacant, inward-looking hours in his apartment, in the shape-shifting anger and sorrow over Uche's arrest and Keiji's continued absence.

Prosper reflected that he had become like a character from his favorite novel, which satirized the lives of Nigerian immigrants. *For all the time he had spent in this country, Uncle Happiness had never arrived.*

Without deciding to, he lay down on the bed, resting his head in Yumiko's lap. She hardly moved, except to brace her hands against the mattress so they wouldn't brush him. He sat up right away and considered apologizing—saying he was tired or upset. She was frightened, he could tell, and he felt he didn't deserve that. They'd fought constantly before the divorce, but he'd never put a hand on her, and in physical intimacy he had never overcome the timidity of their early encounters.

She didn't leave. She explained herself, permitting Prosper the enervating experience of hearing her say what he'd guessed she was thinking. She put it in the arid language of obligation, what was owed and what wasn't, and why. She'd handled the divorce poorly because she was young, and recently she wanted to make that right, but she had already given an admirable effort, and in the future all she really wanted between them, if she could be honest, was Keiji. Regarding Keiji, she believed she was being generous, since Japanese people don't mind if a child is raised by one parent. It was a favor to Prosper, letting them know each other.

She said, We shouldn't talk, because you describe the way Uche treated his wife, then explain why Uche isn't a bad person, and it's impossible to tolerate. Of course he's a bad person. Only bad people lie like that. He deserves to be detained; isn't it fraud what he did to his wife? If you want to live peacefully in Japan, you should help the police put people like Uche in jail. Otherwise, it's completely fair for Japanese people to think Africans are criminals.

When she finished, Prosper asked if she wanted to know why he had come to Sendai. She said yes, she did. He said, It's nice of you to ask, and allowed silence to occupy the moment when he could have explained. He held a smile and would later reflect that it was one of the more difficult he could remember, from a lifetime of smiling to prove that he could.

He took the Tupperware into the bathroom and washed it in the sink. She began apologizing, and he said over the sound of the water, It's fine. It's really fine. Then he said, leaning out the door to place the containers next to her on the bed, "I'm sorry for touching you without asking permission." She nodded, but she'd turned to look out the window and he couldn't see her face to read it. This made everything suddenly impersonal, and he had an image of himself walking past an open hotel room door, seeing a scene like this, thinking of its sadness as something entirely separate from his life. He closed the door to the bathroom and ran the shower. He heard Yumiko let herself out.

Later, Prosper reflected on the moments after Yumiko left the hotel room in the way he reflects on other moments in his life that marked the end of one intention and the beginning of another: as if there was no epiphany involved, only the end of a slow starving process, of trying to live on doubt or hope alone, to avoid an answer he had learned a long time ago. "I just sit inside toilet, no light, just listen to sound of shower, praying, meditating myself. In Nigeria, we say all the time: 'It happens to everybody.' But can you imagine a Japanese man says, 'My mind was paining me, so I just try to put my head down for body of my former wife, and she behave like maybe it's not safe if she's alone with me, and say racist things about my country?' In Japan you can only say, 'It happens to foreigner,' or 'It happens to Nigerians.' Not because I am making mistake, or because anyone is bad person who prefer to be unfair to me. It happens to me because I am from another place."

14.

Four hours after Yumiko left his hotel room, Prosper was back in Tokyo. He took a change of clothes to work, and overnight he borrowed a colleague's residency card. When the sun came up, he caught the first of three trains required to reach Ushiku, where Uche had been transferred. At Ushiku Station he descended the escalator and approached the nearest bus. "To the prison?" he asked. The driver gestured for Prosper to board.

"You knew it's a prison—not an immigration center?" Prosper said.

"The passengers talking. Their faces," the driver said. He looked old enough to be near retirement. There was a small pile of loose, unsmoked cigarettes on the windowsill to his right, with the butts chewed flat. During the ride Prosper watched him adjust the rearview mirror continuously, reorienting it to keep the sun's reflection out of passengers' eyes.

Prosper sat in the back, alone except for a young Indonesian couple who appeared startled at where they'd found themselves, but said they visited the detention center every week. He knew about Uche's file (I'd told him), and his occasional remarks were colored by a revived sense that the conspiracy here, in Japan, against whoever was vulnerable enough it could find him, was a cruder device than its worst equivalents in Nigeria, where at least the consequences of interfering with it were grave enough to justify the cowardice of choosing not to. He noticed that the roads near the detention center needed resurfacing. The gates and signs inside the compound were rusted. The staff collided with visitors in the hallways, not out of disregard but because their boredom had transported them somewhere glazed and private. The guards lingered at the entrance to the waiting room until it was time to call a new number, each of them fidgeting an invisible irregularity in his uniform. Their restlessness revealed an ambivalent awareness of their own authority, Prosper thought, like children who had discovered a flattened worm or a toppled anthill in one of their footprints.

He chatted with an activist sitting near him in the waiting room and learned he had worked to free one of Prosper's friends from jail years earlier. The activist was convinced that Prosper's friend had been railroaded by an overzealous prosecutor who had since received a political appointment in the Ministry of Justice. "Actually, Ike was a drug dealer," Prosper said. The activist stammered, then described several evidentiary issues in Ike's case. Prosper listened patiently, wearing the same smile he had given the intake clerk a moment earlier, handing her his colleague's ID, which listed Prosper sixty pounds lighter than his weight and showed somebody else's face, including a scar where Prosper didn't have one. When the activist finished, Prosper assured him that the entire Nigerian community had known about the charges, and Ike had never claimed to be innocent in the presence of his countrymen. If it was any comfort, Prosper explained, Ike's criminal abilities had diminished by the time he got out, and he followed his short-lived, comically unsuccessful return to drug trafficking with a series of odd jobs that enabled him to meet his current wife, whose parents were now bankrolling his unprofitable hostess club. One of the guards called Prosper's number. Prosper stood, put a hand on the activist's shoulder, and said, *"Itterasshai."* Bon voyage.

In the visiting room, Prosper talked with Uche about his wife and children in Nigeria, living now with Uche's brother in Lagos, and how to get the children back into school with a minimum of tuition paid up front.

After lunch, Prosper changed clothes in the restroom and went back to the intake desk with his own ID, timing it so he could join the crowd who'd arrived on the early afternoon bus.

In the visiting room again, Prosper showed Uche the ID he'd used that morning, and Uche's laughter, leaning in to see the photo, fogged the glass between them.

Uche absorbed Prosper's description of the investigative shortcomings in his file with a patient nod. He asked Prosper to write a summary of the problems I had found and send it to the Immigration Bureau in a letter vowing litigation. Of the lie Junko had threatened to tell their son, Uche said, "Let's pray for her," and they bowed their heads against the glass. Uche gave a long prayer, about seeing Jesus all the time now in his dreams, or seeing his son, but not the two of them together, and asking God, as a sign, to send a dream about both of them at once.

When Uche finished, Prosper said, "If your son may come back, you think it's going to be difficult, if some time have passed?"

"I don't mind. See other way I am struggling, maybe for nothing, but son is never for nothing."

"If mother has tell some lie—"

"I can correct the lie."

"If son become like mother, because she have lie he will also make lie."

"Any reason can be for child who is lying. He can grow up honest man. *A dog in contemplation is mistaken for a sleeping dog.* My father prefer this proverb."

"My father was a clever somebody," Prosper said. There was, in his mind, the image of a compound in the village, with a traditional gatehouse, and under it the mineral-red village soil. The serenity of this scene briefly overwhelmed him, and he waited to feel the humidity he imagined there, as if it were a place he had recently visited.

"My father had business making oil," he added. "Then business smoking fish, business with yam. First in village with car. First with tin roof. War come he lose everything. When I'm small and I have small cousins, we play this Biafra currency he was keeping. Play like it's toy, tossing everywhere."

Uche studied Prosper a moment. He said, "Now he's late?"

"Late?" Prosper said. "Oh, he's late now."

"I have wanted to be a mother," Uche said. "To show how the person I am, his love is not smaller than mother's love." His shoulders lowered, and some animating rigidity went out of him. Prosper was reminded of the illustrated Bible his mother had read to him, of the tranquil expressions of the supplicants in the Gospel of Matthew, who were soothed more by the act of prayer than the miracles that followed. "No refugee, no provisional release," Uche said, and knocked on the door behind him, to alert the guard. "Just send letter, list of mistakes. God will come."

The guard opened the door on Uche's side. Uche waved to Prosper instead of touching his hand to the glass, and went out the door.

At the bus stop forty minutes early, Prosper watched a line of cars pass, moving slowly behind a tractor. A car in the line shifted to the opposing lane to overtake the queue but moved back when the cars behind it began to honk and flash their lights. Prosper said, "If you horn here, they will hear you." Then, pointing at the road, as if what he said next had something to do with the brief disturbance he'd just seen: "One part of the story, I copied it." The part about playing with his father's old Biafran banknotes. "The other part, I made it up."

The borrowed part of the story belonged to one of Uche's former employers in Roppongi, Prosper pointed out. Uche had heard it, or Coup had told Uche that Prosper never knew his father. Uche ended their visit early, Prosper guessed, to spare him further time in the detention center, which had begun to affect him in the same way it affected inmates. "Because I say lie to see comfort, which is how this place will change you."

On the way back to the station it was the same bus with the same driver, but crowded now. The other visitors to the detention center were on their phones, telling friends, family, and employers when they would be back in Tokyo. Some sent texts and emails. Others had the cheap phones available to short-term visa holders, which worked only for placing and receiving calls, so they dialed and cupped their hands over their mouths while they talked, visibly annoying the handful of Japanese passengers who had taken the bus to see Ushiku's Great Buddha, in 120 meters of bronze rising over the rice fields, its fifty-ton palms open in a gesture of peace and welcome.

Prosper was the last to disembark at the station. When he reached the front of the bus, the driver put his hand over the fare box. "It's broken," he said.

"I'm not poor," Prosper said.

"Is it the same friend?"

"Is who?"

"A month ago, maybe two. You stood. He sat there." The driver pointed at the folding seat to his left.

"Yes."

"Fast."

"I don't know," Prosper said. He couldn't say to the driver, Yes—it was fast. Not without knowing if they agreed about other things, too.

"It would feel fast, even if you'd been out a long time." The driver leaned back, draped his legs over the dashboard, and tilted his hat onto his face. "I'm Saburo," he said, speaking into the hat's brim. He used the simple Japanese spoken among friends. "I need to rest my eyes. If I don't see you again, I'll be happy for you."

Prosper disembarked, then turned to look at the bus. There had been something, the day of Uche's release, an interaction between Uche and the bus driver during the ride back to the station. He tried to remind himself what it was, but could only form a picture of two shadow figures in his mind, talking to each other, making the gestures of speech, but nothing he could recall having heard or understood.

15.

East of Kabuki-cho, where the nightlife businesses thin out and revelers realize they've walked past the red-light district, one of the first Nigerian immigrants I befriended owns an after-hours club. When I met him, a regular bar (as his compatriots would call it—a bar that doesn't employ hostesses) was precisely what he aspired to. He intended to attract the Lonely Planet crowd: foreign tourists and curious locals who might want to try a bottle of African Guinness and listen to Fela Kuti. From the beginning, the club was only busy in the wee hours, and most of his customers were Nigerian. When I stopped by, he'd tell me I should have come a few hours earlier, or just the other night, when the club was full of Japanese office workers and European tourists. He never omitted to conclude these speeches by reminding me that he hadn't borrowed any money from his in-laws to open the club.

Once, at a party he held in his apartment, with many of his neighbors present, a tax collector buzzed the intercom and demanded a payment for NHK, Japan's national broadcaster. Every Japanese household with a television is subject to this fee, which is collected by notoriously impolite, commission-earning contractors who go door-to-door extorting it. My friend said, in Japanese, "Just a moment, I'm coming to pay you," and the party went on. Every couple minutes, the intercom buzzed again, and my friend answered, "Sorry about that. I'm still coming to pay you." He did it casually, and at first a few guests chuckled, but the rest hardly noticed. As the tax collector gradually lost his temper, my friend persisted calmly—"Just a moment longer, I'm coming to pay"—until the tax collector was screaming into the intercom and laughter from the party could be heard (as we later learned) by neighbors three floors below. When I think of my friend, I think of him in that moment, precisely who he wanted to be: the charming outsider, expressing his affection for his adopted home by making sport of its inane rituals. When he was saving to open his bar, he must have imagined his customers seeing him in similar moments, thinking of him the same way.

I'd met his wife several times, and his children, the son a teenager, the daughter still in elementary school. It was a love marriage, not the result of visa pressure, and it showed when my friend and his wife joked good-naturedly at each other's expense, or finished their conversations with a gesture of physical affection. During the years I knew his family, my friend's daughter was reaching the age when children mostly take care of themselves in Japan. His wife was rediscovering her own interests and the career she'd deferred to raise their children.

The longer he tried and failed to attract overnight customers, the more he picked fights with his wife about how much time she was spending away from their apartment. One morning they raised their voices, and a neighbor called the police. After sending his children to school, he encountered the responding officers outside his building. They questioned him while his neighbors watched from their balconies and through their windows.

He didn't go home after that. At first he was living in his nightclub. I would stop by at opening time and find him asleep in the DJ booth. Business was slow, he said, and he made this sound like a problem that developed recently. I learned from a mutual acquaintance that the club didn't lose money, but hardly made any. My friend had been able to keep it open because his in-laws were supporting his family.

I worried he wasn't spending time with his children, and gave him football tickets so they could see the team that played at the stadium near their apartment. During the game somebody sitting behind him leaned down and talked into his ear. He never told me what the person said, only that it was ethnically charged, so he stood up and left. His children searched for him until the stadium was empty, and were returned to their mother by the police. A year later the same stadium was in the news because somebody hung a sign that said, "Japanese Only," over the entrance.

His wife had befriended the other Japanese wives of civically involved Nigerian immigrants, and after he moved out, I saw her at a christening celebration for a mutual friend's infant daughter. Her son and daughter were with her. They stayed near each other, apart from the other children, while she circulated. I'd known the daughter as the more outgoing child, but during the party she hardly spoke or moved unless her brother cajoled her. I was watching from across the room and felt I had to notice something. I was sure their father would ask.

I talked to their mother and she asked if her husband, who was no longer answering her calls, would be attending the party; the children wanted to see him, so did she. I stepped into the hallway and called, but somebody else answered, in Igbo, and I could barely hear them. An industrial sound roared over the line, rhythmically, as if the phone had been left on a factory floor.

He called before sunrise and woke me. He said he wasn't coming to the party, but I should tell his wife he'd been there and she'd missed him. I said the party had been over for hours, but he repeated himself, then asked me to put her on the phone. He didn't sound intoxicated; it was quiet wherever he was—I heard cicadas. He said he had to hang up, he was in the bar and it was busy. My caller ID showed him calling from a pay phone.

I asked if he wanted to know how they'd seemed or what they'd said. He said, Who? And for a few silent seconds, until his coin dropped into the phone and the line cut, I waited for him to remember while he waited for me to remind him.

Shortly after Prosper returned from Sendai, I visited my friend's club. It was the first time I'd seen him since he called from the pay phone. He was behind the bar when I got there, but we didn't talk; the woman Prosper thought he'd been with on the day of the earthquake was sitting alone at a table. I sat across from her. She was the only customer.

She didn't notice me, and seemed drunk. She was examining herself in the mirror that ran along the wall of the club, leaning almost out of her chair, catching herself as the chair began to tip. She was dressed as if she'd come from work, in the black blouse and pants a hostess at an upscale restaurant would wear. Her makeup had smudged, rimming her eyes and giving her a painted-on, surprised expression. There was a full bottle of Nigerian beer on the table, patterned with fingerprints, like she'd been picking it up but not drinking. My friend brought a drink for me and another beer for her, but she waved hers away, then waved it away again when he explained I was buying.

"Her first?" I said.

"When she came she's *yopparai.*" Drunk, he meant, and his use of the Japanese word struck me as an aggression toward her. He leaned across the table and took the ashtray. I noticed he hadn't shaved, or shaved his head, and the stubble in both places was uniformly white. He was older than I'd thought. He had always been fastidious with his appearance, and I remembered his wife being young.

He sat on a sofa at the opposite end of the room and switched on the club's television. The noise made me realize there hadn't been any music playing. Behind him, down the hallway, the bathroom door was open; a running water sound was coming from the toilet, and the lid had been duct-taped shut. If he'd been alone when I arrived, I would have thought I'd caught him on a night off.

The woman had leaned onto the table and was resting her head in her elbow. "You need a taxi," I said.

"Waiting for my husband," she said, talking into the fabric of her shirtsleeves. "He's like him."

"The bartender?"

"Refugees." This, for some reason, made her laugh.

"The bartender's not a refugee."

Her purse was on the chair next to her and she went into it, then put her phone on the table. "My husband," she said.

The phone had a password. The wallpaper was a photo booth picture of her with two of her girlfriends. She looked about sixteen in it.

"That's not your husband," I said.

She tried to unlock the phone but wasn't sober enough. I asked if she had children, how old, if her daughter was her husband's, what it was like to have a daughter who was visibly the child of a foreign parent. She was wobbling, I could tell she was tired, and she answered any question she could with a single word or a nod. When I asked her what it was like raising her daughter, she used the adjective that means—in Japanese—far away.

I said, "Were you in Senzoku with a man you met in this neighborhood on the day of the earthquake in 2011?"

She turned in her chair and touched the speaker mounted on the wall near her shoulder, trying to feel with her palm if any sound was coming out of it. "My daughter is mine, from my husband," she said, and rested against the speaker.

With an arm over my friend's shoulder, she made it to the street and into a cab. My friend and I lingered on the sidewalk to watch the queue of taxis at the end of the block, dropping off nightlife workers who'd planned to ride into Kabuki-cho on the last train but missed it. I explained who the woman was. My friend reminded me that I'd first met him in a hostess club he was co-managing and had seen him quit when the other managers brought underage girls upstairs to drink. He'd been so offended by their behavior, he admitted, because it made him wonder about himself. He'd been dating his current wife then, who was seventeen when he met her. His son, he pointed out when I looked confused, was from a previous marriage. His wife was twenty-eight now, his son was sixteen, and he was no longer speaking to either of them because his son had taken his wife's side when he left.

While he spoke, his attention was on the nightlife around us. If a luxury car parked nearby, I watched the interest in his face as he waited to see what kind of people would emerge from it. He interrupted his story to point out the spectacle of a sumo wrestler, in traditional attire, crossing the nearest intersection, trailed by a retinue of hangers-on. In the lull that followed, my friend said, "If Japan is my home, this street is my backyard." I remembered the apartment where my friend had lived with his family, and the adjacent green spaces, including a park with a pond, where his children would sit with their legs dangling over the side of a footbridge and watch turtles swim under their shadows. The apartment had a balcony that overlooked the park, where my friend slept on his nights off, so the sun would wake him in the morning and trick his body into a daytime schedule, the better to enjoy his family's company.

I said: It doesn't seem like you wanted your life back, after what happened.

We talked about how it's different, feeling you've been embarrassed by your wife, if she's Japanese, if you're foreign and your nationality is a constant source of discrimination. Eventually he came to it: Sunrise barely visible when the police arrived; the emergency lights flashing so everyone in his building noticed and watched from their windows; someone called his wife's father, known to all the tenants because he'd been involved in local government and lived nearby; his wife's father showing up and smoothing it over with the police; the familiar feeling, like a pebble in his throat, of having prepared to speak as if honesty were required, when his best remedy was silence; saying to his wife's father, when the police were back in their car but hadn't left yet: Any Japanese tenant's raised voice and the police don't get called, Guess I'd better practice my accent, Guess I'd better practice my conjugation—as brusque and informal as he could make himself sound in Japanese, and ironic, because saying it like that required him to imitate a native speaker; then his father-in-law, a soft-tempered man as long as they'd known each other, pushed him hard enough to knock him against the police car, and they started to scuffle; the police pulled them apart, his father-in-law apologized, but my friend started getting text messages from his neighbors, saying they'd heard he assaulted his father-in-law.

"I should knock every door?" he said now. "Say: It's OK your children are near me, because this my wife's father, old Japanese man, he is the one who have first put hand on me." He'd lifted a fist to pantomime knocking but caught himself in the gesture, in the bitterness he'd begun to revive, and dropped his hand.

"What made him push you?" I said.

"He tell me he don't know." He grunted in his throat, the beginning of a laugh. I could imagine the conversation: his father-in-law not wanting to admit, even to himself, some of his reasons, and my friend not wanting to ask what his father-in-law held against him, or if money had anything to do with it.

I asked about business, the broken toilet, about not playing any music in the club. He said he'd laid off his employees, stopped opening the club in the evenings and learned to live with the after-hours crowd, but the money was steady. I said it was just me asking, out of concern, I didn't care how much money he had or where it came from. I knew the building across the street was owned by the yakuza and bars nearby laundered money. He said he didn't do anything like that. When we finished talking, he locked the club behind us. Until after hours, he said. "If you touch dirty business," he added, "it must be the very last thing, so you take the money and go."

I met Prosper at the end of his shift, at the only convenience store in Kabuki-cho with a place to sit inside—a few tables, a counter against the window that looked out onto the street. I was always meeting people here, people I knew always met here, they were always sitting at the counter because they could watch the street through the window and if anything happened, they'd see it. I once sat at the counter with a nightlife worker and helped him make sense of some business documents, the gist of which was: his savings from a decade of factory work, which he'd plowed into a few containers of used car parts, were gone, and so were the parts. On another night I was sitting with another nightlife worker when his phone rang and it was his brother calling from Nigeria, to tell him his wife had miscarried.

Prosper and I sat at the counter. Outside, it was palm-of-the-hand time—what Igbo nightlife workers call the earliest minutes of daylight. It's a village phrase, part habit, part endearment, describing an adopted home with language native to the heart, on a street where your palm is always visible in the light from the electric signs. This hour of the morning, everyone finds a wall to lean against or a railing to sit on, and they gossip to pass time until the trains start. The slow, sleep-deprived pace of the conversations, the drowsy, knowing nods in place of laughter, make it seem not like gossip but as if everything happened a long time ago, and they were all there when it happened.

I told Prosper about my conversation with the woman he'd asked me to follow. He looked through the window while I explained. His view was blocked by a group of Japanese businessmen who'd paused in front of the store to smoke. One of them, drunk, had sat down against the glass, and when the group moved on they forgot him. He'd nodded off but still had a lit cigarette between his fingers, and Prosper watched its trail of smoke travel up the window. Prosper tapped the glass. The man stood and joined the flow of pedestrian traffic, his right arm bouncing limply off his hip. Abruptly, Prosper squinted and leaned close to the window. When the man had drifted into the crowd, he said, "The arm is not crippled."

"He was lying on it," I said.

"Nigeria," he said, and touched the glass to show me where he'd seen, in his impression that the man was injured, his nation of origin. "Imagine I'm chasing her in the street to ask if she is having my child. In Nigeria, even police may say: Bring the families, let them resolve it." In Japan, he would settle for uncertainty. "What will you ask here that they will answer?"

We walked to the club where I'd encountered her. It was still chained shut. Workers at the clubs nearby told us it hadn't been open for after hours in at least a week. Prosper reached for the padlock and it fell open in his hand. The combination hadn't been set. "Long time he wanted to open regular bar," he said. "You remember?"

As long as I'd known him.

"And the wife is quite beautiful. You met the wife?"

I'd met her.

"I don't know the particulars why he doesn't take responsibility for this place." He stepped close to the entrance and indicated the spot where our friend had taped his business card to the door frame. The name of the club was on it, and our friend's preferred alias, and a cartoon drawing of his smiling face, wearing the red cap reserved for Igbo men of title. "But along the line," Prosper said, "someone have convince him the place is not his own."

16.

Nightly, Prosper woke in the early evening, called Keiji, who never answered, then dressed for work and tried Keiji again before leaving. Prosper could have gone to Sendai. Yumiko would have let him into the apartment, and Keiji would have been there. Then his relationship with Keiji would have something to do with Yumiko, and Prosper sensed that in the long term Keiji would need a father apart from his Japanese family.

At work, Prosper resumed his habit of spectatorship, though fewer and fewer of his colleagues could afford to join in his aloofness. Clubs were closing, succumbing to the Nigerian community's problems of reputation. A growing number of nightlife workers were awake during the day, chasing their former employers in search of final paychecks that never materialized, then staying awake overnight to earn back, on a freelance basis, what they'd thought they would already have.

Dictating to an old friend in Gifu Prefecture over the phone, Prosper was able to have Uche's concerns about his file typed in Japanese. He mailed them to the Immigration Bureau and expressed his frustration with the futility of this gesture by staging a satirical performance (on the street, in the red-light district) of the bureau's likely response, imitating the alarmed sincerity of the clerical employee who would open and read the letter, then the affected decisiveness of the mid-level bureaucrat assigned to review it. With great haste, the letter would be stamped six times and filed in triplicate. A form response summarizing the stamping and filing process (as if it remedied the problems listed in the letter) would go to the return address Prosper's friend in Gifu had suggested: a postal box that received fan mail for a popular boy band.

Prosper and Coup arranged to meet for the first time since their trip to Nigeria, but Coup never showed. He called the next day and declined to give an excuse, saying only that Prosper was better off not knowing.

They met a few days later, and Coup was his usual self to an exceptional degree. When Prosper asked whether the rental property business in Nigeria was worth a look, Coup said he wouldn't know, it was something he did to launder money, and he never paid upkeep on his apartments. In Coup's formulation, he was the victim of this neglect, because it required him to endure the unproductive guilt of knowing, for instance, that a pregnant mother of five was living under a leaky roof on his account.

He went on to claim that he was among the majority of Nigerian landlords, not spending anything on his units. Furthermore, he did not steal cars, but accepted a bargain when it came to him without police surveillance attached. In Nigeria, four orphans were attending private school at his expense. He'd been faithful to his wife since marriage and conquered all his vices—alcohol, cigarettes, betting on football—more than a decade ago. The moment he could trade moral rectitude for food and shelter, he would favor law over profit and allow his blood pressure to improve correspondingly. Already, he abstained from luxury items. He owned one car. His money was in a savings account, in low-risk index funds.

For Prosper, there was no figuring what event in Coup's life had triggered this misplaced outburst, except it was probably whatever had caused Coup to go missing when they'd been scheduled to meet, and it signaled to Prosper that it was time to take another few years away from their acquaintanceship—but not before he'd exploited Coup's evident moral panic by asking him to underwrite Uche's legal bills. They stopped at a bank before parting company, and Prosper left with enough money to buy a modest new car. Uche didn't want a lawyer, but the money could go to his wife in Nigeria.

To avoid the cost of wiring the money, Uche's wife asked Prosper to bring it to a relative of hers who would be visiting Nigeria the following week. The man had worked the same factory job since arriving in Japan thirteen years earlier. His employer had kept most of his work off the books, which he'd agreed to because he didn't want to pay taxes and wouldn't have known how. A few months ago, his knees had locked up from all the lifting the job required. He was without workers' compensation or unemployment benefits and would soon be without work.

He lived with his Japanese wife and their two young children in Kanagawa, where Prosper went to see him. His apartment complex was built in the style of townhomes, the building's vinyl siding patterned after the stones of a castle, and the sign over the entrance said, "Le Café." They sat in the living room, where the sofa and the chairs were angled toward the kitchen because the light through the floor-to-ceiling privacy window that overlooked the driveway was too bright to face.

His children, a boy and a girl, sat on Prosper's lap and read to him from their picture books, in Japanese, stopping now and again to ask if he could read the characters on the present page. Prosper said no, and their surprise, their delight, made them laugh; the only other person they knew who could speak Japanese but couldn't read it was their father, whom they loved. They demonstrated, in the air, how to draw the characters, and Prosper tried to imitate them, without success, so they took his fingers in their hands and showed him.

The man talked to Prosper in Igbo, asking guardedly about money in Nigeria, how far you could get with how much. It had been years, he said, since he last went home. At first the questions were about the kinds of expenditures you might make on a brief trip, then began to include purchases you wouldn't make in a place you didn't intend to live. Neither Prosper's manner nor his tone changed as the obvious presumption suggested itself: Maybe this man wouldn't come back. He'd start over in a country where he spoke the language and could look for a job that didn't require a young man's body.

The little boy on Prosper's lap wanted to bounce, the little girl didn't, and when one of his legs was tired, he switched. When both legs were tired he let the children down and they played on their father's body, one on either side of him, holding on to his shoulders and lifting their legs off the ground, trying to hang on. Their father was tall and broad chested, the kind of father with whom such games are played. He was silent while they crawled and hung on him. He stared at the closed door to the kitchen, which had a small panel of frosted plastic cut into it at eye level, where the flicker of his wife's dark hair was visible as she moved back and forth on the other side. The little girl, who looked younger, stopped her brother and said, "Papa doesn't want it."

His wife brought lunch, a Nigerian meal she'd prepared, and said as she served it, "I'm learning to cook Nigerian food because it makes him feel better, being home all day." When she'd gone to the kitchen to start the dishes, her husband said, "We can't leave any." While he ate, he speculated with Prosper about what she'd done to the food to make it look authentic but taste so unfamiliar. They managed some mirth at first, but the more they ate, the less they joked, until they were silent except for chewing and swallowing and clearing their throats, then sipping their beer to get the taste out. They finished eating, and Prosper left.

On the train back to Tokyo, Prosper remarked that Uche's wife was likely unaware of the risk she'd taken by trusting the man with the money. If he'd told her he was considering a permanent return to Nigeria, he might have altered the facts, perhaps to portray himself as a successful investor, returning to claim the Nigerian life he'd intended all along. Or, if he was the soft kind of person Prosper took him to be, he'd told her the version he wished could be true: that he had discussed it all with his wife, whose love and generosity had solved the trouble somehow. Perhaps in this version his wife agreed to go with him and bring the children. In the man's mind, Prosper said, it had happened already, this other life, imperfect but habitable, where every hardship could be solved as long as something was left to sacrifice.

Keiji called the next afternoon. He had come to Tokyo and run out of money in Odaiba, a reclaimed island in Tokyo Bay occupied by shopping malls, museums, and theme parks. Prosper had never been to Odaiba and got lost trying to find the monorail that reaches it. While he rode the monorail over the bridge, the view of the bay was obscured for a moment by a flock of cormorants that rose to fly alongside the train, frightening the children aboard with their clamor. Prosper listened as the man sitting across the aisle explained to his daughter how the city had paved its waterfronts, forcing all the seabirds to nest on the World War II–era battlements near the bridge. His tone was bright; he was trying to exercise her imagination: All of Tokyo's birds on a tiny island—can you picture it? When the girl looked up from her father's lap, Prosper waved hello and said, "If you lose your home, you can live under a bridge, too." He'd said it in a way that imitated her father's voice, and for a stunned moment the father looked at Prosper across the aisle, while his daughter looked at him looking at Prosper. The train glided into the next station, and the father took his daughter to another car. The other passengers watched him go, and their mouths didn't move, but the murmur was in their gaze.

Keiji had money, but not enough to enter his favorite indoor theme park and buy a train ticket home. He explained how much he wanted Prosper to give him, and why, and Prosper listened, standing with him on the section of the theme park's boardwalk that over-looked a strip of white sand beach, facing the bridge and the city. It was late on a weekday afternoon and students filled the boardwalk, in their uniforms, carrying their backpacks. Prosper watched them enter the mall that hosted the theme park, or come out of it, watched them linger together in circles, peering into their phones. When Keiji finished explaining, neither of them spoke. Anyone passing by would have thought they were strangers, standing near each other by coincidence.

Prosper said, "Let's go," and he meant, *away from here,* but Keiji walked toward the entrance to the amusement park, threading his way through the other teenagers into the shadow of the mall's facade, where it was too dim—and the sun too bright from behind the building—for Prosper to make out where he'd gone. Prosper started walking toward the monorail station, a few steps. Later I asked about those steps, and he said his departure was interrupted by Odaiba's skyline, which was condo towers, glass and steel, the Ferris wheel on one side with its lights changing color in the early sunset, and the hard white glare off the modernist architecture of the Future Museum. When he'd attended college for engineering, he'd carried lab notebooks with him wherever he went, bound hardcovers with gridded paper in them, for sketching his assignments and ideas. After he graduated, when his uncle was arranging his Japanese visa, he filled the notebooks with perspective drawings of the skyscrapers he'd seen in newspaper articles about Japan's miracle economy. The Tokyo he'd learned to imagine, the one he'd thought he would live in, had looked like Odaiba.

On the boardwalk, as he remembered his drawings, another memory followed, of going to church with his mother, with his uncle and cousins, of churches in Benin City, in Onitsha, in the village. Outside the church after Mass had been said, everyone coming out of the fan-cooled plaster interior of the church into the courtyard, where the families' drivers had idled during the final hymn, building a fog of exhaust between the compound's walls. Pulled along by the hand, sand leaking into his shoes, his eyes half-shut against the light, the sweat starting in his pores. Thinking the way a boy thinks, it would be good to go back in the church, now it was empty, let the breeze from the fans slip through his collar, spread his fingertips on the cold plaster walls, tap his knuckles on the pews to hear the dry, echoing sound he imagined it making in the trapped air of the nave. The impulse never left him. As an older boy, then a young man and an adult, he entered churches wherever he encountered them, to be alone in a quiet room until the boy with sand in his shoes and sweat down his collar sits next to him.

He joined Keiji and paid for both of them to go inside. Keiji chose a row of arcade machines played by shooting the screen with a light gun. For a while, that was all. Prosper stood to the side, at an angle that didn't let him watch the screen, only the shapes it projected on Keiji's face. After the better part of an hour, without looking away from the game, Keiji said, "I don't want to be asked those questions," and when Prosper didn't respond, listed the questions he meant, which were the questions adults asked him whenever they had an opportunity.

Prosper was surprised by the intensity of Keiji's resentments. Yumiko had accustomed him to the image of Keiji locked in his room, with nothing he wanted to say about how he'd wound up there, but Prosper could read the past few months, because Keiji had gained weight, and he could read the years, too, because Keiji's hair had grown and lay in a tangle of natural curls, forced to one side in a failed attempt to part them like Japanese hair—there had never been anybody in his life to tell him what to do with it.

Recalling it later that night, while he listened through the wall to a proselytizer reading from a pamphlet in his elderly neighbor's apartment, Prosper could describe this process of sympathy, up to the moment when he understood that there had been a way to care for Keiji as his uncle's family in Nigeria had once cared for him, and he'd missed it. This awareness was physical, like the whole-body gasp of looking down from a high place, and he would have put it into words, but Keiji was already on a nearby escalator. When Keiji reached the bottom he deliberately collided with a woman and her young son. He led with his shoulder, knocking the boy over.

Prosper ran to the bottom of the escalator and pulled Keiji away, catching his arm with one hand and the collar of his shirt with the other. The woman was leaning her body over her son, to shield him. Prosper still had a handful of Keiji's shirt when she looked them over, and he could see her expression enter the moment when she didn't know whether the people she was looking at were related to each other, or how.

Then it was gone, and Prosper said, "Sometimes he gets—dizzy." He'd used the English word, but said it in Japanese syllables, *dee-jee,* and it hadn't made sense. "He gets—" and he started to use the Japanese word for "sick" but realized it wouldn't be any comfort, so he said, "I am embarrassed and sorry." The woman started to say it was all right, but he'd already begun leading Keiji away.

In the restroom, Prosper dropped Keiji's wrist and stepped back so Keiji would have to look at him. He said, "Something happened to you. I don't mean—not like that. Not particularly anything. What happened, it's something. What happened to you every day." Listening to himself, aware of all the Japanese he didn't know, he found he'd been able to say it anyway, had sounded only as confused as he really was.

Keiji's expression didn't show anything, or it showed that he'd lived with his difficulties long enough he could use his face to conceal them. Prosper told Keiji they would have to leave. Keiji didn't talk until they were on the platform at the station and Prosper said, with the train pulling in, "Go, just you." Then Keiji said, "Thank you." There wasn't any surprise in the voice, wasn't relief, nothing to suggest Keiji hadn't already thought he would board the train alone. Keiji's eyes kept busy watching the train pull in, watching the waiting passengers shift closer to the automatic doors at the edge of the platform. That was the way the expression showed nothing. The eyes were always going somewhere else, and Prosper thought how, in Igbo, the eyes play the heart's role in metaphor. The eyes are the organ that feels.

He finished the thought later, in his apartment, and next door the elderly neighbor had forgotten where he was, had started talking to the Jehovah's Witness he'd invited in like she was his wife, but the wife had been in a home for people with dementia since a month ago. Prosper said, "Somewhere I'm knowing the same eyes." The neighbor's door shut and Prosper waited for a knock: on his door, or two apartments over, but none came, just the neighbor's voice on the other side of the wall, asking was anyone still there. "Long time ago I'm knowing them," he said, and gestured at the frosted glass of his apartment's privacy window, the kind of gesture that means it's too far, you won't get there. Then he started telling the story about the rehabilitation camp for Biafran soldiers, the last time he saw his grandfather, whose eyes he had nearly forgotten.

17.

The walk from Shinagawa Station to the Marriott was long enough for Prosper to wonder why he was taking it. He guessed the motives of the other Nigerians he noticed disembarking: related to the buffet and carving station, the Japanese executives who attended embassy events and the credulous hobby investors whose attention had alighted on Africa. Nigeria's Independence Day had arrived in concert with the nation's triumph over Ebola, and the handful of major Japanese companies scared away by the outbreak were nearing the public announcement of their renewed Nigerian intentions. The banquet was fanfare, with a highlife band in matching robes.

None of this he was eager to see, but his feet moved him anyway, in the direction of the hotel, with the outlines of Tokyo's southernmost skyscrapers barely visible in the dark, so the lit offices inside them appeared to float on translucent wires. It was possible to walk like that, not wondering too much what carried him, because the route was familiar. Theodore, one of his earliest friends in Japan, had worked as a cook at another hotel nearby. Prosper had shared a room with Theodore at the first guesthouse he found when he came to Japan, and Theodore was Igbo, but told everyone he was from Mozambique; he liked to have his name pronounced the Portuguese way. He wasn't fooling anybody, and Prosper never developed the impression that Theodore believed anybody had been fooled, but during the years they knew each other, the game—if that's what it was—never changed.

The restaurant manager at the hotel where Theodore worked had grown up in a fishing village on the Sea of Japan and worked on his father's trawler until it wasn't possible to earn a livelihood that way anymore. He was an outspoken conservative, and maybe Theodore was the first African he'd ever talked to, but he was without pretense or judgment toward people he knew and, in his rural way, unreserved about getting to know people. He came to prefer Theodore over the other employees, whose Tokyo upbringings had included none of the difficult, uncertain work that he and Theodore had in common. Every night before the restaurant opened, he insisted on serving Theodore dinner, and Theodore was welcome to invite a friend. Prosper, who was working in factories, depended on these dinners to break the rhythm of labor and sleep. They'd sit, the three of them, and Theodore's boss would talk about winters on the back coast, which involved Japan's worst weather, but the way he described the place, Prosper was certain he would feel at home there.

When Prosper started working nightlife, there wasn't time to visit the hotel, and a year passed before he returned, looking for Theodore, whose phone had been disconnected. When he asked to see Theodore's supervisor, a stranger came to the lobby and identified himself as the new restaurant manager. He explained that his predecessor had returned to Niigata after his daughter was struck by a train. Theodore had resigned a few days later.

Prosper never saw Theodore again, or heard about him, which was unusual; there was always at least a rumor. He supposed Theodore had joined the minority of long-term Nigerian expatriates who accepted deportation after they were caught with expired visas. It amused him to imagine that Theodore carried a false passport from Mozambique and had been sent there to reinvent himself—again—among strangers, in an unfamiliar language.

Accompanied by this memory, Prosper arrived at the basement lobby of the Marriott, where the doors to the ballroom had not yet opened, where the mood was self-conscious and self-displaying: women in traditional Nigerian attire, including the embassy's Japanese employees and the Japanese wives of influential attendees; men dressed according to how they intended to work the room—as the tuxedoed, well-assimilated informant, fluent in Japan's white-collar decorum, or the magnanimous African titleholder, with sparkling filigree woven in his native clothes.

He noticed Coup and Austin immediately because they comprised the exception. Coup's tie was stuffed in the breast pocket of his untucked flannel shirt, as if it were a pocket square. Austin had arrived by plane an hour earlier and sent his bags ahead of him before remembering it was the night of the celebration. He had a T-shirt and corduroys on, and over the T-shirt wore a blazer with sleeves that barely reached the middle of his forearms, the largest he'd been able to find on the rack in a department store near the station.

"The first minister now tell us he don't allow such clothes," Austin said when Prosper joined them, and they laughed together. They had a shared image, Prosper was sure, of the first minister trying to enforce this order. He was a foot shorter than Austin, and none of the embassy staff checking names against the list liked him. He was always calling them at home and putting them on speakerphone, to interpret conversations with clerks and cashiers.

"Vitus don't come," Coup said. "I never see him any other place." The way he interjected—and the degree of preoccupation it revealed—were uncharacteristic. Prosper searched Coup's demeanor for an indication that the comment had been intended as more than gossip. Because it was Coup, there wasn't any, except the comment itself, and Prosper's awareness that Coup laundered money through Vitus' hotel. Prosper reflected that he had known Austin and Coup (and they had known each other) when they were young, and when they were young there would have been more to say, a discussion of what could be done to determine whether Vitus had been detained, but now the time they would have spent saying it passed in silence, while each of them formed the thoughts that once would have formed the words.

An elaborately garbed middle-aged Nigerian walked briskly through the loose circle they'd formed, his drapes flowing behind him, producing their own breeze. He stopped nearby to admire himself in the reflective surface of the door to the ballroom, adjusting the angle of his finely embroidered cap, then the bottom of his matching tunic until it rested just so around the bulge of his stomach.

There was the sound of a camera shutter, and the man turned to see Austin photographing him with his phone. "Looking fine, sir," Austin deadpanned, and the man didn't reply, but the grin he'd practiced for his reflection vanished with such speed—and was replaced by an expression of such cheerlessness—that the change had a cartoonish effect on his features, as if his face had emerged from a spaceship into the exaggerated gravity of an alien planet.

Gradually, certain attendees sought the aloofness of the small gathering that Prosper, Austin, and Coup had formed. They were chairmen and officers of town unions. They were entrepreneurs who claimed they had arrived in Japan just yesterday to attend the celebration, but business in Nigeria was booming, a plane ticket was a trifle: Who needs Japan?

Being the genuine manifestation of their pretensions, Austin's appearance (and the state of his business affairs) became the subject of the group's collective inquiry. A few of the people present had been to Austin's restaurant, or his warehouse, and began to exchange descriptions, weaving into them some estimate of how much profit each business could generate. This was a form of flattery and a way to probe Austin's finances, by gauging his reactions to the numbers ventured. Before his immigration difficulties had cost him the upward mobility he built during his first decade in Japan, he had been a reliable donor to the town union system, and they expected him to resume his largesse.

Watching Austin respond to their estimates with a pantomime of shock, with flattery in return, Prosper had the feeling of looking through a microscope, where very small things, too small to be noticed, occurred on a tremendous scale. Everyone in the conversation was animated, expressing themselves at the volume customary to a public dialogue in Igbo. Austin was less enthusiastic, had to master himself when someone thumped his back or grasped his arm at the elbow. Prosper saw what he hadn't seen when he visited Austin in Nigeria, and maybe what Austin had meant then, when he said a few years would be the same as forever: the fragile nature of his new prosperity, built on two businesses he established within the past two years, and the trouble he could expect from people whose motives and methods resembled his.

The hazing might have continued if not for someone else inviting himself into the conversation, standing next to Prosper with a hand on his back and a familiar voice in his ear: Theodore, and he had aged the way Prosper had, more weight on his body, less certainty in his movements. Theodore handed out business cards that indicated he was CEO of a company with a name that could belong only to a vanity project. He was too quick with the cards, too eager to say he had worked with Toyota, with Komatsu, with Mitsubishi and Marubeni. Prosper smelled *shochu* and lemon on his breath—the scent of a cheap, high-proof canned drink. Wanting to spare him the indignity of the circle breaking up in response to his monologue, Prosper gestured with his fingers to indicate a cigarette break. As they left the group, a voice called after Prosper, thanking him in Igbo. Someone chuckled in reply.

They walked behind the hotel, where Prosper begged off the cigarette. "But you, please," he said, and Theodore lit his. There wasn't anywhere to sit, so they stood, Theodore cradling the elbow of his smoking hand against his chest and holding the cigarette an inch from his lips. To the south, the lights of Tokyo's industrial suburbs—where they had shared their room in the guesthouse—were indistinct behind the glare from the hotel's flood lamps.

"Inside," Theodore said, "I'm embarrassing myself with those big boys."

"I don't think you made problem," Prosper said. "They forget your face if you want next try. You never talk to your boss again, the chef?"

"Why I should talk to him?"

"He was cooking for us, allow us to spend time."

Theodore flicked the half-smoked cigarette against the wall of the building. "I lose my job. So his daughter get suicide. He go home now grieve. When he is ready, maybe government give him money. My money don't come."

"How your kids? You were having two in Japan and five in Nigeria."

"For Nigeria, real one is three."

Prosper didn't inquire. It could have been for sympathy, saying he had five children to support. Maybe two had died, or he'd disowned them.

"I'm having a son," Prosper said. "The boy is thirteen before I'm knowing him. I'm never having father. Brother even, I'm not having."

"They're knowing my face," Theodore said, and pointed behind him, at the hotel. "Those big boys. One is my brother, the one they are calling Austin."

"Mother is same?"

"Same mother, same father, together in same house. We are like age-mates, he is junior by two years. If you tell him my name, he may like to discuss any other thing." He reached for his cigarettes. There was one left, and he inspected it, then replaced the pack in his pocket.

"He don't say he has brother in Japan, since I'm knowing him," Prosper said. "Few weeks ago I'm seeing his place in Nigeria."

"He have?"

"Warehouse, for Ladipo. And the restaurant is—"

"Oshodi."

"It's the place," Prosper said.

"It's better," Theodore said. "It will make me happy to hear it is doing good thing for him." Prosper saw their features were alike, Theodore's and Austin's, and wondered if it came from their mother, their father, or further back, too long ago for them to know whom they resembled, except each other. "That time when our father is soon to be late," Theodore continued, "we are receiving small inheritance, but the main thing is that place in Oshodi, which I am named to receive because I am the senior. The same time, Nkem—Austin—is having his deportation. He have arrived to Nigeria. I give him that place. I said, You are the one who is having difficulty, it's better you have it. Even he want to sell the place, I don't mind." He paused, perhaps deciding whether this was true—whether he would have minded if Austin sold it. "Until now, as you are telling me, I don't know what he have done with it. If I call him, he don't pick. I see him, he may walk opposite direction."

Hearing Theodore's explanation, Prosper could guess what had produced their estrangement. Probably, the property had never been legally surrendered to Austin, and Austin had gambled by building the restaurant anyway, leaving Theodore with leverage neither of them had intended.

"It should be the same time the chef have left his job," Prosper said.

"Near," Theodore said. "Maybe few months after I gave Oshodi place for my brother, the daughter of that chef have lose her life."

That was it: Theodore's kindness, then his misfortune. In the likeliest version, Prosper reflected, Theodore had developed the habit of asking Austin for money, and Austin of weighing these requests against the trouble his brother could cause if he decided to controvert ownership of the property, until Austin tired of the expense and ignored a phone call, then another, and it became clear that Theodore wouldn't use his leverage, which made any further loans (or the experience of refusing to make them) an unpleasant distraction.

"I'm needing a place to stay in Tokyo," Theodore said, interrupting Prosper's contemplation. "I'm staying for work site, it's Tochigi, for factory." He hit the heel of his palm against his temple to show there wasn't any mattress where he slept.

"Maybe final train for Tochigi is very soon," Prosper said.

"You have apartment?"

"Smaller than room we were sharing back those days."

"If I stay with you one week, two weeks, it's no problem?" Theodore touched Prosper lightly on his wrist. The gesture seemed choreographed, and in the small, startling distance Theodore had opened between them by imposing his need on their friendship, Prosper intuited the life of diminishing possibilities Theodore had lived since they last saw each other. Some of it he would learn later: divorce, Theodore's wife disappearing with the children, arrest on a drug distribution charge, time in prison, the loss of his visa, time in immigration detention, provisional release while a disingenuous asylum application was pending, and now the best he could do was a blanket on the floor of a twelve-square-meter room where six laborers slept; awake at five to process recycling, double shifting if he could manage it, in bed at midnight, leaving the site only to make the three-kilometer walk to the nearest convenience store, where he bought phone cards so he could call his wife and children in Nigeria.

"Ten days is even okay," Theodore said.

"It's impossible," Prosper said. "Nor anyone inside," and he pointed where the driveway curved toward the lobby. He'd answered the question as if Theodore had said, *Can you help me?*—as if it were more than the apartment he was asking about.

"Though I'm trying."

"They will see you are trying," Prosper said. He hung his arm over Theodore's shoulder and pulled him abreast.

"It's all right," Theodore said, and Prosper let go, but Theodore didn't increase the distance between them. He began to speak, and the softness of his voice lowered its pitch, so that Prosper, standing close enough their shoulders nearly touched, could feel the air vibrate between their throats.

Theodore said, "That time we're young boys, my brother is having dream. How the two boys, having his same face, they are wearing exactly the clothes he is wearing for sleep—they will chase him. The dream give him sleepless night, so the time we are returning to village side, I go see one woman who know dream tradition. She say anyone who has such a dream, there was twin with him inside the mother, where the twin have died. A thing must be done by parents, to send the twin for rest. If I'm asking my parents, they will chastise me that I don't speak of such thing again. I am thinking how I can make my brother to rest comfortably. In this dream, if the boys are catching him, they will demand money, but he is never having money in the dream, so they can do any evil thing. When he is sleeping I go to him, put small money inside bed. When he is waking up, he will see this small money have remained, he will know the boys have not truly come."

Prosper said, "I don't believe he forgets such a thing."

"I see he is still sleeping well."

And we thank God, Prosper thought. It was the standard Nigerian reply to a report of good fortune, and occurred to him on reflex. He remembered the room he had shared with Theodore, how Theodore had always come in after he had gone to sleep, but he had never woken to the sound. Once, he was awake when Theodore arrived, and lay still, watching him moderate his movements for the sake of silence. In the darkness of the room, Theodore found everything he needed by touch.

"Take," Theodore said, and removed the last cigarette from the pack, offering it to Prosper in his palm, where pale lines of raised flesh had formed around the cuts, from his work, that hadn't been given time to heal. "I must give you small gift."

Prosper took the cigarette with his thumb and forefinger and lifted it in a gesture of acknowledgment.

"Seeing you," Theodore said, and departed.

Prosper held the cigarette toward the skyline, one eye closed, sighting it against the buildings and bringing it closer to his face until they disappeared, one by one, behind it. The band's brass section was audible from the ballroom, warming up. When the music began, Prosper stood with his ear against the wall, listening to the Japanese and Nigerian anthems. "The same song, many years you hear it," he said. "You will tell yourself it have changed." In a moment he will rejoin the occasion inside, and the world of the years he has mentioned.

Decline

|||||||||||||

18.

The forecast predicts Typhoon Etau will reach Wakkanai on September 12, 2015, the first day of field research for Japan's national census, taken every five years. At the city office, arriving workers are greeted by their colleagues' appraisal of their umbrella habits. Kushibiki Hayato receives applause for leaving his in the rack, where he's likely to forget it. He is the youngest employee in the census division and a lifelong resident of Wakkanai, aware that television weathermen are regarded, in Soya Subprefecture, as the esoteric fortune-tellers of a neighboring culture's nonsense religion. Unanimously, his older colleagues agree that the typhoon will weaken, change course, or both.

Already, they have seen television images of a flood in Ibaraki Prefecture, caused by the failure of a storm-battered levee. Media analysis alights on the intermittent nature of maintenance on the levee, though it omits the connection that Wakkanai's city workers make at the watercooler, between the levee's failure and the politicized nature of infrastructure budgets in Japan: It's easier to build a new levee, a new anything, than it is to fix the old.

Sugimoto Akemi, the career municipal bureaucrat who will oversee this year's census, gathers the younger employees at her computer to show them a website that provides visual interpretations of demographic data. Sugimoto, who also works as the sportscaster for Japan's dogsled championships, is regularly caffeinated, a demonstrative talker. She explains why this year's census is significant: It will establish the pattern of age-based population decline in Wakkanai, exceeding economically driven decline, a process the city has attempted to manage for longer than Kushibiki has been alive. By

2060, the city of thirty-three thousand is projected to lose more than half of its population. City leaders insist they can avoid this outcome or, if they have to, scale the solutions they've devised for a city of twenty thousand to work in a city of ten thousand. "People are older than that," Kushibiki says, touching the screen with his pen where a graph shows mortality rates according to age. "When you see them." In Wakkanai, you see them often. Nearly one out of three residents is over sixty-five.

A year ago, Kushibiki was watching daytime television in the house he shares with his parents when the phone rang to tell him he passed the city's civil service examinations. He knew the voice on the other end. It belonged to the civil servant he shadowed when his high school sent him to visit the city office. His teachers had guessed correctly he'd like to be a bureaucrat. The civil service exams were more complicated than the tests to get into college, but he studied for them exclusively. Out of twelve students from his school who took the tests, he was one of three who passed.

Before his school sent him to the city office, his grandfather died, the grandfather who talked with him about how many people lived in Wakkanai after the war, explained that it was once possible to cross the city's harbor on foot, by walking on the decks of anchored trawlers. This wasn't something Kushibiki discussed with his father, who worked at a franchise auto parts store, a livelihood unaffected by the city's decline. While he took the exam, Kushibiki could have reminded himself that the city government was Wakkanai's largest employer, how that was true in more small cities as time passed, especially true in Hokkaido, sparsely populated, postindustrial because there had been, after the war and into the bubble economy, enough land in Hokkaido to industrialize. That's what his classmates told him they thought of, to focus themselves for marching their pencils through the exam booklets. Instead, he thought about wanting to visit the city office when he was in junior high school, with a group of friends, because the toy store in Wakkanai had gone out of business and they planned to ask what the city could do (maybe offer an incentive, maybe somebody else would open one), but they lost their nerve the night before. He remembered that his grandfather had encouraged him to follow through, and wished he had not felt too young to be counted, wished his involvement with the affairs of the city could have begun in a story they shared. Then he passed the exam.

Passed it not long ago. Not a hairstyle ago, which his co-workers make light of, asking him doesn't he have to get back to school, the principal couldn't have meant for him to visit and never leave. It wasn't natural the way he'd imagined it, his first day of work, meeting his colleagues (and everyone outranking him). But listening from the inside to what troubles the city—there aren't enough doctors, and nowhere to get them; the national government funds programs nobody needs—he feels this is what he expected to hear, suggesting it's the gentler part of fate that dressed him in a bureaucrat's costume. On the first morning of his census assignment, when Sugimoto asks him what kind of person he imagines will answer the first door he tries, Kushibiki brandishes an information packet with the city's logo on the envelope, the way an important piece of legislation is presented to a crowd, and replies solemnly, "It happens to be someone I know. They say, *So it's the census?*"

The district Kushibiki has been assigned to survey is north of Wakkanai's old downtown, an extension of it, surrounding an elementary school. Towns in rural Japan expanded in this incremental pattern because they borrowed the *zakkyo* model from major cities, where constructing rental apartments above ground-floor businesses helped to reduce sprawl. In Wakkanai, the closure of businesses in mixed-use buildings has made residential landlords insolvent, reducing the availability of rental units. The first neighborhood on Kushibiki's list has sixty-nine residences in it. Apartments, only six.

After a few houses where no one comes to the door, he begins to watch the windows. When he sees a shadow—believes he sees one—he rings again and holds his ID as close to the window as his lanyard allows. More houses, no answer. Crossing to the next block, he inspects the ID. "It looks like me," he concludes.

At the northernmost house in his assigned area, a middle-aged man is visible through the front window and comes to the door dressed in pajamas without Kushibiki ringing the bell. The house's entrance is on an elevated porch, up a staircase. Kushibiki says through the glass, "National census." The man opens the door without pausing, before Kushibiki completes the phrase.

They stand as far apart as the porch will allow. "It's our first year accepting responses online," Kushibiki says, and gestures with the information packet, but the man doesn't reach. Instead, he replies, "Someone did die," and wipes his feet on the doormat, as if to go inside. His hands are in the stomach pocket of his sweatshirt, which he stretches away from his body, then examines the distended shape. "Recently," he says. "In there."

"I didn't ask if you live here, I'm sorry," Kushibiki says. "It's my first."

"From where?" is the reply, and the distracted gestures continue: rubbing his hand along the clippered shortness of his hair; shading his eyes to look at the street, as if searching for an object in camouflage.

"For the census. The first person to answer their door."

"I wouldn't have wanted to tell you—who died, if they lived here."

"Not your house?"

"My house, yes. After the grieving period, I could tell you."

"This envelope explains how to respond on the internet."

The man takes it. Kushibiki finds the place on his survey sheet for recording a completed visit. The man watches. When Kushibiki's pen starts to mark the page, he gives the envelope back. "Or this," he says.

"I've disturbed you."

"No," the man says. "There's nothing I need to be doing," but he opens the door and moves behind it. "Come on in," he adds, then shuts the door and slides the bolt.

Descending the steps, Kushibiki says, "Please make sense." He is speaking, he explains, to the remaining hours of his workday.

He goes back to a house he checked before, across the street. Its downstairs windows are boarded. The address plate shows a Japanese man's name and a foreign woman's. He peers through a frosted-glass window where the board nailed over it doesn't reach. The view is obscured by a row of picture frames on the sill. A few face outward, the photos distorted by the pattern of the glass.

"Not the neighborhood I live in," he says, stepping away from the window through thigh-tall grass. "If my friends lived here, I didn't know." His comment implies the difference between South Wakkanai and the area surrounding the city's historic port. After the war, Wakkanai expanded as far as land on the cape could accommodate. When the bubble economy brought franchise stores and car dealerships, they built south of the city. New housing took the land nearby.

After the bubble, the part of the city that kept doing business looked like a suburb, and the part of the city that was actually the city had a train station and city office, but no economy.

School-age children would have been scarce in a neighborhood like this when Kushibiki was growing up, and would have been mostly the children of multigenerational fishing families, working on their fathers' and grandfathers' boats when they weren't in school.

A family like that comes home not long after Kushibiki has tried their house. From two houses down, he counts them coming out of their minivan, three generations, carrying groceries, negotiating the piles of tackle near the porch. As he counts, he points with his pen. The minivan's driver puts down what he's carrying and watches back. Kushibiki approaches and they speak on the lawn. When they're done and Kushibiki has walked to the intersection, the sound of a soprano voice drifts from the house, the foreign syllables and careful vibrato of an étude. "It's not the radio," he says, looking where the curtain behind a window billows from one end to the other, implying the breeze from an oscillating fan.

Among the houses that appear abandoned, none are alike. One seems lived in from the outside because the vestibule is tidy and the children's toys on the front lawn haven't lost their new-plastic sheen, but the patches of sedge growing through the cracks in the patio are waist-high. At a house across the street, the lawn is clipped except for a ring of grass and wildflowers around a parked car, which is clean inside and out except for a layer of rust spreading from the place where the hood meets the grille. Later, the neighbor tells Kushibiki her husband mows the lawn but doesn't want to hit the car with his trimmer, and she doesn't know where her neighbors have gone, but it was a year ago at least.

Before lunch, Kushibiki climbs the hill to visit four houses that make up the western boundary of his assignment. The first, set toward the woods, has a detached structure next to it, a garage or workshop. He gets no answer at the house. There's a ladder leaned against a second-floor window and two open cans of paint at the foot of the ladder. He waits and listens, but the only sound is a wind-bell, then the crows talking on the hillside.

He starts marking on his list that no one is home, but moves as he does it across the gap between the house and the outbuilding.

There's a thud on the grass, and he lifts his head in time for the deer—160 pounds at least, with eight points on its antlers—to sprint past, on a trajectory that barely spares a collision. The noise Kushibiki makes, a scream interrupted by swallowing, brings the owner of the house out of his workshop with sawdust down his shirtfront. The deer has stopped partway toward the street, where it watches them. The owner sees it, and Kushibiki says, "Census." The owner starts to laugh, maybe because there's a joke about the residential neighborhoods attached to downtown, that they're home to more deer than people. Kushibiki laughs too, then catches himself, touches his chest, and inspects the fabric of his shirt. "The tip of the antlers," he says, and makes a circle around the spot with his finger. "I think I felt it."

*

In 1980, after Monma Katsuhiko had been awarded the posting of radio operator for Japan's twenty-first mission to Antarctica, but before the two-week journey by boat concluded, his cabinmate composed a song—for a cappella recital, in unison—memorializing how the members of the expedition had been led to their task by a common sense of longing. The song and its title went, *"Watashi no anata."* Literally: "My you," but the *you* was the familiar *you*, to indicate intimacy. "Only you." When the boat reached Antarctica, they found their arrival had been preceded by weeks of unseasonably warm weather and a rare day of rain. "Looks like a construction site," someone remarked. They performed the song anyway, as an ode to the cinder-colored Antarctic mud their supplies had to be hauled through, and the odor of their pit toilets. Over time, the song acquired the self-conscious bittersweetness of replacing the expected with the real.

Monma was reminded of the song three decades later, when a notice appeared in Wakkanai's local newspaper soliciting new members for a choir. The choir rehearsed at a school near the edge of downtown, recently closed for lack of enrollment. Between the end of daylight and the beginning of rehearsal, the parking lot was the choir director's office, where she arranged and annotated sheet music in her car, under the dome light. She drove to rehearsal from eastern Hokkaido, five hours along the Sea of Okhotsk. Her motives were related, the choir's members guessed, to the city of her upbringing, which depopulated after the collapse of its mining industry. ("I'm not a psychologist," she said when their guess was quoted. "I did have a childhood.") The choir had been founded by young mothers eight years earlier, to travel the region performing for children.

Monma had been informed by the director that his limited range fell between baritone and tenor, and although at first the choir members next to him edged away when a song began so he wouldn't drag their pitch, he could sing on key now, in the same timbre he spoke with. An inadvertent glissando lingered between the notes, like the space between syllables, puzzled out phonetically, of an unfamiliar word.

The song at present was about a lost kitten, whose cat sounds had been put to melody. The refrain went, *"Anata no ouchi wa doko desu ka?"* (Where is your place?), and place wasn't house or home, it was place in the broad, metaphorical sense. The choir had the school to themselves, so they left the doors to the room open. While they rehearsed, the custodian checked the fire extinguishers. They watched him enter another classroom, where the wiring for the light switches had failed and he waited for a car to negotiate the hairpin turn behind the school, to check the extinguisher in the flash of headlights.

The "Katsu" in Monma Katsuhiko means "victory," and Victory Boy was named in reply to the bombs the Americans dropped on military and industrial facilities across the river from the Monmas' home in Nakoso, Fukushima Prefecture (present-day Iwaki). During that pregnancy, Monma's mother woke to the night raids, to the sound of the heat from the explosions desiccating the air.

By his second birthday Victory Boy was becoming Occupation Boy, who wouldn't remember the night raids but who, in elementary school, was mortified to learn that students a few years earlier had spent school hours constructing and releasing balloon bombs, which they hoped would cross the Pacific, to cause fires and kill civilians in the country that belonged to the smiling GIs who tossed chocolate from their jeeps. Occupation Boy's taste buds were formed on American condensed milk, and the moral convolution of the adult world became clear, years later, when the announcement was made that Japan would repay the milk's cumulative cost, causing him to reassess a gesture that had made his childhood seem safe and ordered. Japan's diminished possibilities had been dramatized in Nakoso while he was growing up; he once watched an American jeep drive up the steps of a Shinto shrine and park in the sacerdotal space where the deity resided. The radio kept Monma and his age-mates advised of the American-Soviet space race and Japan's Antarctic expeditions, so the latter appeared in the role of the farthest frontier a Japanese explorer could reach.

Before his father's death (Monma was in his early teens), Monma's family hadn't been required to work. Afterward, all of them worked, and Monma's Antarctic aspirations seemed implausible. He was a distractible, butterfly-chasing child who finished high school ranked fortieth out of forty-three in his class. His intentions relied on the Coast Guard remaining shorthanded. When the Antarctic research mission was reassigned to a different branch of the military, he was undiscouraged; the missions would include a Coast Guard radio operator, a development that obviated the need for Monma to select from several types of training.

He was married and stationed in Hiroshima when his fourth application to join the expedition was approved, and his daughters were one and four when he left for Antarctica, to commence the pattern of a career that would often take him away from his family. His reason for leaving never changed, and was a boy's reason: fascination with the parts of the world least claimed, his colleagues in the role of the schoolmates with whom the fantasy had once been shared. Monma told me he settled in Wakkanai for snow and remoteness. A community of retired Antarctic explorers had formed in Wakkanai for the same reason, but Monma was an accidental member: Exhausted from previous relocations, his wife and daughters refused to leave Wakkanai when the Coast Guard changed his assignment.

After the earthquake in 2011, Monma went to Iwaki, where his siblings and their families still lived, running Monma Bike and Motorcycle in the place it had stood since 1903. The earthquake knocked half of it down, and the tsunami carried away tracts of the neighborhood. There wasn't a shop to do business out of, but demand was acute. Fuel was scarce in the disaster area. People rode bikes, or mopeds for longer trips. Everything on two wheels needed fixing.

When Monma and his nephew (who had inherited the shop) were too tired to work, they drank beer, listened to a hand-cranked radio, and pushed shovels along the mud-covered floor of the warehouse, to prolong the rhythm of progress. His nephew described the period between the tsunami alert and the worst of the flooding, which he'd spent evacuating houses around the shop as captain of the area's volunteer fire brigade. Elderly people, he said. Grandparents waiting in the doorways of their homes for their grandchildren, because word had spread at precisely the time when children would have been walking between school and home. Most had overcome their panic and made a decision to stay put, so their grandchildren wouldn't arrive to empty houses and wait in them to drown. Monma told his nephew about the choir, about the children's songs that nurtured the instinct to shelter somewhere familiar: "Where is your place?"

Streetlights weren't on yet in the neighborhoods surrounding the shop, and Monma joined his nephew when he went out with the other firefighters on safety patrol. One night they found a man nodding out of consciousness in his car, his skin the unnatural color of lemon peel, the features of his face constricted by exhaustion so it looked, as he talked, like he was deliberately baring his teeth. He needed hospital treatment, but the car had run out of fuel. Monma's nephew brought as much as he could find. Nobody was sure which roads were open, and the fuel wouldn't last if the man tried more than a few.

After a month in Iwaki, Monma went home to Wakkanai. Passing the illuminated environs of the city's new train station, he remembered his older daughter's most recent visit from Okinawa. She had been pregnant during her visit; it was the first time that Monma and his wife, Etsuko, had seen her since knowing. She pointed out the overwhelming glare of the flood lamps in Wakkanai's suburban car dealerships and said, "If you ignore it, you almost believe you can afford it," words that presently reminded him of the meltdowns in Fukushima and the corresponding national campaign to save electricity, but had been intended to describe what it was like to live in a place that was remote, yet people there were capable of behaving as if nowhere was more important. He remembered the rest of her comments several days later, when he heard a rumor that Wakkanai once considered sending a fleet of trawlers to illuminate Rishiri Fuji—an iconic peak on an island neighboring the city—with dozens of spotlights, as part of a tourism campaign. "Where moonlight is the enemy," she had said.

In 1958, when Steve Tamaki—who wasn't Steve then, but a Japanese name he won't disclose anymore—had turned ten and started his newspaper route delivering the *Hokkaido Shimbun* to 80 houses before sunrise, there were 20 or 25 Wakkanai families, he guesses, who were rich enough from fishing the Soya Strait after the war that everybody knew who they were and which boats were theirs. One of the houses a family like that owned (it's a museum now, and the family owns a gas station) overlooked the street where his route started, two stories, with the ceilings on the first floor built a half-meter taller than other houses, to give the upper rooms an unobstructed view of the boats off-loading in the harbor. Wood for the house had been steamed to Hokkaido from old-growth cedar forests in Akita, but there weren't any master carpenters in Wakkanai, and among the shortcomings of the local help was their dislike of the sanding work required to fit the joists into the elevated first-floor ceiling; it made their arms tired, and they quit as soon as the tongue could be forced—by hammer usually—into the groove.

If there wasn't a ship emptying its net nearby, Steve could pause to absorb the before-dawn sound of the house, which was the creaking of the floorboards over the poorly sanded joists, and the stage whispers of the servants, until the clapper noise of the dumbwaiter door sliding open and shut announced the first tray leaving the kitchen, the master of the house now awake. Steve had started his paper route to save for leather shoes, of a kind the children in the house wore, and from the final house on his route he walked to school, where the other children were fresh from sleep, too many to fit in the classroom and impervious to discipline, and maybe if he hadn't been tired from delivering papers, he wouldn't have been the quiet student, one of a few who finished high school. The rest worked on boats. Some, in their time, were able to own one.

In his inarticulate childhood way, Steve wanted the separateness of waking first, passing on foot through the places adults would shortly claim. He would plan his route to conclude at the breakwater if a train was leaving for Asahikawa, three tons of fish in each car, the ice melting so the cars leaked and the stray dogs from the port formed a bewildered herd, chasing the smell and nipping the steel rail tires.

Or it was Saturday and the hostesses came onto the doorsteps of the cabarets to empty the ash from the braziers. The wind touched the charcoal and made it show the last orange warmth, the ending wisp of smoke, which joined the steam from the hostesses' kimono, from the shoulders, from the armpits, and the place where the collar had worked open, around the sweat-gathering hollow in the front of the throat.

Fish is forever, that's what nobody said but how everybody acted, in the cabarets and the stores, in the cedar forests of Akita. Between school and home his classmates asked to see the pocket money his mother gave him, to laugh and ask what it was, was it money, would a sack of it buy you a rice ball, Here—take this, I don't need it, tomorrow my boat leaves. A trawler's hold, filled once, brought a year's salary in Wakkanai. The trawl nets came up with boulders in them sometimes, from zealous dragging in place of smart dragging.

In winter, boats came back with a half-meter coat of ice, and when Steve was a boy, another boy he knew asked a boat's pilot, in passing, if the ice was any hindrance. The pilot paused his ax in the process of knocking ice from the boat and used the handle to illustrate in the snow how the extra weight could make you capsize. The boy asked how a capsized boat was righted. The pilot pulled the boy by his shirt to the edge of the pier and dropped the ax in the harbor. A boat sinks a little slower, the pilot said. Only a little.

Steve finished high school; some of the classmates who'd laughed at his pocket money didn't knock the ice off fast enough, or were zealous more often than smart, and Steve's uncle in Los Angeles— free after three years of internment—told Steve's mother she could send someone. His siblings couldn't see how, not speaking English, but their how was Steve's why, and his mother told him, "The good son goes," which was a type of resignation Steve was too young to recognize, and mistook it for her saying what she wanted him to do. He was the good son. He went.

His uncle helped him find room and board as a houseboy for a farm-owning family in Pasadena. The money wasn't good. His parents would wire more if he asked, but there wasn't more, not under a loose board in their house, or in a bank gathering interest, there was only the extra shift at the factories that filled the train cars with cleaned or processed fish, three shifts a day, they never closed, and the picture in his head of his mother wearing the thick apron and her eye-lids twitching against the cold while the oil-slick tools worked in her hands. Nights, he worked as a gas station attendant until 2:00 a.m., then studied for three hours, had two hours of rest, worked "like a maid" until his afternoon classes. And again. Days off, he slept.

A letter came every six months from his mother, who hadn't gone to school and whose writing was nearly illegible except, in practiced strokes, the names of family members. The letters went with him to the gas station, where between customers he occupied himself mak-ing a logic out of the places where the names appeared, relative to the other words he could read. Every few weeks he called home, but there wasn't time to ask about the letters. He timed the calls using the minute hand of a clock on a bank across the street. "If I tip you," a driver asked Steve while he pumped, "how'll you spend it?" Steve said, "One minute," and the driver, misapprehending, awaited the reply he'd been given.

Steve befriended a Black doctor in the Black neighborhood where he lived after he moved back to Los Angeles, and the doctor told Steve that they were both colored, that white wasn't something you could half-be, or be somewhere between it and something else, the way Steve had been explaining himself to white people, whose politely dissembled surprise he had begun to notice, who were concerned about him living where he lived but didn't know anybody in their neighborhood with a place to rent. At a certain point, Steve devel-oped an Americanness, to the point his where-are-you-from reply went, "I happen to be born in Japan." And happened to have a life in America, the upwardly mobile type that hardworking immigrants were ostensibly entitled to, until it happened that he was the father of two children who thought his accent was funny, and the leather jacket he wouldn't give up, the way he talked to Black people in pub-lic as if they were friends he'd run into, not strangers startled by his casual profanity. Japan's changing place in the world—the notion a white person would think a Japanese person was anything other than poor—struck Steve as fictional and temporary.

He was sixty-two and his mother was eighty-seven when her health began to decline. He hadn't visited Wakkanai, had never seen a pic-ture of the city since leaving. He put his children in charge of his rental properties and moved home. His brother drove him from the airport. His mother was in the back seat, lying down. They wanted to know which "people" Steve meant, saying, "Where the people?" in English, repeating himself, not appearing to register their replies. All the people, it turned out, and the hundred boats he might not have expected to see, but hadn't expected to see none. The empty roads had been widened, and the land beyond the harbor reclaimed. If a bus passed—and caused him to recover an image of the buses of his childhood, the riders crowded against the windows—there weren't any passengers inside.

His first month home, he avoided leaving the house. He worried he'd get lost and wander into the well-lit strangeness of South Wakkanai, which resembled a miracle mile in suburban California, and reminded him of Japanese boys he had known during the occupation who liked to wear imitation GI hats, who practiced their American accents with the slurs and curses the GIs taught them.

Zero-sai, Steve would say to himself after his mother sent him to L.A., adding the Japanese suffix for *years-old* to the age of his life in America, whatever his physical age. Which was sixty-two now, and he'd forgotten enough kanji that he was illiterate in the city where he was born, but the memories were there, the leather shoes coming down the steps from the cedar house, trailing the scent of tatami and wood dust, or he remembered the scent, which the air had probably been too cold to carry, and now he saw it was his first memory of what work had to do with him, what the boulders in the trawl nets and the steam coming off the hostesses' kimono had said, and the wide, empty streets now, the boarded-shut factories, where the palms he could know in the dark were his mother's had been formed by the damp interior of her latex gloves and the narrow handle of the boning knife. Said it counting backward through the years between until they reached the place where backward counting goes and he put the suffix for years-old, as had once been his custom, at the end of it.

*

Of Wakkanai's buses, the Tomioka Line runs closest to Wakkanai Hokusei Gakuen University. Most passengers board in South Wakkanai and the surrounding neighborhoods, with groceries, with strollers and walking-age children. South of downtown, the city has sprawled on both sides of the highway, in housing subdivisions hardly linked by road. The bus passes through several, in the circuitous manner required, giving a tour of bubble-era architecture in Wakkanai. When reporters filing from rural Japan put color material into serious stories about population decline, they're reacting to kitsch (you could call it)—unironic optimism, made ironic by time and neglect, apparent in the objects of a mass-fabricated lifestyle. The built environment is one of them: uniformed schoolchildren waiting for the crossing guard's signal in an archway formed by a pair of concrete giraffes, which mark the entrance to a public toilet; the old signage—a slice of canned pineapple hovering supernaturally above the can, a disembodied hand making the A-Okay gesture with a strip of kombu pinched between thumb and forefinger—showing through the whitewash on vacant storefronts; a polished-to-gleam barber pole revolving in front of a barbershop where snow shovels (it's summer) are visible through the sparkling-clean windows, stacked against the barber chairs.

The Tomioka Line stops at a government-subsidized rest home, where many residents are regular passengers. They have somewhere they regularly go, or the somewhere is improvised while going. A passenger of the latter habit takes a stack of bus schedules whenever she rides. On her way off the bus, she displays the stack to the driver and says, "I wrap fish."

The older children on the bus explore their groceries, or their homework, or the dazzled expressions of their infant siblings. Out the window doesn't interest them. They came from it, they'll go back. The image of smiling neighbors chatting in a community garden while a monorail cruises overhead interests them; it's a poster taped to the ceiling of the bus, where its text encourages riders to cherish public transit. A few children watch the adults watch the windows. The most articulate description of Wakkanai's captive, habitual optimism I heard in seven years visiting the city was spoken on the Tomioka Line. A girl said to her father, "When the wind moves the grass, the grass looks like it's growing."

From Yokota Koichi's office at Wakkanai Hokusei Gakuen University, he could watch the wind move the grass. The university marks the outer boundary of the city's expansion, sited in 1987 as a junior college when Russian traffic through the port no longer seemed like a temporary phenomenon. Outside his office, time had worn an uneven pattern in optimism's institutional arrangements. Floor-to-ceiling windows overlooked bristlegrass and standing water where there would have been businesses catering to students if enough had enrolled. When Yokota was appointed, there were thirty-five in the incoming class.

Yokota took the lectureship in 2012, a year after his final term ended—the year the ferry to Russia carried few enough passengers that a suspension of service seemed inevitable, the year the idea to build a destination-quality curling rink began to replace ferry subsidies as the city council's source of fiscal self-esteem. He still looked like the mayor, talked like the mayor, hadn't lost the weight a mayor puts on at catered events, or lost the imperative quality of his speech, which mayors develop because the stereotype is true: Compared with other political jobs in Japan, theirs gives the most influence with the least accountability.

When Yokota was elected in 1999, Wakkanai was perched at the fulcrum of its postwar economic adjustments. The region's waters had been overfished when he was a boy, and a 1978 treaty with Russia adjusted each country's maritime boundaries, halving the catch of Wakkanai's fleet. Japan's asset bubble burst in the late 1980s, accelerating deindustrialization in Hokkaido, already the nation's most severe. The Soviet Union, however, was only a few years expired when Yokota's administration began, and disorder in the new Russian government provided circumstances favorable to crab smuggling. Businesses and politicians in Wakkanai were receptive; smuggling had been part of the city's culture since 1945, when victorious Soviet forces forbade Japanese travel to or from Sakhalin (former Karafuto Prefecture), where 300,000 Japanese residents were waiting to be repatriated.

Russian crab smugglers left behind most of their pay, spent on used cars and appliances to resell at home, stimulating Wakkanai's economy at a time when long-term rejuvenation of the city seemed possible. During Yokota's first year in office, the Russian economy recovered from its first post-Soviet crisis with surprising speed. Offshore oil and gas exploration had begun around Sakhalin, and Wakkanai was the shipping hub for Japanese materials and machinery. Domestic tourism to Wakkanai benefited from a growing interest in Japan's outer islands, and officials expected Russian tourists would follow.

Crab smuggling was the first windfall to expire, curtailed by the

Russian Coast Guard. The price of oil peaked and declined, the value of the ruble with it. Sakhalin's petroleum industry stopped building and started pumping, no Japanese equipment required. The affections of domestic tourists were temporary. Before Yokota's final term concluded, Wakkanai's affairs had been discussed, for the first time, along the lines of discourse established by the evergreen object of Hokkaido's economic anxieties—*Yubari (city), spectacular bankruptcy and depopulation of.*

After Japan's coal industry began to decline in the 1960s, Yubari suffered widespread joblessness and depopulation, losing 90 percent of its residents within two generations. The city responded with a coal industry–themed water park, the "clown nose" (a columnist's description) of its failed attempt to manufacture a tourist economy. Yubari's misfortune supplied a precedent for economic prognostication in Japan, where natural disaster governs other metaphors. A town didn't fade away, it was swept away, and not by a gradual process like the migration of a sand dune, which resembled the more accurate notion that people in Yubari had lost their equity, all of their earning power temporarily and some of it permanently, and were leading lives of reduced expectations.

Upon his election, Yokota's slogan for the Wakkanai he envisioned had been "A place where many come," but the first item on his desk was the financial collapse of the city's largest hotel. Later he adopted "A compact, sustainable city," which was equally wishful; land in the historic port area, where the city of Yokota's revised slogan belonged, was mortgaged in a way that could not be discharged by foreclosure, eliminating the possibility of redeveloping it.

Of Yokota's attempts to redefine the city, the construction of a rest home in Wakkanai's new train station was the most ambitious to reach fruition. It was a pragmatist's move. Demand for elder care in Wakkanai was high and would remain high. It was also forward thinking. Prominent gerontologists had persuaded a few cities to locate elder-care facilities downtown, but a facility that offered the full spectrum of elder-care services—including care for dementia sufferers—had never been attached to a train station, the commercial nucleus of a Japanese city.

The project faced opposition from the city council. Yokota was described as the mayor who had marked downtown unsalvageable.

Old people don't spend, his opponents argued, and they're a nuisance to tourists. Yokota declined to steer the debate toward the pivotal, unmentioned fact: The mayor's office had pursued other uses of the space; only the rest home could afford to lease it. To prove he'd explored his opponents' proposals, he would have needed to publicize the underwhelming outcome. At the peril, he felt, of his constituents' sense of their city.

Yokota announced he wouldn't contest the next election. The council voted in Yokota's favor, and the new station was built with the rest home inside. Wakkanai residents went back to worrying about the Sakhalin ferry, subsidized by the city but insolvent anyway. When service was discontinued in 2015, employees of the ferry company stopped wearing their uniforms in public. People had approached them, they reported, with a startling intensity of emotion.

The preferred antidote to this anxiety was the proposed curling rink, and its precedent was a similar facility in Kitami, a city with quadruple Wakkanai's population and a rate of population decline less than half of Wakkanai's. On this matter, Yokota and his former critics agreed: They wanted to build the rink, primarily to attract funding from the prefectural and national governments.

"I'm annoyed by the notion that people in Wakkanai should feel chastened," Yokota once told me. "That a city in decline should dare nothing."

A reporter for a national newspaper, who visited Wakkanai and chatted with Yokota after the rest home opened, once expressed to me his disappointment that the former mayor wasn't familiar with the category of ideas he had accessed when he located the rest home in the station: how a government makes sound decisions when its prior intentions—growing a city, a prefecture, a nation—become fanciful. This was during the time when senior Bank of Japan officials were making candid statements about the limited effectiveness of structural economic reforms in a nation under demographic pressure. The media would have liked a story about a local politician who governed with visionary realism. Yokota, this reporter decided, blustered too much about the costly projects he wanted to pursue; he wasn't the guy. Nonetheless, the ideas the rest home dared to demonstrate, arguably visionary and inarguably realist, inspired larger-scale imitations in other Japanese cities. Most extensively, in Yubari.

Yokota taught a seminar at the university and once invited a representative of Japan's forestry federation to discuss the group's agenda for decentralizing their activities, which would spread the economic benefits of their work into depopulating prefectures. By offering competitive salaries, the federation's members had been able to attract job applicants living in major cities, who were curious about rural life and willing to relocate. When the chairman finished, Yokota explained to his students that fishing cooperatives in Soya also incentivized relocation by offering the families of successful applicants a house to live in and letting them keep it after ten years' work. Few of the families chose to stay, Yokota lamented, because local salaries were among the nation's lowest.

Presumably, Yokota pivoted on this unflattering assessment; he praised the willingness of the cooperatives to act independently, or encouraged his students to imagine how much could be accomplished if local salaries improved. Recalling his comments later, in his office, he came vehemently to his point about the ineffectiveness of Soya's hiring incentives and stopped mid-phrase. "Do you ever have that experience?" he said. "Where an old truth comes back and feels sudden?"

*

Kushibiki finishes distributing instructions for the online census and starts his route again, to assist residents who would prefer to reply in person. The first morning, a gas station attendant informs him that his car (a Suzuki Hustler) shares its name with an adult magazine in America, reducing his sense that its branding captures his diligence, but he counts this peculiar omen against two better ones: A billboard has gone up at the city office advertising the visit of a famous baseball player who will deliver a public lecture titled "Chase Your Dreams," and the echo of this sentiment on the floor mat in the lobby of the city office, which, from the dirty-carpet face of a cheerful seal pup declares, "Welcome to Dreamkaido!"

Dressed in a new suit (the collar leaves a stripe of irritated skin on his neck), starting his rounds in the late afternoon, Kushibiki does not rehearse—as he had on previous days, in his car—the words of his prepared greeting. He leaves the car immediately and walks to the center of the street. "Get the image," he instructs himself, having decided that his habits of perception need improvement, separate from the habit of studying.

The neighborhood responds in its aesthetic way, with the sounds of a studio audience from a television in the nearest house, next door the intermittent intonation of a smoke detector, its battery expired, the occasional noise of a car navigating the streets near the train station, the shuffled-paper sound and snow-globe sense of the breeze moving the trees and the grass and the laundry hung on the lines. It's a clear afternoon, and the low-angle sun from behind Rishiri lights the ocean and the islands in warm colors, in silhouette.

"A little longer and it's autumn," he says. "It's brief here." He has his clipboard and knows the streets now without a map.

Today the pets are home. As he leaves a house where nobody answered, the sound of paws on glass leads Kushibiki to discover that a dog has nearly nudged the door open. In a balletic stumble from the street back to the house, his clipboard dropped, he catches the door with his shoulder and slides it shut, listening for the latch. A few houses later, Kushibiki finds Riki, a Shiba Inu, in his outdoor doghouse, where a sign on the fence reads, "I'm a fatty! Don't feed me," and Riki watches a deer in the backyard graze the rhododendrons; while Kushibiki pauses to absorb this irony, Riki's owner exits his house, enters his car, and leaves.

Since Kushibiki distributed the information packets, enough time has passed for residents to look them over, and he expects the letterhead and full-color printing have made a few points on his behalf, among them establishing his identity as a representative of legitimate authority. In a city where front doors are rarely locked, he surmises, the image of the stranger in the doorway is familiar only from television, where he is an agent of organized crime or—best case—the landlord. At the next house, the most belligerent resident of his district demands to know why Kushibiki doesn't "count boats, not people, save the city some money." It is the closest the gentleman, maybe thirty-five and standing in a vestibule that contains a refrigerator box filled with smaller boxes of historical postcards, has come to granting that Kushibiki might be visiting on the city's behalf.

At most of the houses that appeared abandoned on his first visit, the census packets he left have been retrieved. Without pattern: In

some cases, they were collected from homes that are plainly unin-
habitable from neglect. At a house like that he finds a yard sign for a
local property management firm decomposing in a layer of sediment
that eroded into the backyard from uphill. Houses in Wakkanai rent
for less than it costs to maintain them, if they rent at all. Small apartments in elevator-serviced buildings downtown—when they briefly
become available—are priced as high as they would be in Japan's largest cities.

Because the data he obtains today will enter the official record,
he asks about houses' address plates. Nearly every plate has a family name and two given names, a man's and a woman's. Often the
woman comes to the door and replies in a variation of "I'm her, my
husband passed away." By evening, the question has begun to make
Kushibiki uncomfortable.

At a locked door with no bell, he resorts to knocking. A light comes
on. He knocks again, waits, and has begun to depart when a silhouette fills the privacy glass. An elderly woman, for whom the weight
of the sliding door requires visible effort. When it's open, she puts
her hands on the wall to steady herself while she descends the single
step into the vestibule. The plate on the door has two names on it; as
she makes a false start down the step, then another, Kushibiki whispers, "Why am I going to ask?" She puts on the porch light, unlocks
the front door, and slides it enough to let his voice through. After
identifying himself, he mentions the plate.

"I'm alone," she replies. She is shorter than Kushibiki, and stands
with a stoop, so she says it into his chest. He is looking at the top of
her head under the porch light, the undyed roots of her hair and the
pale colorations of her scalp.

"The other name?"

"No." She tries to straighten, her neck strains, and her eyes strain
toward Kushibiki's, where she wants to direct them, but she can't. She
has to lean against the sliding door to recover.

When she's back inside and the door is shut, Kushibiki shows 7:24
on his watch and says that's as late as he should go. He has sweated
a palm print into the top sheet on his clipboard, cradling it against
the heel of his hand while they spoke. He starts to say something
about city office policy and disturbing people who might be resting,
but stops mid-thought to watch a young foreign woman wearing a

school uniform cross the intersection at the end of the block and continue downhill, her face briefly visible when she passes under a streetlight. "Do you think she's a dancer?" he says. (Exchange students from Sakhalin are typically fine artists.) The question lingers until she crosses another intersection, and Kushibiki, feeling in his pocket for his car keys, says, "Not my district." But his first steps, before he remembers his car is behind him, move him in that direction.

*

"The rest play video games," Etsuko speculated, driving past the place where children sledded on the city's piles of plowed snow. Today it was only the youngest children, escorted by their parents. On Antarctic expeditions, Monma replied, the introduction of internet service had coincided with an increase in emotional disturbances. They were driving to Yuchi, a farming town inside Wakkanai's municipal boundaries where Monma had built a cabin, and where he spent most of his time after coming back from Iwaki. The route passed an elementary school where Etsuko had taught, closed for several years, then a cluster of wind turbines, motionless while a malfunction in a similar model of turbine was investigated. "You forget they're not ornaments," Etsuko said.

The cabin overlooked a creek. Where a small bridge crossed it, leading to the road, you could net fish for salmon in the spring and fill a bucket in an hour, but it wasn't legal, the streambed was too narrow. When the salmon emerged on the other side of the chute, their scales had been stripped in the places where the current dragged them across the rocks.

There were baitfish in the stream, too. He could catch one every minute with a short rod and eat them off his charcoal grill. When his grandson visited, he taught him to bait and place a line. The boy lived with Monma's older daughter and her husband on Iriomote, population 2,347, one of Japan's southernmost islands. He'd never visited Wakkanai, and shivered in the spring weather. In Iriomote, Monma's daughter and her husband worked on a pineapple farm and took odd jobs (he'd been a salaryman but left that life by way of Australia). It had occurred to Monma and Etsuko that among the postwar generation young people didn't wed if neither had secured a career. They declined to apply this notion to their daughter's life, where it didn't seem useful.

"There's something we called the echo phenomenon," Monma explained once, to friends who had come to the cabin to eat the bait-fish off the grill. "Operating a radio in Antarctica, the signals go both ways and cancel. You hear a fragment of what you were saying, but the sound changes. If you don't know, you talk to your echo." For most of the years since he encountered that phenomenon, it made him think of exploring: The farther you go, the more of yourself you find. Now it made him think of his children.

Teni shoku wo tsukeru—somebody brought that up. *A vocation takes you there.*

Young people didn't say that anymore, Etsuko remarked. They had truisms appropriate to their era.

Etsuko had taught elementary school wherever Monma's Coast Guard career took them, bookended by their longest periods without relocating, in the Inland Sea and Wakkanai. Both times, the school districts were small enough that a classroom of elementary students depicted the town—the city, the region—in miniature. More students left home than stayed, Etsuko recalled, "but I don't think they had any idea, when they were young, they would be leaving, and when they left, I don't think they knew they wouldn't come back. Except to visit, except if they failed to get the life they wanted."

Growing up in Tohoku, Monma's friends had been the younger sons of farmers and fishermen, who understood before they could express it that their elder brothers would inherit what their fathers built, and they would need to leave home to find livelihoods. Most of them went to Hokkaido, and if Monma had farmed or fished, he would have known them in Wakkanai. His daughters would have married their sons. Instead, for the fourteen years he lived in Wakkanai before he retired, he was deployed more often than he was home. If he encountered someone his family had known in Iwaki, they thought his Coast Guard stories were put-ons. He'd been the dumb kid, of course he'd come north to work in barns and canneries.

His younger daughter worked in Tokyo. She wasn't married. Maybe her sister knew a bachelor on the pineapple farm, he said. That was a good joke for the people who'd come to taste the baitfish off the grill, whose gainfully employed children phoned their parents in Wakkanai for money because they lived in cities where starting a family wasn't affordable, and political humor in part, because the government wanted everyone having children so the economy wouldn't come unstitched, and if you had a heart, you started a family because the economy, in coming unstitched, had revealed itself as a false metaphor. Somebody said, "She can marry the fellow off the new recruiting posters, you've seen them on the police boxes, for the Ground SDF; handsome, too." The Ground Self-Defense Forces were the part of the military that had been, in Hokkaido, for deterring Soviet invasion, and now quietly absorbed their considerable share of the defense budget while figuring in nobody's plans for Japan's self-defense.

"Outearns him," Etsuko replied. With her eyes she followed the progress of a truck on the road across the creek, moving uphill toward the city. Something was piled in the bed of the truck. From a distance it looked like upholstered furniture, the wind vibrating the upholstery, giving it the appearance of a plucked string. "And men don't adjust overnight."

Beyond the road, the clouds were sparse around Rishiri, which was small on the horizon, suggesting dry air and low pressure. Describing Rishiri Fuji as a scenic barometer became popular, Etsuko recalled, during the brief expansion of domestic tourism, when Wakkanai's economy depended on the islands nearby. No one else remembered this coincidence; they thought the notion had been present in their childhoods. "It would be nice to know," Etsuko said, "whose memory we can trust."

*

In a socially instructive anecdote frequently mocked by bureaucrats at the Ministry of Economy, Trade, and Industry (METI), a traveler in the nineteenth century has reached a lodge in the Japan Alps. Dinner, when served, omits local delicacies in favor of sushi. On taste alone the traveler would guess he's chewing sulfur, which is what the porters of that era would have used to preserve the seafaring creatures on their improbable journey from Tokyo. The anecdote allegorizes a convenient notion in Japan's political discourse: that a renewed commitment to Japan's regional cultures, as opposed to Japan's modern

plenitude (cheap sushi everywhere), can restore places like Wakkanai to the distinct identities that once made them prosperous. Then
people will visit and, often enough to stabilize their economies, will
settle.

Two parties responsible for propagating this notion are impolitic
targets of ridicule. The first is the parliamentary coalition built by Abe
Shinzo and his successors, which maintains that population decline
can be curbed. The second is the group of cities that have been most
receptive to this coalition's claims (these are typically the smallest cities, the farthest from metropolitan areas with growing economies).
The government proposes tax credits for young parents, and it seems
plausible the birthrate might rise. Already, academic research shows
higher rates in the countryside. Rural Japan sees: Move here to solve
the problem. Conservative politicians see: Claim progress, capture
the rural vote.

Bureaucrats responsible for Japan's regional economies have cultivated a policy climate that is quietly contemptuous of the delusion
that the birthrate can be controlled, and of birthrate studies in general, which fail to integrate research from other disciplines: Fertility
is lower in urban areas, for reasons endocrinologists have examined;
social scientists find that better-educated survey respondents believe
their work lives are incompatible with child rearing, irrespective of
where they live (they happen to prefer major cities). Instead, policy makers regard Japan's population decline as an epochal, society-
encompassing development, Japanese modernity's third act, after the
Meiji Restoration and the bubble. As in Meiji, when Japanese officials
visited Western nations to adopt the best elements of their respective
social contracts, bureaucrats have recently gone abroad in historic
numbers, to countries that have faced demographic pressure with
aplomb. They admire Germany, where several cities are anchored by
major employers whose share of the global market is stable. In Japan,
the intensification of urban drift has provided a few cities with too
much of what other cities need: People, and millions more won't fit
inside Tokyo, so the cost of housing rises, and the middle class revises
its sense of what's financially advisable. Children, perhaps not.

If population and industry could be normalized across several of
Japan's largest cities (eight is the number typically described), then

categorical steps might be taken to stabilize the economies that surround them, which include—demographically—most of the country. Of these aspirations (decentralize industry, decentralize population), the first is economic and well studied, but the second, domestic migration, is social, and therefore as responsive to policy intervention, one official put it, as it is to a séance.

The government possesses the wherewithal to provide bespoke incentives to companies whose needs are job intensive if they develop corporate or industrial infrastructure in struggling metropolitan areas. The resulting jobs would reduce dependence on distributive institutions, like hospitals or grocery stores, which sustain local economies without growing them. These policies are intended to benefit cities with 50,000 residents or more. While reporting on the nuclear industry in Fukui Prefecture, I encountered national bureaucrats who were visiting their municipal counterparts in Tsuruga, population 66,000, and in Sabae (69,000), where the local economy depends on the manufacturing of eyeglasses and rayon. I didn't see them in the castle town of Katsuyama (23,000), home of Japan's best dinosaur museum, or in Obama (29,000), the only city on Atomic Broadway to explicitly reject the siting of a nuclear plant. Below 50,000, experts say, a city's economic gravity has dissipated. Under this criteria, Wakkanai's economy extinguished itself decades ago, during the general deindustrialization of Hokkaido, before the collapse of crab smuggling, before the ruble plunged and with it trade between Wakkanai and Sakhalin, before anything had occurred that people in Wakkanai identify as the source of their economic inertia.

In the absence of policy intervention, the fate of cities like Wakkanai is chiefly the territory of credulous journalists and people-pleasing politicians. The primary neologisms involved, "U-Turn" and "I-Turn," describe the enticement of young rusticators, the former born in the countryside and coming back, the latter fleeing the costs and stresses of urban life. To the extent that people who correspond to these rhetorical devices exist, a handful of cities have appealed to them convincingly. Onomichi, where Ozu Yasujiro filmed the motion pictures that established the aesthetic appeal of small-town Japan, is one of them. Not all minor cities resemble, nostalgically, the set of a movie. Observing their chambers of commerce, you'd think they'd watched too many; that's how they expect a happy ending so improbable, where an abundance of young people choose anywhere over Tokyo.

"Ask someone from Wakkanai and they might say, 'If we can keep twenty thousand people.' I find that revealing." (METI talking.) "*Keep* them? Mortality is not administered by the state." Which makes it a shade more difficult to control—only a shade—than the birthrate or domestic migration. Bureaucrats working on population decline are certain that Wakkanai won't have twenty thousand people in it, and in private nobody claims, or claims on anyone else's behalf, that something will have to be done.

*

Steve would have sheltered in his mother's house longer if it hadn't become difficult to sleep. Where he'd lived in L.A., cars passed at all hours, pedestrians shouted to each other from opposite sides of the street, sirens penetrated closed windows, and the walls hummed intermittently with helicopter noise. Aware that people in more comfortable lives wished to avoid these sounds, he never considered the way they made the world reassuringly present, until in Wakkanai nothing replaced them.

In the life he had led as an American, work led to tired, tired to sleep. He tried the work his mother needed, mostly errands to the city office. After the number taking and waiting, he'd ask the city office workers to talk him through the forms. He couldn't tell what surprised them: his illiteracy, or the fact that he asked without indicating embarrassment, didn't lower his voice or lean confidentially across the counter.

In response to his questions, the city office prescribed paperwork and patience. Nobody said, when he remarked on missed opportunities for empathy or pragmatism, "You're not wrong," which was a phrase he'd learned before his English was fluent, and remembered thinking that if you said it in Japan, no one would perceive what it meant: that you could be right without your rightness becoming the final concern.

Perhaps to reduce their share of his attention, city office employees made Steve aware that he could obtain weekly access to a classroom at the old junior high school in Midori, the suburban neighborhood closest to downtown. He taught community English classes there, with his students in folding chairs and the ceiling lights painting a brilliant, gelatin gleam on the vinyl floor tiles, which were different shades where they'd been replaced, and the older tiles were curling away from the floor, but all of them had been mopped and not a missed spot in the room, which looked like the room of a school between semesters, not a school that would be demolished.

Steve cut an unlikely figure in this environment, with his gray ponytail and half the hair he could have swept into it floating in its own static charge. With his leather jacket, worn-out and sun bleached in an all-over pattern, the pattern of a city bus seat and the incomplete shade of a palm tree. With his features migrating to one side of his face when he talked, the way working men of a bygone era talked when they leaned against the railing at lunchtime, to get the words to the man next to you without missing what happened on the street.

At first his students were amused, and if his methods were unfamiliar, this was attributed to the authenticity of his experience. When he learned that the Wakkanai bureau chief of Japan's largest newspaper would be attending his daughter's wedding in Michigan, Steve made him stand in front and practice hugging. Yes, the bureau chief later reported, his American in-laws made him the recipient of several unsolicited embraces.

When Steve decided enough teaching had occurred that his students should be held accountable for their mistakes, his indelicately American way of pointing them out had an effect on attendance. More than once a student quit during class.

Steve discontinued his classes in favor of private lessons, which, he estimated, wouldn't expose him to the incompatibility of Japanese culture with his idea of the language he'd lived in, because there would be only one Japanese person—his student—and you need more than one to make a room more like Japan than it is like a room. His new students cooperated. None asked about old Wakkanai or what it was like to come back. They asked if Americans were like the Americans on TV.

If he couldn't sleep, he tried to find his paper route and walked it. Nightlife in the Wakkanai of his childhood had started at sunset, and nobody left the cabarets until two in the morning, if they left.

Now there wasn't any nightlife, just the wide streets paved flawlessly and nobody on them. Steve had heard about U-Turn and I-Turn, and imagined himself as a young man, transported to the present. He imagined leaving Wakkanai, with no intention of returning while there was work left in him; in Wakkanai, where would he put it? In California, a day was less time than the work it required, work you wanted even if you didn't enjoy it, because you looked at yourself while you did it and saw the difficult, dignified truth. He had only one hobby, there was never any time for it, and he had regarded it as typically American (it was fishing) until he was back in Wakkanai, passing the breakwater on his evening walks. The tunnel through the dome of the breakwater—where he still partly expected to see the old train station—made him think what a funny thing it was, fishing not as a livelihood but as a pastime.

The convenience store in Hokkaido is Seicomart. If a Seicomart has hot food, there's a sign underneath the Seicomart sign that says "Hot Chef" on a red background with a grinning cow wearing a chef's hat. The Seicomart with a Hot Chef in downtown Wakkanai is near the handful of factories that still process fish. Steve passed it on his walks. Chinese guest workers from the factories used it for entertainment: cheap whiskey in a plastic flask, a burger or *okonomiyaki* on a plastic tray, a comic book whose Chinese-derived ideograms they could vaguely follow. Steve watched them shop, listened to them speak Chinese, and felt instinctively that the broken English they used with the cashier came closer to American English than his students could. Even in Chinese they were closer. Even between phrases, just moving in the aisles with the air around them absorbing their drowsy gestures, the nodding-off way they signaled each other, then the tired sound of their voices, which maybe were talking, in anticipation, about the beginning moments of a dream.

Tokiko, August

"I'm not afraid to sleep. When I was young I swam between the islands. I learned to hear underwater. Do you understand the relationship?" Between sleep and listening, sleep and submersion, but Tokiko didn't pause for an answer from the other residents of the rest home, at the table with her, decorating the tissue box covers they'd made from coupon paper. "Your heart, you hear—in your fingertips. And voices, anyone talking above the surface. Your tongue in your mouth makes a kind of noise, underwater you hear it. When you go for air, the voices say what you thought they were saying, by the sounds they made that weren't words."

Endo, the rest home's manager, had crossed the room and laid an index card on the table, lined side up and blank, but on the unlined side she'd written, "Certain things, Tokiko remembers vividly." Out the window behind her, from the top floor of the train station, which the rest home occupied, the rear entrance of Wakkanai's downtown grocery store was visible, its windows covered in wall-length, photographic decals of fresh produce. Briefly, the digitally enlarged image of an eggplant matched the silhouette of a passerby, and Tokiko said, "The eggplant has been left on the vine." With her eyes, she directed the comment to Endo, who, alone among the people listening, didn't laugh, only waved by touching her fingers to her palm, the same hello-in-there wave that fooled me the day before, with a newspaper on the table between us and its headline about a questionable death in a rest home in Nagoya. The experts the article quoted had blamed turnover among care staff, a fault of the industry that had employed Endo for two and a half decades. I asked how she stayed in a profession that did nothing to keep her. "By awaiting a pleasant surprise," she said, and waved as if to greet someone, so I turned to look behind me, where I must have known there was only a wall.

Between Endo and Tokiko, observing the other residents while they laughed at Tokiko's comments was a way to acknowledge the distinction between losing your memories and losing your faculties. The other residents thought Tokiko had lost both, because she described her life before the rest home by repeating an episode from it, adding another, then starting again, the way an actor works on a monologue.

"Swimming to Rebun, I swam naked. I was growing, my siblings were growing, my parents couldn't afford swimsuits. There were urchins to eat; you got tired of fish, but not urchins. The way mainland Japanese thought of rice, we thought of urchins. If the tide was low, the rocks were covered in urchins. From a distance they looked like moss. If you waded toward them, you stepped on urchins in the shallow water; they pierced your foot. I went where the water was deepest, I dove with my eyes closed, I let the backs of my hands drift on the bottom until I touched something sharp.

"I liked bento. If we gathered shellfish, the adults sold them to the fishermen's union and paid us part of the money. I was selling fresh seafood because I wanted bento from the mainland. It was a novel taste. I never knew it was stale.

"If I swam to Rebun, my appetite by the time I arrived made the bento taste better. I thought, *The bento in Rebun is more delicious.* If my parents brought money from the fishermen's union, I swam to Rebun with the coins in my mouth. I could stay in the water and toss the coins to men fishing on the shore. If I watched their rods while they were gone, they bought the bento for me. They didn't know, but I took their lines out of the water. I thought if a fish took the line, I would lose the rod.

"We came to Wakkanai after the war. We had bento all the time. My siblings and I longed for the food we'd eaten on Rishiri, so we stole shellfish from the miso urn where our parents pickled them. The factories were busy processing fish. Every worker double shifted every day. Our hands were stained yellow from digging in the urn, but our parents were too tired to punish us."

The young care worker sharing her shift with Endo began to lower the window blinds. She was eighteen months into the job, seventeen longer than she lasted in the psychiatric unit of an Osaka hospital. At the rest home she took notes when a resident told a story obscure to her, and in this way learned—on the internet after her shift— what she hadn't learned in school about postwar Japan, until it was only a matter of wondering why, understanding what the words referred to, she couldn't think of what to say in response. Hoping to discover a point of connection, she asked several residents to help her remember the lyrics to a folk song she had in common with their generation, "Love Is an Island," which her mother would sing to her when she was a girl. In high school, her parents instructed her to sign up for the vocational course in nursing. Her father drove a charter bus. Her mother was a tour guide on the bus, with a repertoire of folk songs for the passengers, and didn't want her daughter to marry a bus driver or sing to tourists for a living.

Watching the blinds descend the windows, Tokiko said, "It snowed all winter. The rest of the year, there was salt on the wind to turn the stones and grass white, and the houses on the windward side. At night I'd go to the beach and feel with my feet for urchins. Any moonlight, everything turned white, like a photo negative.

"But underwater," Tokiko added, now addressing the care worker, who was adjusting the angle of the blinds and stopped to listen, with the wand in her hand. "If the sun goes behind a cloud, the light changes the water before it changes the air; you notice before it happens." The care worker rotated the wand to dim the room and, humming the jingle that played in the convenience store downstairs, excused herself. Tokiko briefly watched the sunset through the gaps between the blinds, then slid the nearest tissue box cover from the hands of the woman who'd made it, who was, with her eyes half-closed, her head against her shoulder, inhaling deeply and holding her breath, trying not to nod off. Tokiko held the cover toward the window, where the coupon paper caught the light. "I'm not afraid to sleep," she said.

Tokiko, September

The windows on the fifth floor of the rest home faced west toward the city; the height of the building made them level with the place where the sun went below the hills in the evening. The shorter days in autumn, before the days were truly short, left the dayroom and the hallways often lit by Japan's longest twilight, and residents of the rest home began to sleep more, a phenomenon familiar to the home's longtimers as a seasonal invitation to dementia. "A dream is to practice forgetting," they speculated, or, "The greedy remember best." Because a dream might arrange remembered things in wrong patterns; because the mind keeps what it believes it needs.

At residents' request, the television was left playing. One time it was a news segment reviewing its coverage from five years earlier: a story about elderly drivers' objections to the magnet they were required to place on their cars, in the shape of a desiccated leaf. Conversation, when someone started it, accumulated around the two couches that faced each other perpendicular to the television, where people lingered and departed according to their interest in occupying the spot where they were most likely to encounter each other.

"Fewer times a day I hear what the boats are doing," said Tokiko, whose name was spelled Time-Girl in ideograms, the way it's less often spelled now, after the benighted boy emperor's pious grandmother in the twelfth-century epics. "The wind turbines are off for the winter, like every year." They weren't, and hadn't been; they were in a brief interruption, for maintenance. It was true the ferry schedule was tapering off and the horns sounded in the strait infrequently enough to startle you.

"This rest home would have been a pachinko, but the pachinko was already there. There are two new buildings in Wakkanai, the other is pachinko, so it could have happened in the opposite order. People wouldn't want pachinko in the train station, but they didn't want the rest home either."

Pachinko Town (the pachinko's name) shared a parking lot with the station. Tokiko was glad to be near it. Before coming to the rest home, she played whenever she could and used her time in the parlors to smoke, a forbidden indulgence in the homes of her children. The compulsive way she visited pachinko showed the beginning of dementia, and Endo predicted she would forget to keep going before staff could find a way to stop her.

"It isn't loud at pachinko, or quiet when you leave. Everyone listens to something. Do you live next to a factory? Does a dog live next door who calls to other dogs? Pachinko sounds never bothered me. If you're elderly, you play slowly, or smoke but don't play and the staff doesn't mind. I keep my cigarettes in my pockets. I carried sea urchins in my *shimizu,* maybe you don't know what that is, it's underwear as long as your body. I filled it with urchins to swim home. The spines made holes in the fabric. My parents didn't like that, in the wintertime they made me sew them, even my siblings' *shimizu* if they wore out, I was the mischievous one and needed to be kept busy. I carried those urchins next to my skin, but they never cut or scratched me."

On the opposite couch, the woman listening had asked if Tokiko saw any Russians at pachinko, but Tokiko hadn't heard her. "Russians," the woman said, while Tokiko was still talking. "Russians, Russians," and the man sitting next to her on the same couch, who had fallen asleep, woke for a moment. The light, the half of it above the hill, was in his eyes. Closing them, pressing his chin into his shoulder, he said, "Some were there." He had been in North Korea when the war ended. His wife had been on Sakhalin. He was falling asleep again when the woman next to him said, "At pachinko?" He said, "I heard you," and the woman repeated her question, and he said, "I heard you."

Their conversation overlapped the end of Tokiko talking, and she watched television until they stopped. "When the steel balls are passing through the machine," she said, "I prefer the noise that makes. That's my idea of pachinko, watch the balls until you believe you're moving them, and I could feel on my hands—a pressure, lightly—making the balls move. I would find two rocks, and the *nona*, which had long spines, sharper than other urchins, place it on the larger rock, and with the rock in your hand press down and move the rock until the spines split away from the body. Several times until there's enough to fill the inside of one *nona* with meat from the other *nona* you caught, then cook the *nona* you filled over a fire, and that was enough food, only the size of my hand, but enough you didn't eat for a long time after. Rolling the *nona* on a big rock, that was the feeling I enjoyed, the spines falling into the water."

One of the staff came to the couches and suggested to the man sitting across from Tokiko that he could sleep in his room, out of the light. He was getting up, with help, when Tokiko put her hand on the wrist of the younger man helping him. She said, "Soon I'll go to pachinko."

Tokiko, January

During the winter, Tokiko was rarely in her room. The other residents, relying on her presence in common areas, lingered more often. Tokiko, who in the past had made everyone an object of conversation, obliged them quietly now, massaging her knuckles while they talked. A flickering motion in her cheeks was her tongue moving. Together, the gestures reminded staff of the concentration required to assemble something from minute, elaborate parts.

When staff prompted her to talk, she continued describing how she ate when she was young. "We cooked rice in horse clam shells. If I showed you a horse clam, you would think it's not much rice, but we were happy if our parents gave us that much."

She mentioned her family more often. "My brothers went to Sapporo to work every winter. The room where they stayed, the landlords demanded a year's rent to use it for three months or four. My parents told me the room was small and you can't walk to the ocean from there. If I had to sleep in a small room, I would have wanted to visit the ocean, but my brothers couldn't. If you're a child on a farm, you have animals, even if you're poor. I would walk to the ocean and touch it. It probably looked like I was touching an animal.

"When we were living in Karafuto, I took the train from Maoka to Toyohara. The train leaves the valley in a loop, the wall of the valley makes a circle. Leaning out the window, it looks like the train is floating. At the top, looking into the valley, it seems too far away and too small to be the place you came from. My brothers' room in Sapporo must have been like that."

The other residents told me that Tokiko had been born in Shirutoru, present-day Makarov on Sakhalin. Her brothers worked in Sapporo every winter. She hadn't lived in Maoka. That part described the woman who had asked if there were any Russians in the pachinko parlor. She had explained the train ride to Tokiko a few days earlier.

"My daughter visited once. I paid for her flight. She lives abroad. My other daughter passed away. Lung cancer. I married a man from the same village, on Rishiri. In a small community, that's like marrying your cousin. He worked on a trawler. The trawler sank. His body was lost." Only the final memory belonged to Tokiko. Her husband, from Akita, had worked on a trawler that sank, and his body was lost.

"Sometimes you tell other people's stories," a staff member once observed, helping her into the dayroom, where the residents were gathering for calisthenics. All the chairs were occupied. When he brought another and placed it facing the television, she made an effort to turn it, and with his help adjusted it toward the window, which faced away from the sun on the opposite side of the building, but toward the sunlight where it entered the room, and in that way was consistent with facing the sun, as children do when calisthenics are performed at school.

She spoke to herself while the other residents exercised. Her words came out rhythmically, matching the music that accompanied the calisthenics: "If I needed rest, I drifted. The water, I could trust. The first train I rode, I wanted my parents to explain how we knew where it would take us. Putting tracks down, to go one place and never another, I didn't understand. A train looked too heavy, there wasn't water to carry it, a hundred people could push but it wouldn't move. A train is a machine, they told me; it doesn't move when it's resting. If I rode the train, I carried my coins in my mouth. The taste was like a lightbulb shining on your tongue."

When Endo escorted me to the elevator at the end of this visit, Tokiko followed us. The shift was changing. Staff were on the elevator when the doors opened. They greeted Tokiko as they entered the foyer, and she greeted them in return. "Not in there," she said when I moved toward the doors. "It only goes one place." She was smiling, hiding her mouth in her hand while the rest of her face gave away the expression. Endo said, "Should we check your mouth for coins?" Tokiko made as close to a bow as her posture would permit, letting her hand drop out of Endo's where Endo had been helping her balance. "I swam between the islands," she said, without significance or inflection, as if giving an actor onstage his cue. Then she hummed a sound that rose and fell. "That's water," she said. "Calm water on the surface. It was all I could think of."

19.

The nurses' station in the rest home included a monitor linked to a closed-circuit camera system. Endo typically sat elsewhere, in a chair at one of the dayroom's tables, the nearest to the wall clock and the calendar, a coincidence she hadn't noticed until, when she was discussing how a care worker's behavior affects the way her patients experience the passage of time, I mentioned it.

"Living includes the present," she said. "A bath, a meal, light exercise. Feelings, memories, the small freedom of having your body and doing what you want: They are less than separate when you age. A day is a year if they aren't considered. We should try to be considerate."

On the occasions she worked at the nurses' station, Endo's attention to the closed-circuit cameras accompanied shift changes. She watched her staff ride the elevator, in a group because they had carpooled from South Wakkanai, or stopped to wait for each other at the public ashtray and share a cigarette. The elevator would smell like tobacco smoke and coffee, another coffee for their shift and a meal for their break in clear-plastic bags from the convenience store, wearing their street clothes, carrying their work clothes in another bag, the end of their conversation reaching the corridor with them: this week's TV, this month's utility bills, the time one of them visited her son in Tokyo, but her son and his wife couldn't tell her where to find anything in their neighborhood, as if they didn't live there either.

In her childhood, Endo was the oldest of three siblings in a house where their father was half-present, not by choice or defect of character, but by illness. Her family owned a laundry business in Toyotomi, the first train stop with its own town office south of Wakkanai. If the business hadn't been the kind that a woman could run when the family patriarch was sick, if she hadn't been the oldest, she might have become educated, but instead learned early there was nowhere to go except home, and, observing classmates whose lives were similarly circumscribed, discovered the possibility of understanding what a person feels, whom you hardly know and who hasn't told you.

Working in the office of a utility company, then in the town office after raising her children, Endo steered herself among credentialed people by cultivating her habit of empathy. They knew more, but she knew them. The town assigned Endo to the municipal nursing home, an institution affected by the retrenchments of the Lost Decade. From her first day, she cared for people who were too poor and vulnerable to go elsewhere.

The town office bureaucrat responsible for elder care maybe loved his job, maybe disliked it, but regardless behaved in a manner unusual among his cohort, articulating to nursing home employees his dissatisfaction with the shortcomings of government-funded homes. Looming demographic changes, he pointed out, would require Japan to relinquish its aversion to institutionalized aging, an aversion he associated with feudal-era folklore, which described the heads of families abandoning their elderly parents on a remote mountaintop. In Endo, he encouraged the trait she'd cultivated, of carefully knowing people, job applicants in particular, of whom there were never enough.

A colleague recalled that Endo once interviewed a young man who avoided interaction with the employees he was instructed to shadow. Endo could learn from him only that he was interested in recreation and might like to design activities for residents. The rest of the staff thought he suffered from social withdrawal. Endo experienced his shyness differently. Attending a festival his school held a week later, she visited a display he'd made detailing his research about recreation programs in care facilities and found him chatting with interested visitors. She hired him. He remained inalterably shy when his colleagues or superiors were nearby. Reports from residents about his general kindness and gentle demeanor reached Endo regularly.

In twenty years of hiring, Endo preferred employees familiar with adversity. "I needed time to recognize how the pattern—for generations after mine, in less economic difficulty—had changed. Now young people make sacrifices without knowing it. On a daily basis they're comfortable, later their hardships affect them but they're not aware."

A result of successful hiring was: An employee comes from another room because she heard the sound of a chair dragging, to find the tennis ball that worked loose from the chair leg; some of the residents are sensitive to noise. Or an employee asks a resident, who appears disoriented, where she is going, and when she says, "Looking for something," he checks that the hallway is free from obstacles, then moves where it's easiest for her to pass.

Both floors of the rest home in Wakkanai could have been filled twice with the elderly whose families inquired in the weeks before it opened. Only one moved in. The facility had been advertised as "housing for the aged, with services." The rest home was the first of its kind in Soya, and this description struck potential residents initially for its vagueness, then later for the associated expense. The role this model had begun to play in major cities, where waiting lists for rest homes were the longest they had ever been, where a third of available beds were empty because staff were short, and where more than half of the homes were scrambling to increase their cash reserves in anticipation of subsidy cuts—none of this was apparent in Wakkanai, whose elderly residents were accustomed to the expectation that they would cohabitate with their children or qualify for heavily subsidized care. Theirs was the first generation in the history of Japan that would be asked, on explicit terms, to pay for aging and death.

Staff who worked at the rest home when it opened were unsettled by its emptiness. "We discussed feeling as if it wasn't a new facility, as if it had been inhabited and we'd been hired after everyone left." If the rest home's sole resident left the dayroom while the employees at the nurses' station were preoccupied, their next attention to the place where she had been sitting could produce an involuntary, irrational panic, "like she was there when you looked, then vanished."

Endo curtailed neurosis by retraining staff during work hours. They addressed themselves to the hypothetical day of regular work, discussing aloud the logistics of a busy shift, which might begin with a hospital transfer, and maybe a staff member needed to accompany the resident. How you take vitals, treat overnight mosquito bites and minor ailments, prepare medications, manage the schedules of dialysis patients with hospital appointments, and not fall behind, not drop the afternoon's exercise and social activities. Even on days when staff were listless, Endo discerned their enthusiasm; several, if they were tired, had come from Sapporo or Asahikawa on the night bus after visiting a hospitalized relative. The altruism they brought to work was genuinely their own.

The residents who moved in next were brought by their children. They arrived nervous and remained nervous until newer residents arrived, then adjusted to their environment self-consciously, as if by seeing their anxious behavior outside themselves they had noticed it was unbecoming.

Staff asked Endo how long she thought rooms would remain available. Endo averred that she didn't like to predict anything; it was hard enough to see to the end of the year. At the end of the year the comment was mentioned, and the irony of its prediction coming true—the last room filled almost exactly a year after the rest home opened, with a resident whose children, Endo was surprised to learn, didn't mind having their father at home. He knew someone who lived in the rest home and liked it. He had asked to come.

*

The bank sponsoring "Chase Your Dreams" has given away all the tickets by the time Kushibiki thinks to request one, but he hears from friends that the celebrity baseball player started his lecture by screening a documentary about himself, after which he explained that "the most important part of success is failure" and temporized from one truism to another for the remainder of the event, periodically interrupting himself to sign *shikishi* boards onstage. *Any Capricorns in attendance? Have an autograph.*

This is near the end of his census taking. In his bag, Kushibiki has organized his materials to retrieve them by feel. Before, he would look at his notes if he heard someone coming, if it was nighttime and a light came on. Now he watches for someone to darken the sliding door from behind the privacy glass, and moves into a position that lets him speak to their ear if their hearing is poor. He says, "Excuse me—census!" at the beginning, so it's the first intrusion anyone will notice, not the sound of the door to the vestibule sliding open, or the motion-detecting light switching on. The questions that are sometimes unpleasant to ask, about the other name on the address plate, or—in a home where neglect is apparent—whether anyone is employed, he has learned (from Sugimoto) to ask as if they were the least sensitive.

Sugimoto, one of her employees discovers, searching the internet for commentary about Wakkanai's population decline, provides the city's website with an exceptional array of statistics. When the task of collating the census data begins, she examines the geographic distribution of households that responded online. "In the city, the rates were low," she tells her staff. "You expect the opposite. The data may show that multigenerational families farm and fish in the countryside, and downtown has become a place where the elderly choose to live." Even in decline, Wakkanai might develop the characteristics associated with urban drift, except for the presence of youth.

The room where data collation takes place is wallpapered with Census 2015 posters and Soya-specific accoutrements, including a pennant to fly on your boat featuring Census Boy, a diapered baby clutching a pencil. New for 2015, appearing opposite Census Boy on the posters: Little Future Girl, with a smartphone where Census Boy has his pencil, pointing to the sky with her other hand (because the census is priority number one, and not because, as some have suggested, prayer is recommended in advance of its findings). Kushibiki works to the sound of the crows who have occupied the park behind the building. "They're saying we've made too much of it," he quips. "That's my intuition."

His intuition omits the question of how many people will U-Turn or I-Turn to the countryside, and whether the neighborhoods he surveyed will keep their schools open. He imagines instead the oldest residents of Wakkanai needing care somewhere else, until Wakkanai is a place where you get little of what you need if it's unusual or important, and the people who decline to leave don't need it or can't afford it, so they age unobtrusively in place, invisible in their good luck, or their quiet poverty, or until one runs out and becomes the other.

Before the end of Yokota's second term, when the city's economic relationship with Russia was in the process of dissolving, the ranking customs official in Wakkanai dispatched himself to a warehouse on the proverbial outskirts of town, where his subordinates had discovered a cache of heavy machinery from Sakhalin. Startled by the quantity of equipment, the official summoned the warehouse's Russian proprietor and demanded to know where it had come from. Immortally, the apple-cheeked Russian, who in some versions of the story had continued imbibing while the customs agents awaited their supervisor, replied through the bureau's interpreter, whose dutiful, deadpanned translation retained the original's brusque patois: "Y'otta know. S'yore job." It was later concluded he understood Japanese, on the basis of his waiting—until the translator had finished—to belch.

If this anecdote (or news of the incident that inspired it) had reached Yokota, it would have portended nothing he didn't already know. The construction phase of oil and gas development around Sakhalin had ended, and there wasn't any incentive to relocate surplus machinery. Nothing in the Russian's warehouse could be usefully liquidated, so it might as well be seized, and why wouldn't the unfortunate Russian—stranded in Soya with no smuggling or selling to occupy him—raise a glass and crack a joke? The raid had released him from a job he was desperate to lose.

Yokota's realism was a recent trait, acquired gradually. In 1980, he inherited his family's car dealership. After the Soviet Union fell, his phone would ring, and people who spoke a few words of Japanese (how they found him, he never bothered knowing) would ask about his stock. Occasionally, someone bought every used vehicle on his lot. Having witnessed demand like that, he responded with equanimity when the Russian Coast Guard began apprehending crab smugglers during his first term. Freight service to Sakhalin had been established specifically to provide the licit, longitudinal basis of continued trade.

Licit isn't free, and smugglers were accustomed to exporting goods in the holds of their vessels. There was never a time, even during periods of growth in Sakhalin's economy, when the freight service was consistently well subscribed. The shipping agency offered

Russian-language services, but the genuine regional bilingualism that might have encouraged Russians to continue shopping in Wakkanai didn't exist. If it had, what would the Russians have bought? Without the informal trade in used automobiles, which was restricted by Russian customs authorities in the mid-2000s, commerce between Sakhalin and Wakkanai looked like a few dozen Russian visitors boarding the ferry with cheap kitchenware, athletic clothing, and secondhand electronics in their luggage.

Yokota was chastened. In meetings with prefectural officials, he detected an old prejudice that reappears when the trough is almost empty: Coastal communities come last, because, the folk belief goes, they eat what they catch. People asked why Wakkanai couldn't do what Sarufutsu had done. Sarufutsu was a village of twenty-eight hundred, the next town over, one of Japan's poorest jurisdictions until a 1971 municipal collaboration with the local fishermen's union committed the town's budget to scalloping; two decades later, it was one of Japan's wealthiest villages.

Smaller, it occurred to Yokota, might provide advantages. Elsewhere in Hokkaido, the economies of modestly sized cities were regressing deliberately, migrating their corporate, retail, and municipal infrastructure toward historic downtowns. Downtown property in Wakkanai was difficult to redevelop, but in the rare instances when developers obtained it, apartments in new buildings sold or rented at Sapporo prices, most often to the elderly.

He instructed the city office to conduct a needs assessment for elderly residents and took a quiet political detour to South Wakkanai, where people lived who put their names on waiting lists for downtown apartments. South Wakkanai was a suburb, but the majority of new development occurred there. In conversations with business owners, he asked a provocative question: What if we could make South Wakkanai Station the transit hub? It might amount to a courageous embrace of the city's beginnings; South Wakkanai Station had been built before Wakkanai Station, in the era when Wakkanai was a whistle stop on the route to Karafuto.

He listened to his staff describe what elderly residents told them: They wanted to be near the life of the city, not sequestered from it. On the conference circuit Japan's mayors travel, he learned that several small cities were redeveloping their commercial districts to attract retirees. Yokota pursued other projects (funding to replace the city's vehicles with an electric fleet; negotiations with wind farm developers), and all were of a type, fashioning a smaller city with modest obligations and a diminished reliance on fickle forms of revenue, but the notion of bringing the city's infrastructure to the people most affected by its absence—people in South Wakkanai and the surrounding neighborhoods—was the notion that occupied him. Coming to his office at the car dealership to indulge his thoughts and conduct his own research, he used twice the heat in one month he'd used all winter the previous year.

Before he could ask his constituents to accept that the city's downtown had become ornamental, he needed the business owners who would benefit to organize. A few were willing, the rest bewildered. Without an enthusiastic response, he allowed the idea, convinced it had been his best, to become an object of detached contemplation. He awaited an equally elegant notion, one that showed milder intentions toward the city that people had lived in, still remembered, and no one living among them could choose to disregard.

*

When city bureaucrats rotated jobs in 2014, the director of Wakkanai's General Affairs Division transferred to the Sakhalin Bureau. His subsequent travel to Sakhalin helped him grasp a notion he had previously experienced in half-complete form: the difference between borders and frontiers. Historically, Japan had expanded northward, Russia eastward. Wakkanai and Sakhalin existed at the respective edges of their former empires. You could go from one to the other, noticing the continuity of the landscape, feeling you had set foot in neither country.

The director's father was born in Karafuto. The Soviets kept him on Sakhalin after the war ended to train arriving Russian workers at an industrial freezer. He never described this experience to his son, and until adulthood, the director (whose name is a literary rendering of ideograms that mean "border") presumed his father was born in Wakkanai. He learned otherwise unintentionally, from his family registry. He assumed that his father had thought an incompatibility would arise, between the family's awareness of its origins—on the periphery of Japanese identity—and his son's sense of himself as Japanese. But Wakkanai had been like that: on the periphery, not all of it Japanese. The director's earliest memories included tasting Coca-Cola and hearing jazz records at the bowling lanes attached to the American espionage installation on the cape, before it closed in 1972. The Sakhalin Bureau was the periphery, too, where his desk had a poster on the wall next to it advertising a subsidy of fifty thousand yen, up to five times, to launch a business exporting (anything) to Sakhalin. That was all the city government could recently afford, in addition to sending a few crates of comestibles, with the notion of getting Russians hooked on rice balls.

When he'd been in the Sakhalin job three months, the director walked me down the *shotengai* and pointed out restaurants where Russian trainees on an exchange program studied small business management. Most were closed, in contradiction of posted business hours. "Seasonal," he said. Then, changing his mind, "I'll look into it." Outside a store selling movies and music, he showed me a poster from the 2010 revival of *Through Summer's Frozen Gate,* a film that dramatizes the group suicide of nine young women—Japanese postal workers—during the Soviet campaign to retake Sakhalin. Three of their colleagues survived and were repatriated through Wakkanai. "We're standing in their footsteps," he said, and meant it literally. "That's one thing about living in a city with no future. We never evict a memory."

The same week, I was introduced to the chairman of the central *shotengai*'s commercial union, a second-generation pet shop owner who was elected when plans were being drafted for Wakkanai's new train station, and promptly defied his constituents by supporting Yokota's decision to build the station with a rest home inside. Ozaki (his name) didn't care whether residents of the home would shop downtown. He wanted, as Yokota did, to develop the infrastructural basis of a more compact city. Sprawl had made Wakkanai car dependent, and car owners had no reason not to shop somewhere pleasant, somewhere else.

He launched a nonprofit to solicit grants from the prefecture. The first, he spent on a branding campaign, beginning with the publication of a travel guide, which included pictures of fashionably attired young tourists standing in a meadow, sipping coffee at the counter of a conspicuously rustic coffee cart, assembled specifically for the photo shoots and making later appearances in the spreads the nonprofit was able to land in the travel sections of lifestyle magazines. He had been inspired by the work of Toyota Masako and her Onomichi Vacant House Restoration Project, which restored neglected buildings in the town where Ozu Yasujiro had filmed, on the Inland Sea. Aware that mortgage debt often prevented restorations in Wakkanai, Ozaki preferred to enroll business owners in management courses, taught by entrepreneurs who could impress upon them the contemporary standards of retail tourism. His older constituents sneered at the suggestions these courses made—interior remodeling, curated inventories. Either the city was doing business, or business was elsewhere and the reason was not aesthetic.

Once, I rode in a taxi and the driver went the long way around the *shotengai*. He claimed that store owners were taking the plate numbers of occupied cabs and flagging them down later, to ask why their businesses hadn't been recommended to passengers. I told the driver he needed a better scam. When we stopped at a traffic light, he said, "It's a ridiculous story." On the other side of the light, a tangle of scrap metal from a condemned apartment building sat in a parking lot, in the shape of a house-sized bird's nest, and a flock of grosbeaks had occupied it; the whites of their wing tips flickered inside the pile. "But this is just the kind of town where you'd believe it."

*

Four years passed before Monma's next visit to Iwaki, to attend the reopening of Monma Bike and Motorcycle. The shop had been operating out of a warehouse, where thirty months of rent had equaled what it would have cost to buy the same property before the disaster, before the city's absentee landlords returned for properties that hadn't flooded, to put them on the rental market at scarcity rates, abetted by the surge of supplies and laborers arriving for reconstruction.

Monma could stand with his eyes shut in the new garage and give himself the illusion of being in the old building, where the wind this time of year, in Indian summer, blew the same way through gaps in the joinery, sending threads of cold air across the floor. The bikes and motorcycles that weren't ruined in the tsunami or given away afterward were here. And his youngest brother, who had been two when their father died and knew his face only from pictures, resumed his habit of sitting in the garage before the rest of the family woke. Except for his recollection of his previous visit, Monma could look at the shop and see a renovation, one you would perform in any city, under tranquil circumstances—because it's a pleasure for a place to feel new.

During the day, he drove with his brother and his nephew on errands across the prefectural border, where Iwaki's twin city had become one of its suburbs. Chain stores were laid out on the gridded streets that once formed the city, and the expressway ran parallel to the local road, which was frontage now. Before coming back, they visited the cemetery where members of the Monma family were buried who had died before Monma's grandfather relocated to Iwaki. It was the first time Monma's brother had seen several of the graves.

After returning to the shop, they sat in the office and watched a segment the prefectural television station had produced about Monma's nephew evacuating the neighborhood in 2011. It shows him rushing to don his fire brigade uniform, circling the neighborhood in an emergency vehicle with the siren on, warning residents over the loudspeaker. The siren and announcements are poorly dubbed. Monma's nephew and the other volunteer firefighters play themselves, and their body language doesn't match the editing. In an interview, one of the firefighters recalls noticing the tsunami out of his peripheral vision, and before he recognized what it was, his foot was shaking too violently to work the gas pedal. Describing it, he pauses to blink away a facial tic, and the effect is to remind you that the event occurred, not the dramatization.

A customer arrived to pick up his motorcycle, and afterward Monma joined his nephew on the sidewalk, where he was changing the tires on one of the mopeds belonging to the post office. Monma's nephew talked about the ongoing privatization of Japan Post, the only delivery service that operated in Iwaki during the tsunami's aftermath. If Japan Post had been privately owned, he said, they would have suspended deliveries like their competitors, creating a shortage of medical supplies.

Monma asked if his nephew ever heard from the man who needed hospital care, after they sent him across the city with their last container of fuel.

It took Monma's nephew a moment to recall. "Never," he said.

"He didn't give his name?"

"I could find out, if I knew where he came from."

If the man lived in temporary housing for tsunami victims, Monma pointed out, he would have moved several times by now. The people he knew would have moved. There might not be a place where a description would be enough.

"I'm not sure the face I remember is the face of the man I saw," Monma's nephew said. He was looking at the tempered glass they had installed as the shop's front window, to resist future tremors. Not at any reflection. It was dusk, and the lights were on in the shop. At the new showroom and new motorcycles, and the plastic shrouds his employees had removed from the motorcycles, piled shoulder-high. "He wouldn't remember our faces either," he said. "It has been that long."

*

When he wasn't teaching his English class anymore and discovered himself, during night walks, straining to believe that the neighborhoods he passed would wake in the morning and appear inhabited, Steve attended his middle and high school reunions. He intended to encounter his classmates who had told him, before he left, how they wished they could go with him. He thought they might understand what it was like coming home.

Instead, they gave advice he hadn't sought, about acting Japanese so people wouldn't find him coarse and bewildering. The transformation time had worked on these people, who had once been young and more interested in what they didn't know than what they did, Steve could only interpret by assuming it was dishonest. "Isn't it you, wanting to go to America?" he asked someone, and the reply was,

"Everything you can buy in America, you can buy in Japan." He showed Steve a Swiss watch, purchased in Tokyo during a trade conference. "The lobby has shops," he said. "I never left the hotel."

Repelled by the possibility of adopting his classmates' preoccupations, Steve went back to work, as a custodial employee at the hilltop park surrounding the Northern Memorial Museum, capturing beverage cups and tourist pamphlets before the wind carried them into the strait. Visible ethnicity aside, Steve could distinguish the Japanese tourists because they clustered near the bus to silently indicate they'd finished with the park, and if they talked, they made polite comments about the museum. Chinese tourists occupied themselves roaming the park's perimeter, as if to make a grid search for sources of amusement. Without understanding what they said, Steve sensed their comments were candid.

Besides being a job, and Steve liking what jobs imply about the people who do them, the custodial work was appealing because it provided a panoramic opportunity to contemplate the city. From the hill, the vacant lots and disused buildings were too distant to notice. Only a trained eye would know the port was empty for its size. Instead, the pattern of traffic indicated the absence of urban activity: Some cars moved north, to the cattle farms; most moved east, to South Wakkanai. Few entered the streets that terminated downtown. The noises the traffic signals made to assist the blind carried up the hill and gave Steve the impression of a conversation taking place among unattended machinery.

Because they were accustomed to a landscape that had been leveled for rice planting, Japanese tourists from outside Hokkaido sometimes experienced vertigo after visiting the museum's observation deck. Using binoculars to follow the valleys, they lost their sense of the horizon. If that happened, the person would come to the parking lot and face the bus. It made them dizzy to look downhill.

Telling his mother about it, Steve said, "They're used to that different picture, so they don't remember," as if their unfamiliarity with Hokkaido's landscape was a matter of forgetting. It was the nearest idea he could express in Japanese that described the experience of not possessing, to a sensory extent, what your surroundings demanded. In this way, he was able to understand what had become of the curiosity and youth his classmates lost, overnight or over fifty years—neither he nor they could say which. "For Wakkanai, time is the past. Never saw little bit of little changing every day, only remember the time in your memory. Even Wakkanai is your home, it used to be you never been here."

*

The twenty years she worked at the nursing home in Toyotomi, files describing wait-listed residents arrived on Endo's desk—then the people they described arrived at the nursing home—after the onset of dementia. In some instances, no other conditions were noted. Other times, incoming residents weren't ambulatory, or competent using a wheelchair, and Endo could guess from the files whether losing their freedom of movement had precipitated their loss of lucidity, or the loss of lucidity had led to an injury.

At the rest home in Wakkanai, many residents arrived in good cognitive health. A separate floor of the home was reserved for advanced dementia sufferers. Endo didn't rush to fill it. She knew that couples would enter the rest home together, then husband or wife would deteriorate, and the last comfort would be their sharing of a building, the ease of visiting between floors—if a bed were available.

A journalist who visited the rest home told Endo that residents of similar facilities in Europe and America were fearful of losing memories that stirred their emotions. In Japan, they feared losing practical skills and worried their caretakers would privately blame them for allowing themselves to need care. One resident in Wakkanai would push the call button and forget why she did it. When staff entered the room, she stared at the ceiling light and wouldn't stop—a self-imposed punishment—unless staff assured her they were stopping by routinely.

Residents from Sakhalin, if they hadn't evacuated before the arrival of the Red Army, were the group most likely to discuss which memories they would be glad to relinquish. Russians arriving to settle on Sakhalin had lived at first in houses abandoned by Japanese evacuees. When these were full, they cohabited with Japanese families. Compared with Russian homes, Japanese houses were cold. Their only heat came from kitchen stoves, which the Russians left lit, with too much fuel inside. Homes burned, and burned the neighborhoods around them. Most fires were accidental. Some were set clandestinely to spite the Russians, some set by Russians in reply. Rest home residents had seen neighbors—Russian and Japanese—jump from their windows and not always survive, but the unwelcome memory mentioned most often was the morning after a fire, the carrying odor of smoke and the stains of dead embers on exterior walls, and the feeling of not knowing why the fire had been set; if out of malice, to whom—what person, what nation—the malice belonged.

In Toyotomi, a woman had tried to leave the rest home every evening, to go home and cook dinner for her husband. Endo took the woman to her room, asked her to stand in front of the mirror, and showed her a photograph of her husband, who died in middle age. This is your husband, Endo said. It's you in the mirror.

Endo thought the woman would remember. She didn't. She learned it, as if Endo had brought the message from the scene of his death. After that, Endo accommodated the fictions that residents' minds formed.

A woman who was ninety thought she was a child in her family's curry shop after the war, waiting for her mother, who left every morning to salvage tin in the firebombed neighborhoods. When she said, "Everything washed before she returns," Endo gave her towels to fold. She was placid with the task, which Endo could prompt her to repeat by unfolding the towels.

Younger staff were confused by Endo's reluctance to contradict disoriented residents, and thought the memories of the fires in Sakhalin were exaggerated versions of a less traumatic experience; when Endo asked why, they said, "If something terrible had happened, we would have learned in school." Occasionally, they approached Endo to say a resident was trying urgently to express a need, but wasn't making sense. Endo found these residents were in bright moods, using their natural diction, language that went out of circulation after the war, or developed, this far north, apart from standard Japanese. The first time this occurred, Endo reminded the resident that the younger staff couldn't understand the old dialect. He said, "These words happened." He wasn't referring to a specific event, but she was certain he was right.

Her certainty related to an intuition she had developed, when residents in Toyotomi wanted her nearby in their final hours, that her presence displaced the environment of disquiet they remembered—of your mother digging tin, the ash stains on the sleeves of her *yukata;* of your father making you stand on the balcony, and your punishment for crying out during the air raids was watching the light of the bomb strikes; of noticing how the fire in a burning building illuminated a polished object in the arms of a Russian woman who had climbed through an upstairs window, then she jumped and you saw, not a samovar, the object among your neighbors' possessions you wanted to touch, but an urn, the lid off and the ashes falling on the lifted hands of the people who had gathered below to catch her.

Sachiko, August

"The war was ending. I was visiting my sister in Karafuto. Her husband was in the military, posted there. Their son was two. The last boat, I came back. After that boat, if you wanted to leave you were smuggled. The harbor was crowded. People brought everything they owned. My sister and her husband couldn't board with their belongings. I tried to board without them. I had my daughter with me, on my back. I couldn't put her down. There was no way to predict when the crowd would shift. I thought I wouldn't be able to board, but people passed me up the line with their hands, because I was a young woman with a small child and no possessions."

Sachiko moved her wheelchair parallel to the dayroom's window, which began at the height of her shoulder. "Festival week," she said, and tapped the glass. The street was busy with tourists who had come to see Japan's northernmost train station and would take the ferry to Rebun or Rishiri tomorrow. Having exhausted the attention they could dedicate to the station's gift shop, few appeared certain where to go. "That one is walking in a circle," Sachiko said. "I've seen him three times."

A police cruiser had pulled a car over opposite the rest home. Throughout the morning, Sachiko said, she heard the police ordering drivers to the side of the road. "The city is building a curling rink," she said. The cruiser with its lights on gave the children something to look at, and there were a few, standing with their parents in front of the station, most of them carrying cones of Soya soft serve. In the rest home, too, where they'd come to visit grandparents and great-grandparents, running the halls, colliding with the shins and knees of delighted residents. "On holidays it's like this," Sachiko said. "People don't live here. They're from here."

She wore a dark beret tilted forward in the flapper style, to overlap the brow-length bangs cut into her wig, and a dark shawl, covering her to the armrests of her wheelchair. In pale makeup with immoderate blush, her appearance matched the description staff liked to give, of an actress in the drawing room of her high-rise apartment, in a lifetime of receiving distinguished visitors.

"I am without my friend today," she said, "who works at a *ryokan* and takes me on walks. She went to Sapporo. Her husband is hospitalized. There is no appropriate place to be ill in Wakkanai. Our grandparents taught us to gather herbs. When I told my friend, she said, 'Teach me to gather flight vouchers.'

"Before the boat left that brought me back, my sister and her husband were able to come to the fence and toss rice balls. That's what we ate. Other people were hungry. Immediately, everyone had lice. I cut my daughter's hair and threw it overboard, but they were everywhere on our bodies. She struggled to fall asleep. When we arrived in Wakkanai, I said to my family, 'You shouldn't embrace us. We have lice.' They embraced us. News came from Karafuto that people couldn't leave. We wondered if my sister was safe. Two years passed before she came home. During those years I remembered the way the forest and the village looked, where my sister lived in Karafuto. I imagined my sister carrying her son on her back in the countryside near that place. I thought, 'She can't put him down.' I felt sorry for her."

Sachiko's family was from Urausu, near a coal-mining city in central Hokkaido that depopulated when the industry went into decline. Her daughter had married a man from Wakkanai. When Sachiko's husband died, she entered the rest home to be near her daughter, who had visited rarely in the two years since.

"I don't like the vehicle service here, to go somewhere in your wheelchair. The windows are tinted. Some people don't want everybody to see them in a van like that and think, 'That person is disabled.' Instead people think, 'Oh, a shadow in the shape of a disabled person.' I would prefer if you could find me. Someone can tell you they saw me in the car. Maybe the window was open, and I told you: There's somewhere I'm expected."

Sachiko, September

"My sister lived in Karafuto during the war. I was never able to visit. When she came home I asked what it was like where she had lived. She didn't want to describe it. Only, 'There was a river, low in the summer, on the banks smooth rocks the size of a hand. Every year a child would drown, we forbade the children to swim.' The boat that brought her home was the last. Their son was two. Belongings weren't allowed, to make room on the boat for more people. The departure was delayed, then delayed again. The passengers became hungry. Her husband traveled overnight to reach the port. It was forbidden for soldiers to leave their posts. I imagine he arrived at the port with food to give them, and he learned they had gone. Or the train stopped somewhere partway, and he learned the boat had departed, then waited for a train in the other direction, in his soldier's uniform, at a station where a soldier was not permitted to be."

Briefly, Sachiko was alone in the dayroom, listening with her eyes shut. The humidifiers in residents' rooms exhaled intermittently, with a plosive sound human enough it was never in the background; the vinyl flooring in the hallway stretched when a wheelchair or walker or feet moved along it, a sound like the cracking of skim ice. A resident who had recently begun taking a dementia-prevention drug came to the couch near Sachiko's wheelchair. He said, "It's an aquarium." A staff member asked if he wanted to go outside (panic attacks were a side effect of the drug), but he was describing the muted television, not the room, and he was smiling. He sat on the couch. The television played College Dating Party, and the captions scrolled over the contestants' faces, concealing them. In the window beside the television, the wind from the cape blew the clouds across the hills at a pace appropriate to a time lapse.

"In Urausu, we learned the boats from Karafuto had stopped. When I reached Wakkanai, I was told that one more boat would arrive. I didn't know if my sister was aboard. When the boat became visible, I went to the harbor and joined the crowd. Women were waiting for their husbands and children. I pushed those women past me, toward the pier. My sister was on the boat. She had lice. My nephew had them. She had cut his hair. I put my hand on my sister's head, like this, and pulled her against my shoulder. She began to weep. My nephew had been weeping during the trip. The skin around his eyes was red. When we returned to Urausu, he was a quiet child and spoke in whispers. He was imitating us. We whispered when we discussed whether his father was safe.

"Two years passed before my brother-in-law came home. I wanted to tell him about meeting my sister in Wakkanai. How tired she had been, carrying her son, and her son's face looking over her shoulder without any expression. I was relieved the first time he blinked. His face had been like a face on a dead boy's body. But they were living together again. It would have been wrong to disturb their happiness by asking them to join me in a painful memory."

Sachiko, January

Once, when her daughter visited, Sachiko asked to take a walk on the path that overlooked Wakkanai. She had wondered why the rest home was built to face the city, not the sunrise. The view from the hillside, when they reached a place to see it, showed geometric consistency: The longer sides of the larger buildings ran parallel to the water. The city's shape, Sachiko decided, was a matter of unrevised history, beginning when plots were purchased and buildings constructed in relation to the harbor. "There were hardly any boats on that day," she told me. "I couldn't tell where the harbor ended and the ocean began. Then I thought, 'What a strange place to live.'

"My sister's husband was sent to Karafuto during the war. She stayed in Urausu, to give birth to their first child. My nephew was two when she took him to meet his father. Before she disembarked at Otomari, crew members were informed that no other boats would return to Hokkaido. Only theirs. No one knew when it would leave. She went to the fence that separated the boarding area from the pier and found our aunt. They decided she should stay on the boat.

"The ship was held at port. The passengers became hungry. Our aunt remained in Otomari to buy rice balls and pass them through the fence. The passengers were unable to bathe. Lice spread. My sister cut my nephew's hair, but the lice can live on your body. It was difficult for my nephew to sleep. He became ill, in his chest. My aunt was searching for a doctor when the boat was able to depart for Wakkanai.

"In Urausu, word never came from my sister. We became worried after we learned that voyages between Karafuto and Hokkaido had been suspended. We prayed that she had found her husband and they were safe with their son. We prayed no battles would occur in Karafuto.

"During the warm months in our village, the river makes fog overnight. On the morning my sister entered the village, I was the first to see. I recognized her face. I saw the face of my nephew above her shoulder. I didn't know she was carrying him. In the fog his face appeared to float. She was coming from the direction of the mountains, where nobody lives and the land is uncultivated. I felt the responsibility of waking my family and telling them what I had seen. They would believe my sister had died.

"She said, 'Don't embrace me.' I embraced her. My nephew was asleep. His face was near mine. His breath touched my face. I heard the unhealthy sound from his chest, like a paper ornament was inside. I knew they were alive.

"I'll say what she told me: After leaving the train and walking, she worried she had passed our village. If she could climb high enough to see the landscape when the sun rose, she thought she could recognize home. 'When I came to a high place, each village's farms resembled the others. I felt, *This is a strange place to live.* I became afraid I would drop my son and forget he was with me. I decided whatever town was nearest was the place that fate intended. I would live there with unfamiliar people and never tell them my name. Entering the village, before I knew the village was mine, I saw you.' That was how my sister spoke.

"My nephew recovered. He slept more than other children. He was quiet. We wondered if the fever he developed on the boat had affected his brain. No word came from his father until two years later, when a letter notified us that he would return. I never described to my brother-in-law the day when my sister came home. It would have disturbed the peaceful life he provided for his family.

"When my nephew became an adult, I told him the story. He often visited Wakkanai and tried to recover his memory of that time by walking to the place where the boat from Karafuto arrived. On those days, he called me from his hotel room. He said, 'I went to the breakwater.' I told him, 'I understand.' He stayed in the hotel near the harbor, in a room with an ocean view."

The sound in the hallway adjacent to the dayroom was the sound of deep breathing, produced by humidifiers, which concealed the sound of deep breathing, produced by people asleep in their rooms. (Sachiko had described it that way.) Two of her could fit in her wheelchair, and two of her wheelchair could fit abreast in the hallway, with room to walk between them; they had been designed like that, the hallway and the wheelchair. Not to make you feel small—that was unintended (her description, too). Now, sitting at the end of the hallway, with a pattern of shadows projected on her face by the window blinds, she remembered an article from yesterday's newspaper: The Soya Line north of Asahikawa had been selected for development as a tourism corridor before the Olympics. The same part of the Soya Line, the article noted, had been labeled revenue deficient by JR Hokkaido. Three stations, scheduled to close, averaged less than a passenger per day.

"How will you write them in English?" she asked—the Ainu names of the stations, whose ideograms, selected for phonetic reasons, were unusual in Japanese.

I could transliterate the stations' names, I said, write the Japanese words in Roman letters, or name them in a way that would make Americans find them less foreign, the people living there less separate from themselves. Nukanan was almost New Canaan, a biblical name an American town already used, and the image it suggested in America was not unlike the fertile, uncultivated plain the original Ainu implied, or the fertile, uncultivated plain where the train briefly stopped.

"Only," Sachiko said, "I think they intend to dump there." They did. To test storage methods for nuclear waste. "They'll keep the name," she added. "You can still use it."

Pulling her wheelchair closer to the window, where this morning she had mistaken the view of cars snowbound in a parking lot for a view of boats in the harbor, she said, "I didn't imagine it? The dumping—I read it?"

"You read it," I said. "You imagine things?"

"My sister and nephew—" It was the beginning of a thought she didn't complete. Her expression became serene. She said, "It's the story I won't forget."

20.

In the third term of his administration, Yokota was described to the public by a journalist emeritus from Japan's leading newspaper, who had been allowed to choose his final assignment before he retired. Wakkanai was the bureau closest to home (Toyotomi was home). His first trip to Rishiri and Rebun came shortly after his arrival, during election season. On Rebun, in the town near the ferry terminal, he drove off the ferry behind a parliamentary candidate's sound truck. "This is the largest electoral district in Japan, geographically. If the truck wants to cover every town, there's no time off. On Rebun, you can see both ends of town from the pier. That's how small it is. When the truck's front tires touched the street, the driver must have floored it; the truck went through town a hundred kilometers an hour. Some of the buildings near the water, they're no bigger than a sound truck. The truck went past like that, and the amount of air it shifted, it shook the frames of those buildings. When I passed through they were still shaking. The person on the public address system was shrieking the candidate's name. You couldn't tell what the name was. It was unintelligible. I thought about it for weeks, but I couldn't get in the newspaper."

An event like that discloses directly what newspaper coverage labors to suggest. In Soya's case: The relationship between national government and everyday lives is not reliably rational, or is reliably irrational, or makes a study of how these statements become indistinct. Under what circumstances had someone formed the impression that residents would rather be screamed at through their shaking walls and windowpanes than have their town passed over?

The ANA Hotel Wakkanai, the architectural and budgetary equivalent of screaming at city residents through a wall, had been built by Yokota's predecessor, at a price that dwarfed previous government expenditures in Soya, on a jetty reclaimed for the purpose of building it. Financing was provided by the third sector, a wave of public-private partnerships intended to bring government money management in line with Japan's ascendant corporate model. The collapse of Japan's asset bubble notwithstanding, third sector projects were frequently unsuccessful, and profligate in rural areas, where municipal partners were intoxicated by access to financing of historic proportions. Among the handful of third sector projects Wakkanai had undertaken, the ANA Hotel, appallingly unprofitable when it opened in the mid-1990s, revealed the city's least plausible aspirations, derived from the prosperity accumulated during its relationship with crab smuggling.

The hotel's looming insolvency landed on Yokota's desk early in his first term. Six years of restructuring were required before the hotel posted an annual profit—slim, but helpful in the search for buyers, which occurred during a period of stagnation in domestic tourism spending. The category of tourism that benefited Wakkanai (travel to Japan's outer islands) was among the hardest hit.

When two potential buyers emerged, the city rebuffed the party that reportedly intended to submit the highest bid. In negotiations with the other party, favorable terms were exchanged for a handshake agreement that a few lean years would not affect their long-term intentions. Yokota's constituents discerned the pragmatism of his approach and returned him to office the year the sale was finalized. But the intervening years without a potential buyer, the modest price the hotel ultimately obtained, the debt the city kept, and the weakness of its negotiating position provoked civic anxiety, including, to Yokota's surprise, the widespread use of the Y word (Yubari). In a city where population had declined every year since 1975, the cutting of losses could no longer be accomplished independent of politics.

The city had chosen the site next to the breakwater for the ANA Hotel because it gave the twelfth-floor bar and restaurant an urban panorama, and a view of Sakhalin on the opposite side. Yokota knew elderly fishermen in Wakkanai who recalled with evident pride their imprisonment on Sakhalin in the late 1940s, when they had been caught smuggling Japanese families off the island. The same pride, in its contemporary form, recalled the livelihood that illegal crabbing around Sakhalin had recently provided. But what could be good that didn't last? Or lasted only until you depended on it? In budget meetings, one of Yokota's employees made a habit of saying, "After you eat what you crave, the taste of your mouth doesn't change."

The same employee had gone overboard from his father's trawler when he was in junior high school. He was preparing the net when a swell came over the stern. Falling backward, he reached for the beam of the net and hung for a moment, long enough to avoid going out to sea with the swell. It was night and raining; the searchlight wouldn't have found him. He lost consciousness when he entered the water. He woke in the boat. Because his childhood had occurred on boats, he never thought until the wave struck that anything was risked when he boarded, or anything on the other hand was gained, or the two existed in mutual relation. The work was familiar; it was nothing, it inevitably occurred.

Yokota was the mayor of a city that had once been loud: harbor traffic, industrial facilities along the waterfront, festivals on weekends and holidays, karaoke through the walls of the hostess clubs. Land near the ANA Hotel had been the place where American servicemen from the espionage installation would barbecue, where local children wandered close, pretending they wanted to practice their English, but the motive was hunger, the type produced by amazement, at the size of the feast Americans prepared outdoors and ate standing, with their mouths open to laugh while they chewed.

The city was a habitat for quiet now (a professor had told him the cape's natural geography formed what scientists call an acoustic shadow), and Yokota reflected, visiting the twelfth-floor bar, where he was often the only customer, that it provided a superior view of all the things he had promised to keep, because the promise had seemed familiar, and even if he hadn't made it, the difficulties that followed would have inevitably occurred.

*

When the Ministry of Internal Affairs releases the results of the 2015 census, they neither fulfill nor contradict the forecasts the media has disseminated. Instead of a million and a half fewer souls, the headlines report it's a million. This number is conspicuous and inaugurates an anticipated century of decline. It's also fudge: The tally is 962,607. Rounding up ripens the prescience of last year's coverage, filed from towns where more houses are abandoned than inhabited, from schools where students no longer enroll, from islands whose role in Japan's territorial status quo is under threat from the end of permanent habitation. The five-year decline is the first in the history of Japan's census. Set against expectations, it is underwhelmingly incremental.

This matches the intuitions Kushibiki expressed to his supervisor. The gap between the percentage of Wakkanai residents aged sixty-five years or older and the corresponding percentage nationwide has widened. Japan is an aging nation. Wakkanai is aging faster.

When I run into Kushibiki downtown, he tells me he was recently assigned to the office that manages children's programs. When he was in junior high school and made a plan with his friends to ask whether the city office could address the closure of Wakkanai's toy store, their inquiry would have been forwarded to the division he's working in now. "I thought of that when I learned I had been transferred," he says. "I wondered if I had mentioned that story to my colleagues and someone made a record of it. But it's a coincidence."

He hopes a future transfer will place him in the city's fiscal division, where fate and bureaucracy will share a desk when more of Wakkanai's residents are ill and fewer than ever are young, when the city is the region's largest employer to a greater extent than it already is and people reassure themselves by wanting to build what you build when you already have a curling rink. "A city is a budget," he explains. In Wakkanai, in an era claimed by decline, I am not certain he would believe me—or I would believe myself—if I said, "Not everyone sees cities that way."

He does not discuss census canvassing until I inquire, but his comments suggest it occupies him. "I don't wonder if young people will move to Wakkanai, start businesses, telecommute. I understand the city would benefit, but people already live here. I met 1 percent of the city, taking the census. Some were old, living alone, didn't have money. They didn't treat their difficulties like interruptions."

He adds, "I imagined my grandfather living in the houses I

visited—after that first house." Where the person who answered the door came outside seeming not entirely lucid, then told Kushibiki someone had died recently and the household was in mourning. He imagined his paternal grandfather, who died after Kushibiki decided to take the civil service exams but before he learned he passed.

That grandfather had fished for a living. At home that evening, Kushibiki's father, Hitoshi, recalls it. "I helped. It was my family's business for generations. Waves had an effect on my stomach. I never saw a future in it." He took a job at the auto parts store where he still works. "I would have wanted more education if I knew about the bubble." He's fortunate anyway, and aware of it. The chain he works for is the largest in Japan, and the store stayed open after Russian patronage declined.

Kushibiki is an only child. A description of his family could appear in the human development chapter of a biology textbook; their mannerisms are identical, and natural to the dimensions of their home. They live in tract housing in Midori, the neighborhood that connects downtown and South Wakkanai. This is where, as Hitoshi describes it, his son spent two years of afternoons, evenings, and weekends behind the locked door of his room, studying for the civil service exams. "I told him to play sports with his friends, because friends are harder than jobs to come by. He was hurt, though, when he failed the one exam. It would have hurt more to fail two."

The exam he failed was for the National Tax Agency, the only job that would have required him to leave Hokkaido. "I didn't think that was his first choice," Hitoshi says. "Afterward he said it was."

During an interview I taped, talking to Kushibiki about the exams, he described studying for three of them, then said he had passed. He didn't say how many and never mentioned failing, but later in the same interview he claimed not to understand why most of his classmates wanted to leave Wakkanai. When his father mentions the tax exam now, he starts to interrupt. A glance is exchanged; it's not clear if Kushibiki knows what his father is likely to say, or if he minds hearing it.

"In a fishing town, my father fished," Hitoshi says. "In the bubble era, I went from high school to work for a corporate franchise. You didn't want to work for the government. Even the government wanted to work for corporate Japan. My son works for the government. Today, that's the better job. The lesson seems to me: Your job tells you more about the world you live in than it tells you how to live."

Kushibiki, sitting cross-legged on the floor (he still lives here), has the jumbo lint wand his parents keep handy for dog hair. He is picking fibers off it, making his reaction to his father's rhapsody at once known and hidden in the way only a teenager (or recent teenager) can.

"I didn't have a first choice," he says. "Sometimes I thought I did."

"They rotate you through departments," Hitoshi says. "There's a first choice you haven't discovered."

"Working with children is important," Kushibiki says. "I'm on the boat." He gestures with his forearm, his wrist, and his fingers to make the motion of waves. He savors his cleverness a moment, hiding a grin in his shoulder. He says, "I see a future in it."

*

After eleven years running it, Takagi Tomoyuki, director of Wakkanai City Hospital, retired into the lesser responsibility of managing the city's geriatric care facility. He kept his office at the hospital. When I visited him there, he was not familiar with the American idiom *I'd rather be fishing,* but had hung portraits of himself holding a variety of wildlife you get that way. "I'd rather," he said, adopting the idiom's structure, if not its spirit, "you were a surgeon." The hospital had fewer than it needed. He knew because he'd spent a decade addressing the problems imposed on Soya's health-care system by a severe shortage of physicians.

Few Japanese doctors are willing to move their families to far-flung or second-class cities; the willing promptly exhaust their opportunities for professional advancement, ease their consciences with the time they've put in, and return to Tokyo. To the extent that someone like Takagi would "pay what it takes" out of the discretionary parts of his hospital's budget, that hardly matters. The recruitment of doctors is an elaborate process, founded on professional prestige and institutional relationships. Rural hospitals can't compete.

"The shortage of physicians gets worse," he said. "The doctors who staff the clinics retire. The hospital struggles to attract specialists. See what I'm implying? Hospitals become clinics. People are becoming accustomed to the idea that a hospital—what you or I think of as a hospital—is an urban amenity. People who are less educated, less wealthy, people with fewer resources, they take what they can get where they live, which puts hospitals in the business of providing care that obscures the human consequences of a region's decline."

Before settling in Wakkanai, Takagi was a military physician, and for a time he was assigned to Antarctica. Now he was a reverse commuter, visiting Sapporo twice a month to see his family, who had not volunteered to remain in a city where trains depart without anyone aboard and movies screen to empty theaters. He still gave an impression of military discipline. His carefully groomed mustache suited the severity of his speech, and in his office the clutter was arranged—without exception—at right angles. His dignity, I knew from the other Antarctic explorers who lived in Wakkanai, made him a preternatural victim of humorous accidents, and it was easy to picture a gust of wind destroying the meticulous order in his office, to imagine the blur of frantic gestures he would make to hold everything in its place.

When I met Takagi, he asked if I'd gone people watching at Wakkanai Airport when the nonstop from Tokyo arrived. The flight was prohibitively expensive, but seniors could book it with a subsidy. "You'll notice the lightness of souls," he said, "and the process that dissolves them." The next week, I watched the arrivals lobby fill with unaccompanied elderly passengers. Before they reunited with the friends and families who came to meet them, an interval of quiet passed through the lobby. The elderly went to Tokyo for medical procedures. In the airport, their loved ones waited to see how they would disembark—healed or unchanged. A man wearing an eye patch handed his son an envelope and said, "I couldn't spend it, knowing your circumstances." The son, wearing the billowing trousers used by Japanese construction workers, rose to twice the height of his hunched father and led him by the hand onto the waiting municipal bus, where two airport employees were assisting the driver with a growing line of wheelchair passengers.

A woman pushing her aunt in one of the wheelchairs wore a clipping of silvergrass on her blouse, in memory of the crews of fishing vessels lost in the Soya Strait. Leaning to her aunt's ear, she said, "I know they would have done the surgery here if they could, but it still feels as if you were turned away."

In Wakkanai, the human consequences Takagi described are addressed in part by the Council of Social Welfare (CSW), a nonprofit organization operating nationwide, used by local governments to administer loans to small businesses, to plan festivals and cultural events, and in Hokkaido to fill gaps in municipal snow removal budgets. In 2015, the prefectural government partnered with CSW's Wakkanai branch to implement a pilot program of antipoverty initiatives.

These initiatives targeted the entire subprefecture. Koziya Yoshiaki, director of the Wakkanai branch since 2008, hired seven new staff. There wasn't time to provide training, so he hired on temperament, with preference given to interviewees who were indelicate enough to say that the problem of poverty deserved a better program than a local nonprofit could offer.

Koziya was known in Wakkanai for opening his office to children suffering from social withdrawal, even if they were truant. "I get impatient," he told me. "In your country, churches help the vulnerable. We're Buddhists; we wait for disaster, then help the victims. The average person—the average social worker—doesn't see a problem with this."

When the pilot program became public knowledge, cases appeared, corresponding to Wakkanai's economic hardships: store owners who had spent their savings staying open after tourism declined, machinists and mechanics who outlived the city's fishing fleet, and what was left of the city's laboring class—people from the last few canneries, working for poverty wages on a contract basis. And cases came from out of town, including thirty day laborers from Kansai who bought one-way tickets to Wakkanai because a longshoreman told them the port was hiring (it wasn't), then came to Koziya for return fare.

I had known Koziya long enough to exhaust the pretense of having intelligent questions to ask him. Out of perversity, I asked if his clients included any I-Turns. They were rhetorically significant residents of Wakkanai, and I had never met one. As a matter of fact, he reflected, there was one active case.

This I-Turn, call him Hank, was unusual for deciding to remain in

Wakkanai; people in his position typically leave. He was not unusual for deciding to ride to the end of the line. In a given year, Wakkanai experiences a handful of nonresident suicide attempts, people who drive to Japan's northernmost point on Cape Soya and wade in. Some people come to Wakkanai to live anonymously, as far as they can get from the source of their despair. Koziya's staff meets them when they're out of ideas.

Hank was one of these. A friend tricked him into guaranteeing a bank loan, then disappeared, leaving Hank, fifty-three and unemployed because his eyesight was failing, with a debt beyond his means. He came to Wakkanai and found a part-time job processing fish. He was fired when his deteriorating eyesight interfered with his work. In the next breath, his supervisor phoned CSW.

"Everyone is so thoughtful," his caseworker deadpanned. This after she visited Hank in the hospital, where he'd been staying for two weeks, getting his eyes treated and learning to manage his diabetes, which he hadn't identified in six years of vision loss but she guessed when he spent his grocery vouchers on chocolate cereal.

When he lost the job, he lost his place in the dormitory the employer provided. CSW helped him find accommodation and registered him as a resident of the city, which he asked for, saying he didn't have any living relatives. The hospital said his vision could be expected to improve to an extent that would permit him to live unassisted. As yet, he couldn't distinguish a door from a window but insisted on taking care of himself in the hospital, doing his own laundry, for instance, after which hospital staff discovered everyone's clothes were mixed up and called his caseworker. Hank's misbehavior occurred on days when his roommates received visitors and he didn't. The hour and a half his caseworker spent with him, absorbing his incongruously exuberant way of criticizing himself for upsetting the hospital staff, she couldn't necessarily afford to spend. She had ten cases. Even her first—a year and a half going—was a month away from its conclusion.

She had never lived anywhere except Wakkanai. Before working at CSW, she was employed by a funeral home. When Koziya hired her, the pilot program hadn't been announced. She managed the office while he kept hiring. The phone rang rarely enough that when it did, the noise startled her. After the start of the program, she volunteered for elder-care cases. "To meet them in time to know them."

For privacy reasons, I didn't meet Hank. After her visits, his case-worker debriefed me in the parking lot. Hank wasn't aware of these debriefings. I never had the chance to tell him how, whatever the perversity of my asking about him to begin with, he had fulfilled his duty as the first I-Turn I'd known in Wakkanai. I had verified that some would visit and a few would stay, in a city becoming more itself through its citizens' neighborly gestures.

*

A resident of the rest home, in the final weeks of his life, remembered the town where he'd lived on Karafuto. There had been a paper mill and deposits of bituminous coal. The dangerous job of mining a new vein was assigned to indentured Korean laborers. As this resident's ailments progressed, he frequently asked, "What happened to their bodies?" In his lifetime, he had explained almost nothing to his children—middle-class adults now, near retirement age—about life on Karafuto. ("Only the time he went back," his son said, "on a government trip, and in a museum he saw the border stone from the 50th parallel, with a flower on Japan's side and an eagle on Russia's.") Contending with his partial descriptions and deteriorating memory, they weren't certain whose bodies he meant.

When he was a boy, dozens of Korean laborers had died in an accident. Their bodies had been sealed in the mine. Oral histories from former residents of the surrounding villages indicate that children were protected from the details. He might have learned as an adult. Or the rumor was all he knew, and he no longer wished to restrain himself from asking.

He died with his questions. Endo would say the death deserved gratitude. His odd behavior gave the family notice, and they were present during his final moments. She would say the death deserved attention. His generation was the only sufficient observer of its own passing, and his questions were an attempt to observe.

Endo had been in her late thirties, working at the nursing home in Toyotomi, when she was present for the first time as a resident suffering from dementia gradually neared death. Until the resident's family members arrived, only the setting's familiarity offered comfort; it would have been worse for the resident if he had entered the nursing home recently. A face you know is the face you want to see, Endo thought, and couldn't discern, in this resident's final days, if he knew who Endo was or why she stayed in the room.

"Your mind is ahead of your heart when you witness the process of dying. If emotions come first, that's harder. The impulse to leave—or look away—never diminishes. Not everyone feels this, but I did: The rest of the job is a way of preparing to be present when someone dies."

In Japan, most elderly die in hospitals. Most would prefer to die at home. Hospital deaths are expensive, and home deaths often require family support. As Japan's elderly population increases, deaths of each type will strain the resources of families and the resources of the state. Several reforms have attempted to manage these burdens by increasing the percentage of deaths that occur in the least common setting: rest homes. Residents of rest homes are most likely to avoid premature or unnecessary hospital transfers, researchers have determined, when supervisory staff specialize in end-of-life care. If staff matching this description were not so rare, Japan's largest generation of senior citizens would receive better care (and better-funded care) before death, and more would die in a familiar, comfortable place.

The rest home in Wakkanai is not a palliative facility. In the event of an acute concern, an ambulance is called. Endo stays with the resident until the doors of the ambulance shut. This happens outside the train station, in front of the Seicomart, lit twenty-four hours, where the night buses board. If Endo isn't working, staff take her place. "I remind them: You may be the last familiar person they speak to. Demand from yourself a convincing word of comfort. You may be the last familiar face. Demand to see your expression, when you're forming it, as they will."

A visitor to the rest home once surprised her by asking, in a conversation about employees offering comfort to the dying, if she knew the aphorism among theater professionals, *It's not important the actor cries, it's important the audience does.* She has remembered that. "The depth of responsibility for what is felt, the acknowledgment you aren't the one who feels it—in Wakkanai, I find myself thinking I haven't done my job because I'm rarely present in the moments when this is true." The worst deaths, she says, occur during holiday periods, when experienced staff use their vacation time and premature hospital transfers become inevitable. The staff who transfer the residents are strangers (part-timers, recent hires), and holidays remind the dying, even if they suffer from dementia, to miss the presence of loved ones. More often than Endo would have thought, residents enter the rest home who have outlived a son or daughter. A bereaved parent might have forgotten their bereavement, and their final moments—if they occur during New Year's festivities—are more likely to remind them.

The poverty of wartime had made Endo's father a frail child who became a frail adult, frequently sick when she was young. She was twenty-six when he died, and thirty-three when her uncle became frail in a similar way. She had gone to work at the Toyotomi town office the same year. When she agreed to work in the town's nursing home, her uncle told her he'd like to live there, but she was never able to arrange it. His final words to Endo were, "I wanted you to care for me." When she was with a dying resident of the nursing home for the first time, "I had been looking for somewhere his words would fit."

"I would like to die in the winter," she explained once. "It's quietest. The deaths I've seen, I'd want a gentle death. No pain, or hardly any. If it helped me not to be afraid, it would be all right to forget or imagine things. And the people I love, I would want them with me. I would want them to feel, for the moment it takes, that the place where we are is our home."

*

Monma Etsuko (that's Pleasant Girl, married forty-eight years to Victory Boy, and "our names," she said, in the hospitable, arch voice of a schoolteacher, "are always the irony of our times") would have saved three thousand yen taking the once-a-morning local train on the Soya Line, from its northern terminus in Wakkanai to its southernmost stop in Nayoro. The ride would have been longer, the train a single car, seating twenty and never full in winter, with a vestibule holding the old fare box on either end where passengers could watch the landscape. She paid to ride the express. "Those three stations scheduled to close, it's inauspicious stopping there. On this errand." To visit her husband in Nayoro City Hospital, where he had been waiting for a transfer to Asahikawa for three weeks, with stents in his veins, taking blood thinners and avoiding furniture with sharp corners.

Wearing her walk-in-the-countryside clothes, Etsuko gave an impression of levelheaded restlessness, a trait she developed when she was a girl and preferred leaving the house to climb the hills, to walk in town in the temple courtyards overlooking the Inland Sea. Town was Fukuyama, and when she was old enough to ride the train by herself it was Onomichi, Japan's visual metonym in postwar cinema. *Twenty-Four Eyes,* filmed on the nearby island of Shodoshima and released in 1954, dramatizes two decades in the life of an elementary school teacher who labors to protect her former students from the effects of escalating militarization and, inevitably, war. To Etsuko, the teacher's role in her students' lives and the metaphor of the island became objects of *akogare,* a yearning given its intensity by the nearness of an ideal impossible to fulfill.

Etsuko taught for three years on an island not far from Shodoshima, classes the size the film had portrayed, of children on whom it was gradually dawning that people left home. In her rare idle moments, she found she pictured herself and her students on a smaller island, a few meters in circumference. The image had the quality of eternity, of nothing that would happen, and no reason it had formed.

"Wakkanai isn't unlike that." Isolated, and she would never stop teaching; the city was perpetually short of help, and it was in her nature to provide it. "Thoughts are like children that way. Later you realize you always planned on something you couldn't imagine."

Her current term, teaching part-time in the city's after-school program, would end in two weeks. She planned to visit Okinawa for her grandson's school enrollment ceremony. Her husband would have accompanied her, if not for his health, if not for the wait to address it.

When regular screening discovered cancer in Monma's prostate, his private insurance paid for a hospital room in Tokyo so surgery could occur immediately. He was in Wakkanai a week later, convalescing at home, when his leg began to ache and his breathing became shallow. After tests at Wakkanai Hospital pointed to a pulmonary embolism, he was transferred to Nayoro by ambulance. The paramedic riding in the rear with the Monmas stopped Etsuko at the door to ask whether her husband would want extraordinary measures taken to revive him if the embolism caused cardiac arrest. It was a three-hour ride after that, under the advice that Monma should lie still and draw even, moderate breaths.

At 4:00 a.m., Monma was stable in intensive care. Etsuko had been awake twenty-three hours, twelve in hospitals and ambulances. The hospital staff in Nayoro, who were aware of how far she'd come, could not provide a place to rest or to sit and close her eyes until it was time for the morning train. The hotels they recommended, in walking distance of the hospital, offered no hourly plans, and none were willing—upon explanation of the circumstances—to let her sleep past checkout. By the time she learned as much, there were four hours left in the night she would have to pay for, so she did, and woke recalling the exhausted expressions she'd noticed in the waiting rooms of the hospital's new wing.

On subsequent visits, she learned the faces of the people she had seen, arrived from subprefectures like Soya, where hospitals no longer provided comprehensive emergency care. Everyone talked, everyone spent their pensions to visit their loved ones, who were waiting for care farther south, everyone explained to each other what sacrifices this required in lives like theirs, which were the lives you expected in rural Hokkaido.

I saw Monma in Nayoro the night before Etsuko came to visit him. The city is partly Wakkanai's mirror image, and in other ways they appear as peculiar relations, as if there were such a thing as brothers once removed.

Nayoro is, like Wakkanai, the northernmost city in the subprefecture that contains it. Kamikawa Subprefecture is administered from Asahikawa, and in practice Nayoro represents the geographic limit of Asahikawa's distributive economy; ideally—politically—Wakkanai represents that limit. The two cities are equally populated. Both have sprawled. Yokota Motors in Wakkanai sells cars; in Nayoro, a business with the same name operates a fleet of delivery vehicles, and its owner has never met a Russian. In neither town has the *shotengai* kept its businesses open, but Nayoro has a downtown department store. There are more convenience stores in Wakkanai, an indication that its retail economy depends on transient laborers dispatched by the construction state.

Separated by vinyl curtains, three other people shared Monma's hospital room and kept him awake talking in a way he recognized from the early weeks of his Antarctic expedition, when the intimacy of shared privation was novel. (Hospitalized for his prostate surgery in Tokyo, where the population density was 117 times higher, he'd been given a room to himself.) In the unit where he was staying, the reading room contained a sofa upholstered in turquoise vinyl, a bookcase of manga, and its own gas heat. Other people rarely used it. Monma found it peaceful, watching the snow accumulate on the window, listening to the intermittent, metallic knocking of the heater and the ghost sounds of the wind in the building's seams.

We watched movies about Antarctica on a laptop. The part in a documentary when the forklift driver at McMurdo Station describes the common traits of travelers who reach the edge of the map, Monma asked to watch again, and composed a Japanese translation: "The dreams the world is dreaming, those dreams are born in Antarctica." He talked about his daughters, whose interests and abilities were a thousand times his own, but whose era demanded mostly boredom. "When Wakkanai depopulates," he mused, "it might look like Antarctica. Then people can dream again."

Passing a television on my way off the unit, he pointed out news footage of the snow festival in Asahikawa. He said, "Very popular, that stuff," and because of his curtness it surprised me when he said *stuff*, not the four-letter equivalent. It took him a while, shuffling on his achy leg, to accompany me to the elevator, most of the way following the red line on the floor that led to the ICU, where he had spent five days. The ceiling of the lobby for the hospital's new wing was ringed in an abstract pattern of birds in flight, some milled from wood, some from metal. We were alone in the room, which was filled with more turquoise sofas, arranged in parallel rows like seats on a passenger vessel. "See?" Monma said, apparently in relation, by nearness in time, to his comment about the snow festival.

At the station, I gave an arriving passenger directions to the hospital. I guessed her destination from her behavior and belongings. Overhearing this, the gate agent tried to sell me a ticket to South Wakkanai, and when I said no, Wakkanai Station, he replied, "Oh. I thought probably you live there." What passes for a compliment in Nayoro. Hospitals, it occurred to me, were the place you went to blend in.

Etsuko was taking a photography class at the Wakkanai library. Waiting for the express train to Nayoro, she spread her photos on the café table and selected as her favorite a picture of Lake Onuma's frozen surface, textured by lumps of snow the wind formed—"snow tumbleweeds," she resolved to call them. "Our assignment had a theme: Little is shown, everything is present."

The plan was, get Monma from the hospital and take him to Asahikawa, where they would meet their daughter and spend two days together. Their daughter was able to reach Asahikawa because low-cost carrier flights were cheap this time of year. She hadn't remembered the snow festival or its effect on the cost of hotel rooms.

If you received a document, Etsuko said, explaining what you should be prepared to spend when you're hospitalized away from home, that would be one thing. Or if there were a support group to attend, where necessary information passed between like people. Instead, everything was discovered, everything came by surprise. Monma's surgery hadn't been scheduled yet. Etsuko booked accommodation presumptively, based on tentative scenarios, then canceled when further delays arose. In Tokyo, he would have undergone surgery weeks ago. But at what point is it a matter of recognizing the body is not a chattel—a matter of what is humane—not to self-transfer in a state of medical distress across a distance of a thousand kilometers, where loved ones can't reasonably follow?

"For every two doctors that retire in Wakkanai, one young doctor arrives," Etsuko said. "People don't trust young doctors. Some were good, the government sent them on a program, but the program didn't last. Teachers, it's the same. Some success, persuading people to be teachers, persuading teachers to work in the countryside, then a program ends or a policy changes. In difficult districts, young teachers become mentally ill. The students need help with their home lives, but teachers are trained to teach. Doing the right thing, but it's not what's needed and it isn't enough, over time it affects your mind.

"I'm uncertain why the dogsled races occur to me," she continued, "but when our daughters were young, they entered our husky, and every year he wouldn't pull. The spectators tried coaxing, so he crossed the ropes, he thought they wanted to play. You couldn't do that anymore, interrupt a tourist event. You would be guilty of something. Of getting in the way of that business Wakkanai has now: the business of Wakkanai, no matter what happens or how much time passes, trying to be Wakkanai."

*

When Heartland Ferry sent the MV *Eins Soya* to Sakhalin for the last time, Yokota didn't note the date, and later didn't remember it. A hundred and twenty-one passengers joined the outgoing voyage. The largest group were elderly Japanese born on Karafuto, traveling together on a tour of the island's former Japanese settlements. They sang while the boat departed. "Furusato" was the song's title, meaning "homeland," written with the character that attaches when the subject has died or vanished.

I counted half as many passengers on the return trip. Heartland's CEO was aboard and sat within earshot while a Russian shipping agent spoke admiringly of a Japanese acquaintance who collected maps of the Karafuto-era rail network on Sakhalin, then used them to locate trestles where brass plaques had been affixed at the time of construction. The plaques sold to antiques dealers in Tokyo for several times the cost of an excursion to Sakhalin. Coming through immigration, the same gentleman pointed at the trilingual sign that read, *Gaijin / Non-Japanese / Russkiy.* "Look," he said to the man in line behind him, also Russian. "We were a special category."

After Heartland Ferry withdrew, a committee of Wakkanai's leading citizens was constituted to develop a plan for maintaining the route. In time for the high season, a joint effort with the Russians was agreed on, to be operated by a Sakhalin freight company. It would carry passengers—eighty at capacity—but no vehicles, no freight, and would run for three weeks, weather permitting (the boat wasn't intended for rough seas; the Russians had leased it from Singapore). More ambitious plans had been discarded. The city couldn't afford them.

Quoted in newspaper coverage, Yokota's spin was the same as his successor's: The route had been retained, which would make it easier to attract investors. "A boat is some steel in a clever arrangement," he reflected. "To watch some people mourn the *Eins Soya,* you'd think they had lost the idea. I kept my idea—of Russia."

His first trip aboard the *Eins Soya* occurred on May 1, 1999, his first day in the mayor's office and by coincidence the day of the ship's maiden voyage. He was accompanied by a trade delegation. Waiting to disembark in Korsakov, Yokota and his colleagues watched a ship being loaded for departure to the Kuril Islands, where Japanese citizens were forbidden to travel. "The buildings were alike," a member of the delegation remembered several people remarking. "You couldn't notice them the way you'd notice buildings in Japan."

The built environment was familiar to Yokota from the first trip he had taken to Sakhalin, in the 1980s. His hotel in Yuzhno-Sakhalinsk had periodically been without water (which was never hot) and electricity. A golf ball had been left in the sink, to plug it if he wished. The bucket next to the toilet, the concierge explained, could be used to fill the tank. He thought of Japan in the intermediate phase of postwar recovery but was aware the comparison was flawed: Fishermen he knew who had come to Sakhalin for decades told him it had always been like this.

During the years between the Soviet Union's collapse and his election to the mayor's office, Russian patronage of Yokota Motors increased at a rate that supported his interpretation of global events: The world's other superpower was joining the economic system Japanese products dominated. What hadn't been present when he visited Sakhalin—reliable public utilities, proper plugs for your sink, a life of the material quality that Japanese citizens enjoyed when the postwar years had passed—would be provided at a profit, much of it sold, shipped, or stored by businesses in Wakkanai.

Arriving on the first Heartland Ferry to Korsakov in 1999, Yokota waited "an hour that felt like two—maybe it was," to clear immigration, on a wharf that Hokkaido's prefectural budget had paid to build. This might have distressed him more if he had been familiar with a joke (and the home truth it expressed) told in the Sakhalin towns the Japanese had built: For best results with infrastructure, lose a war with Japan. People paid handsomely for houses connected to Karafuto-era sewerage.

The streets of Korsakov, when Yokota was permitted to enter, were unpaved and potholed. Stray animals gathered around the steam pipes that ran from the harbor into town. Every few minutes a car alarm would sound, and to Yokota the noise momentarily had an accusatory quality, as if the cars he had sold to Russia were asking, *And what would you say we are doing here?*

I have imposed this foreshadowed sensibility on a few recollections Yokota provided later. When he entered this scene, his sense of destiny was not perturbed. The moment he cleared the customs barrier, he was with Korsakov's mayor, who took him to city hall. A reception was held, a banquet by Japanese standards, typical hospitality—of the overindulging variety—for Russians. No one had warned him to put water in his vodka glass, and he was drunk by the end of dinner. A Russian attendee remembered Korsakov's mayor asking to meet the captain of the *Eins Soya,* to which Yokota, comically grave with the effort of suppressing his intoxication, insisted, "The captain, it's me. I'm the captain."

Earlier, in the best stage of the mood that vodka provides, he ascended a few toddling steps onto a podium at the front of the room and became, on his first day in office, the first mayor of Wakkanai to deliver his first public address in another country. "We are building an environment of mutual hope," he said. "We people with a place and history in common, not a language or government. They will come to Korsakov, to Wakkanai, to know how it looks: the ideas and institutions that belong there. The feeling it gives to share these assets across national boundaries, to expand them so the lives they affect are continuously improved. To recognize our difficult past in the successes of our better present."

As he said it, he could see it.

*

It was time to close, and somewhere in the restaurant a television had been turned on to discourage lingering. Steve watched the waitstaff with amused detachment—in his case, an American habit displayed in American moods—while they waited for the message the television implied to settle on the other diners, a few tables of young people mostly drinking. Listening to the occasional loudness of the television audience's reactions, failing to gain the waiter's attention with his culturally extraneous hand signals, he said, "I'm watch a lot of

TV. Today's one, from Hokkaido University. They play sounds, like a fan in the bathroom, one of them. Even the computer signal"—he made a falsetto hum like a synthesized sine wave—"it's a kind of research. Sound never change, but your mind think it's kind of cycle, like wave in the ocean, hear it stopping, starting." The weather in Wakkanai had turned cold, and the park on the hill was closed. The city had reassigned Steve and his colleagues to other jobs. Recently, landscaping the grounds of the city's fitness center. Steve had been in Wakkanai six years now, and in the last two months his mother had begun to need the care he had come to Japan to provide.

The city had arranged the construction of the dining complex where the restaurant was located, shortly before Steve came home. It was, he thought, the only recent instance of the city procuring downtown land for commercial redevelopment, and he remembered his nephew (editor of Soya's local newspaper) describing the negotiations. The lobby's centerpiece was a topographical model of Soya and the southern half of Sakhalin. The local radio station broadcast from a booth nearby. A souvenir store sold trinkets opposite a ceiling-height replica of Wakkanai's old train station. "Guess it's for tourists," Steve remarked on his way out. "Local people using it, come here for dinner. Nothing else to use."

He lingered in the lobby until the young patrons of the restaurant left the building. On the sidewalk, he waited to see which way their taxis would take them. Quieted by the cold, four of them stood facing each other, shaking the falling snow from their hairstyles and dress shoes. Breathing into their hands, they made a fog, the way the men would on the docks when Steve was a child, so dense he had wedged through their bodies once to warm himself, and discovered there wasn't a fire between them.

Working for the city and giving English lessons, Steve came home nightly in time to bathe his mother and put her to sleep, he slept after that and the schedule repeated. Every few weeks she asked to go to the hospital. "Bath is like minute hand, my life," he told me. "Ask for hospital, it's the hour hand." Often, a doctor told Steve, an aging family member who feels she has become a burden will find excuses for hospital care, which is formal and feels less like dependency. Steve would pretend he'd called, pretend the hospital was full but they had given him the number for the mortuary.

"Tell them they're crazy," his mother would say, "I want to hear," and waited to hear him say it.

Steve learned what she remembered from the years he lived in America. Vividly, the trip that she and Steve's father took to visit him in Los Angeles during the early 1990s, a quarter century after he left. She recalled that there were fewer crowds in America, and people talked to strangers. Your father, she said, thought Americans were his enemies, but the landscape and their lack of affectation charmed him.

She remembered a handful of encounters that revealed the boundaries between Japanese Americans, among whom Steve had lived, and the expatriate emissaries of Japan's soft power, the wives and children of corporate managers assigned to California. Steve's mother chatted with these expatriate families in the supermarket, comfortable in their presence because they represented the prosperous Japan she had grown accustomed to; the Japanese Americans, including her own relations, seemed to her a defeated people, whose flight in poverty had not ended.

Steve's first day working at the fitness center, he was directed to join his colleagues wrapping trees for the winter. Many of the trees he wrapped had already died and more would die soon; the building is on the harbor, too close to the saltwater spray. The landscaping, he suspected, had been designed in Tokyo, at a national agency that contributed funding. The funding was probably renewable. Wrapping the trees, the city could keep jobs, if pointless, if part-time and short-term.

In his previous job, at the park on the hilltop, Steve had become weary of his bosses putting off the repair of a dilapidated fence and dialed the contractor responsible, who pleaded poverty. Steve called his nephew next, and the local paper ran a name-and-shame story with an eyesore photo of the fence. Contractors working for an acquaintance of Steve's arrived the next day, and in parting told him the repair was two or three hundred dollars—a couple guys working for an hour.

"My kids, they don't interested coming here. Times they get retirement, maybe want to see place the family from, even I'm alive can't tell them, 'Oh, everybody working, I was young, it's a working place. Lot of fishing, other thing going on.' People here waste time, blanket on the dead trees, take the government money, don't even fixing the small fence. That kind of a city, you never believe the kind of city was before. People knew about it, lost. I'm old, just working a way of life. Even do the work, doesn't matter. Time you're doing work, the work already lost."

During his assignment at the fitness center, Steve's habit of saying, "I happen to be born in Japan," became a habit of saying, "I wish I'm born American." It might have followed that he was looking forward to getting back. He frequently discussed—with me, with his siblings—the question of when his mother would pass. "Guess I'm go back," he would say. His tone had an abruptness in common with bluffing: Guess I'll have to, better not let me. I had always found it strange that he never mentioned the mother or mothers of his children, and finally asked if he had ever been married. He said, "I don't want to talk about it." He wanted to talk more about what an insult to the notion of livelihood it had been, the city wasting his labor on patronage.

It was a long time later he said to me, "It's the place I'm American," and meant Wakkanai. First, he had to acknowledge that Wakkanai was his home, neither chosen nor inherited exclusively—partly each, wholly his. He had to absolve the people in the city of their role in permitting it to vanish, and himself for not having sensed, from a distance, what had occurred. Longer than he expected, he had to carry his mother to bed. Her mind entered its final decline, and she asked him, lying down with her eyes closed and the lights off, where he had gone until he was old. When he answered into her ear, she reached to touch his chest. With her hand there, she said, "Very few hearts beat this way." He left and pulled the door shut. She said, "Don't forget," to herself, or to the impression his presence had left in the room, and he waited behind the door, listening to her repeat it until she drifted off.

Reform

||||||||||||

21.

When the earthquake occurred on March 11, 2011, Iriya Takeshi, a civil engineer employed by the Japan Atomic Power Company (*Genden* in Japanese), was in a meeting at NISA, Japan's nuclear regulator. While he sheltered under the tables of a conference room with his NISA counterparts, he glanced at the earthquake data from Miyagi Prefecture appearing on their phones. "Slow system," he quipped. "Old data." Born in Sendai, Miyagi's largest city, he had heard from friends still living there when a temblor struck two days earlier. Earthquakes in Miyagi Prefecture are rarely felt in Tokyo.

As the data refreshed, the Richter scale rating of the earthquake appeared and the regulators' faith in its accuracy began to seem reasonable. The initial rating would likely increase, and Iriya realized he would no longer be able to estimate the radius of the shaking. Already, the severity of the earthquake had reached the upper limits of the thresholds used in his profession.

Through the windows, he could see Hibiya Park, where office workers evacuating from nearby buildings had started to gather. Nothing about their behavior was inappropriate, but he wished they would disappear. "Not Miyagi," he discovered himself thinking. "Not now." The world's best seismologists had worked for the past fifteen years to predict Japan's next major earthquake. Iriya was familiar with the resulting maps. Comparatively speaking, Miyagi was a low-risk zone.

Iriya had lived in Sendai until he was six, before his father's work required a series of relocations. When he was thirteen, his family arrived in Tokyo to stay. In the suburban neighborhood where they settled, there were only a handful of children whose lives were not the product of careful, class-conscious planning. At school, his classmates greeted him with the presumption that he had learned, from changing schools so frequently, to fight. He hadn't, and soon wished he had.

The high school belonging to Waseda University was the first school he attended without expecting to leave. He went to college at Waseda and stayed at Waseda for a graduate program in civil engineering, which wasn't among the disciplines that well-ranked students were encouraged to prefer, but he was captivated by buildings. Specifically by institutions—buildings whose value is in their purpose, necessarily designed for longevity. The Ministry of Defense was his first-choice employer. He passed their exam and a job would have followed, but was surprised to learn that Waseda had nominated him to the Tokyo Electric Power Company (TEPCO). He could only attribute the nomination to his occasional, courteous indulgence of a rambling professor's anecdotes about his son's career in the electricity industry.

From new hires, TEPCO solicited three preferred assignments. Iriya listed urban infrastructure first, natural gas second, and struggled to select the third, before deciding nothing suited him, and writing—on an impulse—nuclear. Later the same day, a colleague told him the nuclear division was desperate to attract civil engineers, who found the work tedious and passé. The company was certain to place him where need was greatest. Iriya remembered moving cities and changing schools, and realized he had not escaped the childhood rule of going where you must.

After the earthquake on March 11, he walked to Genden because it was close enough to reach on foot, stopping when he felt an aftershock. He thought to buy groceries, but every store he passed was overrun. He noticed—without remembering how far his office was from his home—that lines had formed outside bicycle shops.

He reached Genden ninety minutes later and found his colleagues watching the news. Japan's meteorological agency had issued a ten-meter tsunami warning. The group he supervised at Genden was responsible for earthquake and tsunami engineering, but his team members, like him, remember how little there was for them to do once the disaster began. He prepared to leave the office. He was aware of the reported epicenter of the earthquake and the projected path of the tsunami, but his former employer's plants in Fukushima never entered his mind.

In 1994, when Iriya learned he would be assigned to TEPCO's nuclear division, he assumed he would work in Fukushima but was eventually sent to Higashidori, a site TEPCO shared with a local utility company, where TEPCO was planning to build a new reactor. In 2000, TEPCO loaned Iriya to Genden, which was preparing to construct Japan's first advanced pressurized water reactors (APWRs), Tsuruga 3 and 4, in the hub city of Japan's Atomic Broadway region, where fourteen commercial-scale reactors crowd the coast of the nation's third least populous prefecture.

Japan's regional electricity monopolies, which own Genden, had founded the company to test the profitability of nuclear technology. The nation's first commercial reactor was a Genden reactor. The first domestically constructed reactor was a Genden reactor. When the industry wanted to find out whether a Japanese reactor could refuel faster than its American counterparts, Genden did it. Using the best technology available, Tsuruga 3 and 4 would establish a new safety standard for nuclear electricity. When this goal was discussed with regulators, one possible achievement was omitted: Genden would become the first company to deliberately double the number of reactors operating within two hundred meters of a large earthquake fault, widely understood to be active.

Iriya was assigned to this problem: the Urasoko fault.

For the duration of their use of the Tsuruga site, Genden had maintained that Urasoko wasn't active, dating its last event to fifty-five thousand years ago, narrowly in the realm of regulatory compliance. Iriya presumed that Genden, in anticipation of applying for permission to build Tsuruga 3 and 4, would revise their Urasoko estimates and attribute this revision to improved data discovered during the siting process. He was prepared to design seismic countermeasures robust enough to withstand an Urasoko earthquake. What other approach was plausible? For decades, academic geoscientists had known the Urasoko fault was active. No informed individual, absent a conflict of interest, would continue to corroborate Genden's prior claims.

Many laws and policies forbid misleading Japanese regulators, but none forbid the construction of a nuclear reactor near a major fault system (in Japan, Tsuruga 2 is nearest). In a regulatory or legal environment, this would be difficult to forbid. *Fault* refers to any place where the earth has fractured during a seismic event; a fault could be less than a meter long and prone to displacements (movements) of a few centimeters, or kilometers long and prone to correspondingly extreme displacements. Depending on your interpretation of *major,* you could run out of places in Japan to build safety-class facilities. The Urasoko fault, however, is thirty-five kilometers in length, indisputably capable of producing an earthquake to rival the event that affected Fukushima.

Iriya was awed by the task he was about to join: using the best technologies available on a commercial scale, engineer the world's first earthquake-proof nuclear plant. So close to a fault so large, nothing less was enough.

During his assignment to Genden, many of Iriya's colleagues saw the necessity of relinquishing the company's deceptive assessment of the Urasoko fault. And, like Iriya, were shocked when their superiors overruled them. Genden's leadership had been mentored and promoted by the executives who presided over the assessments in question, at a time when their findings were dubious but plausible. "You want me to bury my friends," one lamented. To call them incompetent, or liars.

Genden reiterated its fraudulent safety claims, abetted by the ambiguous science of seismic risk assessment: If a fault does not pass directly beneath a nuclear plant, its proximity and its most recent movements—however frightening—reveal little about the hazards it presents. NISA's seismology experts, unwilling to sacrifice their agency's reputation by disclosing that Genden had fooled regulators for decades, participated in this process of obfuscation. In a political environment dominated by concerns about Japan's lethargic economy, the collapse of regulatory oversight at Tsuruga went unnoticed, awaiting a future disaster that would illustrate its relevance.

Iriya's zeal, if not his idealism, endured. When TEPCO brought him home after four years, he quit to take a job at Genden and was castigated for embarrassing TEPCO—for quitting the major-league club, his supervisor complained, to play for the farm team. TEPCO's perspective placed Genden, Japan's smallest nuclear utility, at the insignificant edge of the electricity business, a business defined by the places where demand was strongest: Tokyo for the service economy, Kansai for its urban and industrial centers.

Iriya had spent too much time in the field to accept the notion that the plants were one asset among many, easily managed from afar. Nuclear safety succeeded or failed on Atomic Broadway—in Tsuruga—and Tsuruga's reactors belonged to Genden, the only utility dedicated exclusively to nuclear energy. Among Genden's employees, many of whom lived in the communities surrounding the plants, Iriya never encountered the illusions that shaped workplace culture at TEPCO, where generating electricity was considered a trifle compared with selling it.

And Genden had the APWR project, which was compromised from a regulatory perspective, but for precisely that reason needed the best engineering from the best people, and still presented an opportunity to design a structure that would alter society's relationship with natural disasters. "Most people wonder if they're capable of changing the world," Iriya later reflected. "I thought Genden could show me."

Having arrived at Genden with his TEPCO pedigree, Iriya was permitted to devise the trajectory of his work. Colleagues remember his subsequent accumulation of affectations. One colleague compared it to the way a precocious child would behave after changing schools for the first time, having discovered the possibility of self-invention. Iriya frequently showed up late, dressed in a natty, colorful way, with the clothes askew as if he'd slept in them. A former supervisor summarized, "We let him act young."

When I met Iriya after the Fukushima disaster he was the product of this permission, of benefiting from it for a decade in pursuit of his world-altering project. He was in his mid-forties, short and svelte in well-fitting tweed and wrinkled ties, his hair wetted in a variety of unconventional shapes. He was a natural field researcher, able to modulate his voice to be heard through—rather than over—the noise of dewatering pumps at a dig site, and had a way of turning any environment into an opportunity for research. When a Christmas tree appeared among the semiarid plants in the atrium of Genden's office building one December, he scrutinized it before continuing to his office, promising me as he passed that he would find out "where it was grown and why it's here."

Some of his wondering about himself remained from a rootless childhood, but some had been replaced. Early in his career, Iriya began attending school reunions. He married a former classmate. They had a son. Genden grew up, too; steady pressure from the company's investors and creditors resulted in streamlining and a concurrent change in office culture. It was more like working at other power companies, and Iriya was more like his former colleagues at TEPCO, mostly satisfied with the life that provided.

He was preparing to leave the office on the day of the earthquake when a colleague advised him that one of the Fukushima plants had lost backup power. Iriya's expertise was too specific to civil engineering to assess the significance of this comment; he assumed it was a courtesy extended to him because he was a former TEPCO employee. He was nearly out the door when it triggered another thought, or a memory: that he was not an anonymous person with a job he enjoyed and some prepared language to explain why it matters. Earthquake-proof. That's what he was building. He worked in the absolute center of the nuclear industry's relationship with disaster.

He sent a message to his team before leaving: Because the siting of Tsuruga 3 and 4 has required us to develop geotechnical expertise that other engineers don't possess, and because a nuclear plant has been affected by a seismic event, whatever happens next might involve us more than we can imagine.

22.

In 2004, the year Genden reiterated its deceptive assessment of the Urasoko fault, Shimazaki Kunihiko, a Tokyo University seismologist, was collaborating with several fellow academics on a draft report evaluating the tsunami vulnerabilities of Japan's nuclear plants. He and his colleagues were members of a cabinet-appointed committee whose work, notwithstanding its momentous intentions, was routine in the context of Japanese bureaucracy. Many members of cabinet committees become "government-contacted academics" and through repeated collaboration develop a business-as-usual approach in one another's company.

Though quiet by temperament, Shimazaki had been elected to leadership roles in influential academic societies, where he had been responsible for bringing several disciplines of geoscience closer to matters of public safety. At a meeting of the tsunami committee in February 2004, Shimazaki presented his analysis of mitigation calculations provided by TEPCO and approved by regulators. The stipulated ability of the Fukushima plants to withstand a seventeen-foot tsunami, he concluded, was less than half of what might be required. Despite Shimazaki's reputation for thoroughness, consideration of his critique was postponed indefinitely.

After the Fukushima disaster, Shimazaki recounted his frustration with the tsunami committee to *The New York Times*. The article included a picture of Shimazaki in his office, seated at his desk with his back to the camera, the walls bare, and in the corner of the frame several cardboard boxes (he had relocated twice since accepting emeritus status). "Either Professor Shimazaki imposed the conditions of that photograph, or the photographer had an instinct for his mood," one of his colleagues remarked. A career in service of safety had made him, however unwillingly, a party to the Fukushima disaster.

The concerns he raised in 2004 connected him to several earthquake experts whose relationships with government panels had been ambivalent. One seismologist hired by Japan's Nuclear Safety Commission (NSC) called the nuclear plants on Atomic Broadway "a suicide bomber wearing grenades around his belt." A prominent geomorphologist, Suzuki Yasuhiro, declined a 2008 invitation to help apply new seismic safety guidelines to existing plants. So long as it was outside the realm of legal or bureaucratic possibility to declare a site seismically unfit, he would not participate. Perhaps none of the plants needed shuttering, but you cannot regulate, he concluded, in the complete absence of leverage.

Periodic safety failures at Japan's nuclear facilities have ensured that criticisms like these receive widespread attention. In 1995, a dramatic cooling breach and subsequent cover-up occurred at Monju, the government's state-of-the-art facility meant to demonstrate the commercial viability of fast breeder reactors, which produce more fuel than they consume. Four years later, a radiation control failure at a uranium reprocessing facility killed four, irradiated hundreds, and prompted evacuation measures affecting hundreds of thousands. By the turn of the century, the list of accidental radiation releases, equipment failures, and related nondisclosures (or belated disclosures) by Japanese nuclear utilities was substantial. Political resistance to nuclear energy was organized and zealous. The nuclear industry's response was well financed and institutionally mature.

In this setting, "neutral" often loses its meaning. I once visited the office of the labor union that represents nuclear plant employees in Fukui, to negotiate access to union officials. By virtue of the lifetime employment system that applies to most full-time jobs at utility companies, the unions and utilities are friendly, and the union representative wanted to know whose side I was on—for the industry or against it.

"Neutral," I said.

"I'm neutral, too," he replied. "Which neutral are you?"

The notion that science itself is neutral—and so are scientists—is vulnerable. Japanese nuclear professionals, if they work in management, often characterize their academic counterparts as economically naive: They don't know that a resource-scarce archipelago can't compete in a global marketplace without nuclear power. And nuclear engineers regard academic expertise as inferior: The industry does more safety research than academia and funds it better.

Academic geoscientists would object to this characterization on several grounds, among them the difference between academic culture and engineering culture. The goal of academic research is to learn what's true, without agenda. The goal of engineering is to construct a building or solve a problem. The latter profession rarely produces a temperament suited to the possibility that some problems cannot be solved, not even in the foreseeable future, not even by the collective efforts of many brilliant engineers. "To some engineers," Shimazaki once remarked, discussing the many Genden employees who contributed to misleading characterizations of the Urasoko fault, "every problem is a test of ability, not humility."

To Iriya, the task of engineering a nuclear plant safe enough to operate in an unsafe location remained compelling. I have said he was trying to make Tsuruga 3 and 4 earthquake-proof. Had someone on Iriya's team used that term—"earthquake-proof"—he would have found it troubling. There is no such thing as an earthquake-proof facility. There is only the question of how much earthquake a structure can withstand. Iriya expected that Tsuruga 3 and 4 would be among the most earthquake-resilient structures in the world, perhaps *the* most resilient, but would not have conflated this with invulnerability.

"Earthquake" is a term that summarizes the diverse effects of a geophysical phenomenon. One of these effects is ground motion (shaking). The intensity and characteristics of this shaking are considered foremost in the design of a facility's earthquake countermeasures. Another effect is displacement (faulting), which refers to the occurrence of fractures in the earth. The two are plainly related (faulting produces ground motion), but faulting constitutes the rare exception to Shimazaki's criticism of the intrepid attitude of engineers. Few would claim there is a proven countermeasure for the direct effects of displacement. If the ground underneath ruptures or changes shape, no structure can be relied on.

The Urasoko fault is large and dangerous. Smaller faults, less immediately apparent, abound in seismically active regions, some of them directly linked to larger faults like Urasoko, some independent, and in many cases the nature or existence of that relationship isn't clear. Among faults large enough to cause an earthquake, some cause them. They are seismogenic. Others are not; their movement is triggered by earthquakes occurring on other faults.

In 2006, Japanese regulators began a review of the way these nuances were addressed, in assessments of possible ground motion and faulting alike. Most experts who participated consider the 2007 Chuetsu earthquake, which gave the resulting regulatory changes their urgency, the genuine initiating event. The earthquake occurred offshore from TEPCO's Kashiwazaki-Kariwa plant, the world's largest (seven reactors). Ground motion at the plant was significantly higher than anticipated. When regulators cast a backward glance, they were chagrined to discover that their predecessors had approved several TEPCO requests to reduce the amount of ground motion the plant should be expected to withstand. These reductions violated the plant's original siting requirements and enabled TEPCO to reap enormous profits.

The earthquake was followed by a prolonged back check, as regulators would come to call it. Utilities had already been required to check the accuracy of their resilience thresholds on a periodic basis. Now regulators became directly involved in the assessment of those thresholds, especially the NSC, whose role as an independent extension of the Cabinet Office explicitly empowered it to audit other nuclear regulators.

The most influential voice on the NSC's committee of seismic risk experts belonged to Okumura Koji, a Hiroshima University professor widely admired for his contributions to the applied science of paleoseismology (the dating and characterization of prehistoric earthquakes). Industry scientists couldn't criticize his expertise; they envied it. Among the misrepresentations that came under Okumura's review, none were as brazen as Genden's claims that the Urasoko fault wasn't active. Okumura's assessment of Genden's research was not charitable, nor were his conclusions about the seismic safety of the Tsuruga site.

To Iriya and other Genden employees discomforted by the company's nondisclosures, Okumura was a heroic figure. "We wanted the regulator to intervene," a member of Genden's compliance staff recalled. "Okumura's criticisms restored us to the standard we first intended for Tsuruga 3 and 4: World's Safest."

It never occurred to them that Okumura's findings might jeopardize the construction of Tsuruga 3 and 4, or the continued operation of Tsuruga 2, the most profitable reactor in Genden's fleet. Forbidding the operation or construction of a nuclear facility on the basis of a seismic safety review was not—as Suzuki Yasuhiro had observed when he declined to join the back-check process—how the regulatory system worked. After Fukushima, when urgent questions arose about the paucity of past enforcement actions against Japanese nuclear plants, Okumura and his colleagues argued that decision-making authority was not within their remit. They provided expert analysis, which was integrated with the utility companies' submissions and the work conducted by career regulatory staff.

Shimazaki held similar notions about his involvement in nuclear safety. He had supervised projects that affected public policy (including the National Seismic Hazard Map) but rarely considered the downstream influence of his work on the decisions of regulators, until the Fukushima disaster made the painful suggestion that his influence had been minor, or amounted to nothing. When he was photographed for *The New York Times* after Fukushima, he had nearly escaped "that windowless room," as one of his co-researchers described it—the limited authority granted to scientists during moments that ultimately matter. "Are you glad to get out?" his colleague wondered, "or do you dwell on the role a scientist should play?"

In the summer of 2012, Shimazaki learned he would be nominated to the vice-chairmanship of the Nuclear Regulatory Authority (NRA). Because the nomination was unexpected and his interest in the nuclear industry previously existed at an academic remove, he understood the proclaimed mission of the new regulator only as thoroughly as any citizen did: Establishing the agency was a direct response to regulatory failures identified by parliament's investigation of the Fukushima disaster, and the NRA's watchwords were already in wide circulation—Japan's new regulations, lawmakers promised, would be the world's strictest. Until those regulations were enacted, nuclear plants were effectively forbidden to restart.

As soon as he learned of his nomination, it was leaked to the press, and the resulting uproar in parliament—where appointees leaked by the nominating party, as a matter of procedure, cannot be confirmed—left Shimazaki and the other nominated commissioners, whose names were also leaked, in a prolonged state of suspense. The press was charitable to Shimazaki, characterizing him as a stranger to the industry whose expertise was nonetheless relevant to regulating it, a desirable combination of attributes for the leadership of an agency whose purpose was a break from the past but whose workforce would be inherited from it.

Less kindness was shown to the nominated chairman, nuclear engineer Tanaka Shunichi. After the disaster, Tanaka, a Fukushima native, had returned to his home prefecture and volunteered to assist with decontamination efforts. The Democratic Party of Japan (DPJ), briefly in power, was selling him as the expert who could operate on both sides of the division that existed among his commissioners, two of whom had never worked in the nuclear industry, the other two of whom possessed urgently needed expertise you could develop only in an industry career. Tanaka was an academic and nuclear professional alike, and whatever partisanship the latter might have required, the DPJ contended, had been erased by his experience of the disaster. In early comments to the media, he apologized to his former colleagues for the draconian measures he planned to take.

The media wasn't convinced. Tanaka had been a lifelong resident of the nuclear village, as critics called it, a group of research institutions and corporate patrons that conspired to maintain industry control of nuclear safety. No matter your prefecture of origin or the depth of your grief in response to the disaster, you did not develop radical ideas about the nuclear industry over the course of a few months if you owed your career to it. In the press, Tanaka's name was attached to the distressing possibility, so soon after Fukushima and the resulting consensus about the need for reform, of a reinstatement of the status quo. In an impolitic attempt to take responsibility for his past, Tanaka publicly acknowledged that his professional affiliations had been textbook pro-nuclear.

The Liberal Democratic Party (LDP), which remained in the longest of its few and brief periods out of power, was happy to exploit Tanaka's vulnerabilities during his confirmation hearings, already contentious after the leaking of the nominees. Polls showed Japanese voters overwhelmingly in favor of abandoning nuclear energy, a goal the incumbent party was more likely to pursue. LDP leadership, derived from prefectures where the nuclear industry was heavily vested, were struggling to defend their aggressive plan for restarting the nation's reactors. If Tanaka could be made to appear pro-nuclear, the gap between the parties' respective platforms would appear to shrink.

So it occurred that the party most supportive of the nuclear industry played a leading role in confirmation hearings that tested Tanaka's commitment to turning his back on it. In a manner none would have predicted, Suzuki Yasuhiro's complaint about regulatory leverage entered parliamentary debate and public discourse: Was Tanaka prepared to do what regulators had failed to do before Fukushima, to determine that a plant was unsafe and require its decommissioning?

Closing a reactor permanently on the basis of faulting hazards had become a legal possibility shortly before the disaster, when the NSC issued new guidelines for applying the active fault rules drafted during the back-check process. These rules had been adopted to provide a formal codification of the obvious: Don't build nuclear plants on earthquake faults. The guidelines went further, expanding nuclear utilities' obligation to study minor faults near their existing reactors, and connecting this obligation to a rule that had been drafted with major fault systems in mind: No reactor should operate if equipment critical to its safety might be affected by displacement from an active fault, and faults will be considered active wherever the possibility "cannot be denied."

Because the new guidelines would require investigating a wide variety of faults, small and large, seismogenic or not, some experts raised concerns. "If you look for every crack in the earth and evaluate it with the full force of a field investigation, your risk assessments lose context and perspective," argued Yamazaki Haruo, a geomorphologist whose research focused on the relationship between fault maps and earthquake damage patterns.

After the Fukushima disaster, concerns like Yamazaki's were no longer a priority. NISA and NSC geoscientists swiftly acknowledged that their risk assessments had fallen phenomenally short. "We were ashamed and terrified, pondering what else we had missed," said Toda Shinji, a NISA expert. "We went back to work to find it."

By the time Shimazaki's and Tanaka's nominations were submitted to parliament, Toda and his colleagues had identified six nuclear plants where the possible presence of active faults required immediate assessment. Tsuruga figured prominently in these discussions, which the media compared with recent revelations about design flaws at the Fukushima plant, willfully overlooked by regulators: Multiple reactors had operated astride possible earthquake faults for a collective total of 141 years, and until disaster struck, few experts objected. Almost a year to the day after the disaster, the government's industrial research institute published a report reviewing Genden's Urasoko data, concluding that the fault was longer (and possible earthquakes therefore more intense) than the company's current assessments claimed.

Iriya remembers that he was too astonished by the media's response to pay attention to the substance of the report. The first paragraph of the article published by Japan's largest wire service announced, "An active fault running under reactors 1 and 2 at the Tsuruga nuclear plant in Fukui Prefecture is much longer than previously thought and could trigger a 7.4-magnitude earthquake, larger than earlier projections, according to a team of government-affiliated researchers."

From the perspective of Genden's engineers and the wary few academics who had cautioned against the active fault guidelines issued in 2010, the media's interpretation of the report was foreboding. Minor faults adjoining Tsuruga 2 had been confused with the Urasoko fault itself. The fictitious notion of an enormous fault passing beneath two nuclear reactors—when it was actually hundreds of meters away—could cause the public to lose its sense of proportion in matters of seismic safety.

Discussion of faulting hazards dominated the final phase of Tanaka's confirmation hearings. Reactors 3 and 4 at Ōi, a Fukui plant owned by the Kansai Electric Power Company (KEPCO, or *Kanden* in Japanese), had been allowed to resume operation despite the nationwide moratorium on restarts, in anticipation of summer electricity demand. Ōi was among the plants selected to undergo active fault review, and the fault in question showed enough evidence of displacement that a review of the site was considered, like the Tsuruga review, urgent in the first degree. In a session with the Steering Committee of the lower house, Tanaka was asked what he would do if reviewers concluded that an active fault was present under a safety-class structure at Ōi.

"Shut the reactors down," Tanaka replied.

By threatening Kanden, Tanaka threatened the nuclear industry's restart agenda. Eleven reactors on Atomic Broadway belonged to Kanden, the nation's most nuclear-dependent utility and the largest after disaster-stricken TEPCO. None of Kanden's plants were implicated by the engineering failures of the Fukushima disaster, and the company intended to restart twice as many reactors as its nearest competitor, on multiple sites where active fault investigations were pending.

"When Tanaka promised to discipline Ōi, that was the first time his confirmation hearings were in the press and the articles didn't mention the possibility that he was partisan on the industry's behalf," a member of Tanaka's staff recalled. Public concern about seismically vulnerable nuclear plants was deeper than skepticism of Tanaka's background.

While the confirmation process dragged, the five nominated commissioners were appointed to a cabinet committee meant to facilitate their introduction to one another. It was decided that they would travel together to Fukushima, to view the plant. The press wasn't notified. Shimazaki in particular was grateful that no public-facing statements would be required. According to one of his colleagues, "He found it jarring to be made part of the safety myth. In his position, you know the limits of what science can determine, and you discover that society's desire for certainty is nonetheless intense."

On a previous trip to the disaster area, Shimazaki had visited Okawa Elementary School in Ishinomaki, where seventy-four children had died on the day of the tsunami after responsible adults followed emergency procedures that wrongfully prohibited evacuating. Commentary about Okawa has focused on the malign role that policies can play if they displace better instincts. For Shimazaki, the example of Okawa was instructive. The National Accident

Commission report, which served as the impetus for the NRA's establishment, named two priorities: removing regulatory responsibilities from the cabinet ministry whose primary aim was economic growth (METI), and streamlining the multiple-agency system that had previously been responsible for nuclear facilities. All of this would occur in service of safety culture, with the explicit acknowledgment that regulatory standards exist downstream of the principles and behaviors that produce them.

The NRA's new parent agency, the Ministry of Environment (MoE), presented its own complications. Its mandate included the reduction of greenhouse emissions, a policy poorly served by importing more fossil fuels in response to the idling of nuclear plants. More significantly, the MoE had engaged in a decade of mutual antagonism with Japan's nuclear industry, and the ministry's senior bureaucrats were displeased to learn that their other responsibilities would be dwarfed by decontamination activities at Fukushima and regulating the industry in perpetuity.

The NRA would need to develop a new safety culture using staff inherited from its predecessor agencies, under a purely cosmetic plan for merging them. Former NSC and NISA staff would conduct regulatory activity, and research tasks would be assigned to the Japan Nuclear Energy Safety Organization (JNES), an arrangement identical to its pre-Fukushima antecedents. The only difference would be the agency's leadership, which placed Shimazaki in the second most significant position among a total of five that distinguished the NRA from its predecessors.

Of his second trip to the disaster area, to visit the destroyed nuclear plant with the other nominated commissioners, Shimazaki has nonspecific memories: somber silence during viewings of the damage, and the astute, earnest questions his colleagues asked during briefings. He remembers in greater detail the bus ride from the plant to the train station when the tours and briefings concluded. He took a window seat next to fellow commissioner Fuketa Toyoshi. In five years, Fuketa would be the NRA's new chairman. Shimazaki would be absent from the final phase of the restart process, with the press saying his departure from the NRA signaled a return to pre-Fukushima complacency.

On the bus, Shimazaki and Fuketa discussed the clarity one gains by going somewhere in person. Neither could describe what had changed about his intentions during the trip, but both agreed their intentions had changed. Shimazaki had been reminded of Okawa Elementary School and told Fuketa about his visit. Their conversation began when the plant was visible through the window. Shimazaki watched it while they spoke. He was telling Fuketa about Okawa, but his mind wandered. *Not better, not perfect—different.* He couldn't guarantee the safety of Japan's nuclear industry, but he could change the way it was regulated. Whenever the opportunity arose, he would discard the systems his predecessors relied on.

The people involved couldn't be the same. The experts whose contributions to NISA and the NSC were vindicated by the disaster and the experts whose mistakes were revealed—both would be replaced. He wouldn't include too many voices, all talking at once like the members of the tsunami committee in 2004, so no single voice was afraid of its mistakes. Only a few experts would review each safety concern. They would be appointed on a self-nominating basis, to eliminate considerations of prestige. Anyone whose expertise was appropriate would be hired.

Five years later, when Fuketa had become chairman and Shimazaki was prepared to reflect on his vice-chairmanship, he remembered this moment on the bus and considered its relationship to the work that followed. "There was no logical relationship," he concluded, as if the absence of this quality confused him. There was only the ruined plant out the window of the bus, the part of him that was present in the conversation with his colleague, describing the memory of the elementary school, and the other part, his mind at work on the difficult task of determining what makes and changes culture.

"There is no logical relationship," he reiterated, having gone over it out loud, speaking now in the present tense, no longer confused.

23.

Before he was promoted into management at Genden, Hoshino Tomohiko was assigned to the maintenance department at Tokai, supervising the plant's turbines. In Japan, control room personnel begin their careers operating turbines because the system provides a tidy metaphor of an entire plant. Hoshino, the mentee of a prominent professor of mechanical engineering, was assigned to turbines for seasoning, in the expectation that he would ascend to the circle of managerial employees who could communicate operational subtleties to company leadership. Maintenance staff at Tokai quickly forgot he was a company man. "A nuclear reactor doesn't know it has shareholders," one recalled. "Hoshino understood that."

Hoshino was born in the Tokyo suburbs where Iriya's family eventually settled. His parents had grown up nearby, before the area suburbanized. A natural athlete, he ran track in middle school, then disappointed his high school coaches by neglecting his training to hang out with friends in the band room and play Rolling Stones covers. The high school was one of a handful of all-boys public schools in Japan. Athletics were mandatory, and Hoshino couldn't enjoy track as an obligation the way he enjoyed it as a choice.

He was a weak swimmer, and in high school the physical education classes distributed swim caps on the basis of ability. The lowest cap was bright red. Resentful of the possibility that his ineptitude would be publicly displayed, he began visiting the pool before the swimming curriculum started. The following semester, he tested into the strongest group. During the annual class trip to Shizuoka for ocean swimming, Hoshino's group swam an hour-long circuit past the breakers. The first of these trips was also Hoshino's first time swimming in the ocean. Any associated loss of confidence did not occur to him until the first swell pulled him away from shore, and he discovered that his feet no longer touched bottom. He was able to finish but returned to the beach carrying the memory of his fear more than the achievement of overcoming it.

When I met Hoshino, he was in the handsomely attired stage of his career that no longer involved winning the loyalty of maintenance workers. Corporate managers in Japan are taught to distill the image they cultivate until a single word captures it. For Hoshino, *Meticulous.* Once, a TEPCO colleague left a pen in his office. He wrote with it every time I saw him for three months. The Hoshino of the Rolling Stones cover band, who felt no behavior should be mandatory, supplied the original motive. He was fond of the pen because it was considered poor form to carry items emblazoned with TEPCO's cheerfully radioactive pre-Fukushima mascot. He wished it to be known that there was no room in his busy mind for the hot air of corporate censorship. He was management's engineering liaison for Tsuruga 3 and 4 and responsible for Genden's role at Monju, the Japan Atomic Energy Agency's fast breeder reactor.

After the National Accident Commission published its report, Genden's executives asked Hoshino to prepare for an unpredictable series of interactions with regulators, whose interim safety standards had been poorly received. Before allowing the Ōi plant to resume operation, NISA had subjected it to a series of hypothetical stress tests, criticized by experts for their similarity to the spurious risk assessments that preceded the disaster. "Fukushima is where we tear them down," one journalist commented. "Fukui is where we restart them."

The NRA launched in mid-September 2012, after Prime Minister Noda Yoshihiko invoked emergency powers to confirm his appointees without parliamentary approval. Shimazaki immediately commenced meetings with staff inherited from the NSC and NISA. The priority item was active fault review. Reviewers would determine whether faults existed that could cause displacement directly beneath safety-class structures at Japanese nuclear plants. Other regulations related to faulting hazards would be omitted. Only the direct displacement rule was strict enough—zero displacement permitted—to prevent the utilities from claiming their engineers could mitigate the risk of a serious accident.

The decision to assess six plants, also inherited from the NSC and NISA, presented Shimazaki and his staff with the most ambitious agenda for original research a nuclear regulator had ever attempted. To fulfill this agenda, Shimazaki sought Tanaka's support for the development of a new regulatory process, based on the notions that occurred to him during the bus ride after the commissioners' visit to Fukushima, of smaller panels staffed by experts whose histories with nuclear regulation were limited. Tanaka assented.

A week later, Tanaka announced that the stress-testing program that had returned Ōi to operation would be discontinued. The chairman of the previous regulator had expressed disgust with the tests before his departure, and the governor of Fukui, despite his prefecture's economic interest in the resumption of nuclear operations, had declined to endorse them. Tanaka assured reporters that the NRA would not inherit its predecessors' reliance on risk assessments conducted by the utilities. Stress testing would be replaced by a commitment to primary research, beginning with the system of expert panels Shimazaki had proposed. "He made Shimazaki's model, which was really a pilot program to address a single regulatory issue, the heart of the new institution," one reporter recalled.

Under renewed questioning about Ōi, where two reactors were operating on the basis of stress tests the NRA had discontinued, Tanaka explained that he did not have the authority to retroactively decertify the plant. Adjacent to this statement, however, he reiterated his commitment to idling both reactors immediately if an active fault was discovered. Shimazaki, absorbed by the administrative challenges of establishing a new regulatory system, was not aware that his neutrality, credibility, and ambitious research agenda had constituted the load-bearing component of Tanaka's promises.

To staff the active fault panels, Shimazaki issued invitations to every professor holding membership in one or more of the four academic societies that coordinated relevant research. Any willing respondents would be hired, except under specific circumstances: Recent acceptance of research funding from the energy industry was subject to mandatory disclosure and in many cases would be disqualifying. An age limit was imposed, slightly higher than the mandatory retirement age at Japanese universities. Any expert who had previously participated in the regulatory review of a nuclear facility would also be prevented from joining.

The Society for Quaternary Research objected to several criteria and throughout the nomination process submitted names that violated them (Okumura Koji, for instance, despite his former work for the Nuclear Safety Commission). The Geological Society of Japan objected to the age limit and noted that it would exclude a scholar considered the dean of active fault research in Japan. His name was therefore included in the Geological Society's nominations. The Active Fault Society, expected to contribute the largest number of panelists, nominated him as well.

Because it had not been Shimazaki's intention to exclude this scholar, who preceded him as president of the Active Fault Society, the age limit was clarified in subsequent solicitations: *up to and including* seventy years of age.

Following the completion of panel assignments, Hoshino and Iriya began reviewing the research histories of the panelists. The disqualification of experts who had previously participated in nuclear regulation led to the inclusion of scholars unknown to the industry, only four of whom, in addition to Shimazaki, were assigned to the Tsuruga panel, but all of whom would conduct a peer review of the resulting assessments. Most were geomorphologists, specialists in mapping and topography whose discipline was responsible for the majority of active fault research. Their participation in nuclear safety reviews was a relatively recent development, beginning with the back-check process in 2006.

The first phase of the active fault review would address Ōi and Tsuruga. Hoshino had joined Genden when Tsuruga 2 was under construction, and associated the reactor with the peak of the company's bubble-era engineering achievements. This was the first time he had prepared to argue the seismic safety of the site without the Urasoko fault in mind. The rules were clear: The panelists were looking for faults under safety-class structures. Urasoko, no matter how large and close it was, did not pass directly under the plant, and therefore could not—according to the NRA's own regulations—become the object of the proceedings.

The Tsuruga and Ōi panels did not share members, except for

Shimazaki in the position of chairman. Each included three geomorphologists and one structural geologist. The presence of the latter was briefly comforting to Iriya's team, because the geologist's expertise would be better suited to the trenching investigation Genden planned to conduct (a field survey of excavated faults), but Hoshino discovered that the geologist on the Tsuruga panel had given speeches to organizations that lobbied for the development of renewable energy. Of the three geomorphologists, one had been allowed to participate despite recent involvement in industry-funded research. Another, Suzuki Yasuhiro, could be counted on, in Hoshino's opinion, to use the panel as an opportunity to advance an antinuclear agenda. Hoshino consoled himself by noting that Kanden's position appeared more difficult. The Ōi panel included the seventy-year-old geomorphologist the Geological Society had insisted on, Okada Atsumasa, a former president of the Active Fault Society. Because the Active Fault Society was dedicated to expanding the use of cautionary science, its leadership figures were regarded as likely antinuclear partisans.

Although the Ōi and Tsuruga reviews would occur in parallel, Shimazaki couldn't attend in two places at once, and Ōi was selected to commence first. Geomorphologists had identified a fault (F-6 on their maps) near the Ōi 3 and 4 reactors. Layers of sediment connected to F-6 had moved within the 130,000-year threshold the NRA had adopted for the classification of active faults beneath safety-critical equipment. It was unclear whether that movement had been caused by an earthquake.

The first meeting of the Ōi panel was held on November 4. On the day of the meeting, Tanaka spoke with the panelists and instructed them, off the record, to reach their conclusions with alacrity. It was not fair to Japanese society, he argued, to maintain suspense about the safety of the nation's sole operational nuclear plant.

The press attended the meeting in force, and one of the panel's geomorphologists, Watanabe Mitsuhisa, argued that the available evidence justified a shutdown pending further investigation. Watanabe's presentation included photographs of the area surrounding F-6, with animated, brightly colored labels that read, "ACTIVE FAULT."

Shimazaki had never conducted a public meeting under press scrutiny, and the same was true of all his panelists, with the exception of Watanabe, who previously collaborated with watchdog groups.

The momentum of the discussion turned quickly toward the need to determine an appropriate response to the present evidence. This surprised the panel's geologist, who had assumed it would be obvious to everyone—experts and reporters alike—that further information was needed. Critical features of the site had been destroyed during construction of the plant, and KEPCO, despite its enormous regulatory and public relations vulnerabilities, had been slow to conduct new research after the active fault review was launched.

Okada Atsumasa, narrowly included on the Ōi panel after the clarification of the age limit for experts, surprised the press and KEPCO alike (both presumed he was antinuclear) by objecting to the rapid progress of the meeting. The evidence was scant, he declared, and the type of rock they were analyzing was known to produce the false appearance of seismicity.

"It was important it was Okada who spoke up," another active fault expert reflected. "If you lined us up with guns to our heads, geologists, geomorphologists, seismologists, and you said, 'Tell me the one person in the world who can determine if a fault is active,' you would hear one name from the list of excluded experts—Okumura—and two from the panels: Shimazaki and Okada."

The press was ill-prepared to digest the technical content of the disagreement that arose between Watanabe and Okada, and continued to rely on the narrative that formed during Tanaka's confirmation hearings: A plant was operating in a regulatory vacuum, and for several months it had been an open question whether that plant was sited on an active fault. They could now add that they had seen one professor of geomorphology (and government-appointed active fault expert) declare imminent danger while another insisted that haste was the enemy.

Several reporters described the meeting as if the panel were genuinely split. They positioned Shimazaki on Watanabe's side of the argument because he had allowed it to go forward, and assigned the swing vote to geomorphologist Hirouchi Daisuke, because he had said there was "no evidence it is *not* an active fault."

Shimazaki's staff recall that he was grateful for Okada's intervention. In the Japanese bureaucratic context, premature assessments often become irreversible. Shimazaki had seen it: on the tsunami panel in 2004, and in the regulatory history of the Urasoko fault.

Okada was an expert of sufficient stature to inspire other active fault panels to embrace their expanding burden. The official logistical intentions of the active fault review—six plants in the same number of months—wouldn't survive this adjustment, but it was hard to imagine any expert among the panelists who wouldn't choose a more thorough agenda.

Okada, Shigematsu Norio (the panel's geologist), and Hirouchi spoke privately after the first meeting. By virtue of Tanaka's instructions to proceed quickly, they surmised, as well as his statements to parliament about shutting the plant down if the presence of an active fault were verified, the panel was not being asked to pursue the unrestricted, independent research Shimazaki originally described, but to make a determination about whether Ōi should be permitted to continue operating. In light of mounting public concern about active faults at nuclear plants, and the Ōi site in particular, this determination could alter the course of national politics. An election was immediately forthcoming. The approval rating of the incumbent party (DPJ) was historically low; Japanese voters remained widely opposed to nuclear restarts, arguably the sole issue favoring the incumbents. An active fault finding, leading to the idling or planned decommissioning of a nuclear plant, would constitute powerful public evidence that the DPJ was prepared to discipline the nuclear industry, consistent with the electorate's wishes.

Although it would require them to disobey Tanaka's instructions, Okada, Shigematsu, and Hirouchi agreed to engage in willful, scrupulous indifference to the social, economic, and regulatory impact of their work. They would pursue their task according to the ethics of scientific research, obtaining the correct and complete answer with sole regard for the integrity of their methods. Okada would put his reputation and seniority at stake by making himself, when necessary, the public face of intransigence. He would not allow the panel to move toward a conclusion until its research obligations had been satisfied. This would permit Shigematsu to quietly expand his field surveys.

As Okada anticipated, recriminations about the panel's reduced pace were swift and emphatic. Newspaper editorials condemned the panelists for commencing their work with a public disagreement, then moving incrementally to replace it with a conclusion. Political leaders published similar criticisms. Electricity companies, KEPCO included, predicted significant harm to consumers and shareholders.

Despite appointing himself the target of much of this disapproval, Okada—an intensely private person whose career had occurred in an introverted community of specialists—was irritated by the behavior of the press: their descriptions of a panel bitterly divided, their inability to grasp the routine nature of two experts disagreeing, and their calls for haste in a matter that required the careful analysis of physical evidence. He blamed antinuclear sentiment incited by fellow panelist Watanabe for the bizarre threats that reached his university email and home phone. Watanabe had made statements Okada considered scientifically irresponsible. He told the media, for instance, that the softness of the sediment in the fault gouge indicated the seismic nature of its most recent movement. He failed to mention that the mineral composition of the sediment had been analyzed and the results did not indicate seismicity.

Among colleagues, Okada was willing to articulate his low esteem of Watanabe and the reporters who heeded him. "Which created the sense that if you did not repudiate the involvement of the press, you had subverted your science," said a member of another panel. "But the regulatory system is designed to integrate academic science with the public interest."

Shimazaki was scheduled to visit Tsuruga in late November. The first review meeting would occur shortly after. Hoshino observed the Ōi meetings with interest but presumed the meetings about Tsuruga would be less contentious. There was not, at Tsuruga, a similar level of agreement about the specific geomorphological features that required evaluation. Genden's historical records of the site were better, and their plan for contemporaneous research was rigorous.

Shimazaki visited Tsuruga on November 25. Hoshino had arrived the day before to direct preparations. Genden's trench investigation was under way, and Hoshino wanted the walls of the trench cleaned so Shimazaki could assess them independently. Before returning to his hotel, he walked the path that circumnavigated the trench, pausing along the way to anticipate what Shimazaki might examine.

It was blustery, overcast, and raining intermittently when Shimazaki arrived on the following day. He forgot his jacket on the bus.

Rather than go back or wear a Genden jacket, he wrapped a towel around his neck and instructed Hoshino to proceed.

Shimazaki chatted about the weather and thanked the Genden staff—all warmly dressed while he was not—for indulging his interest in their work. Except for these pleasantries he was absorbed by his initial surveying of the trench. Hoshino and Iriya felt his manner departed from their previous experiences of senior bureaucrats. Shimazaki was avuncular and ingenuous, an avid listener whose few comments demonstrated an exceptional ability to absorb information. "I believed that Shimazaki was the man for the job," Hoshino recalled. "We hoped—how could you hope otherwise after Fukushima?—that our new regulator would be too smart for the industry to fool."

Their esteem of Shimazaki tempered their wariness of the other panelists when they visited a week later. By then, Genden had asked the geotechnical consultants trenching the site to research the panelists' reputations. In addition to Suzuki Yasuhiro, the consultants considered another geomorphologist, Miyauchi Takahiro, an antinuclear partisan.

After the first day of visits, the consultants expressed surprise at the impartial nature of the panelists' comments and inquiries. Reasons for caution remained: At a preliminary meeting with Genden earlier that week, Suzuki had delivered a letter explaining that the panel was not required to await the conclusion of Genden's investigation before issuing its findings. During the tour of the site, one Genden employee had reported that Suzuki looked displeased when NRA staff stipulated to additional debriefings. Most of Genden's consultants nonetheless felt the panelists' initial investigation boded well for future cooperation. "We were in a new world after Fukushima," recalled the Genden employee assigned to liaise with the consultants. "We believed in better regulators, better review, better safety."

The first review meeting was scheduled for December 10 and 11. The panelists would summarize their ongoing research on the tenth. Genden could respond the following day.

The panel's structural geologist, Fujimoto Koichiro, had not been available on the date of the other panelists' visit and instead surveyed the site on December 8. Hoshino returned to Tsuruga to facilitate. Fujimoto's presumed political bias had been a subject of wariness, and because his research interests were idiosyncratic, Genden's consultants were dismissive of his expertise.

After touring the trench, Fujimoto remarked that Genden's research had been ambitious. The tone of the comment was conspicuously sober, Hoshino noticed, and it occurred to him, recalling what he knew about Fujimoto's background, that Genden's research was the kind that Fujimoto had always wanted to conduct. Fujimoto had started his career supporting the development of geothermal energy. Following the oil crisis of the 1970s, nuclear became more attractive than renewables. Fujimoto became a scholar, not an industrial scientist. He attended conferences where his friends who worked for research institutes funded by electricity companies presented data-rich findings, obtained with unlimited resources. To Genden's consultants, he was the archetypal overlooked man, who would seize the authority the panel granted him and wield it vindictively. But his admiration for Genden's research was genuine; Hoshino could hear it. This investigation would define both of their careers, and provide an opportunity to be remembered for institutionalizing the lessons of Fukushima.

"He didn't want a fight," Hoshino reflected. "He wanted an achievement. We had that in common. I thought: We'll go as far as we can on trust."

24.

After hosting Fujimoto on his belated tour of the site, Hoshino returned to Tokyo, where Iriya was preparing for the panel's first meeting. It was customary at Genden to anticipate significant regulatory events by working overnight. Neither Hoshino nor Iriya felt the meeting qualified. The other three panelists had visited the trench a week earlier and would barely have begun to interpret the resulting data. The panel would present the concerns that led the NRA to launch its investigation. Genden's engineers had been promised the following day for a presentation of their own. The public would obtain a concise and transparent statement of each side's progress, but neither side had reached a conclusion.

Besides which, reporters were more interested in Ōi, Chairman Tanaka's rhetorical candidate for decommissioning. Hoshino's counterparts at Kanden told him the controversy at the first Ōi meeting shouldn't make Genden nervous. Recent activity on the Ōi fault, they had discovered, was the result of a non-seismic landslide. The media could say the panel was deadlocked, but the mundane facts would emerge.

On December 10, Iriya's team spent the morning updating their computer simulations, a longitudinal task barely related to the review meeting. The probable course of the meeting was discussed in a routine briefing with upper management, and Iriya was given broad discretion to structure his responses to regulators' questions. The other spaces at Genden's table would be occupied by Hoshino and Genden executives, who could intervene on management's behalf in the event of an adversarial exchange.

Before the meeting commenced, an NRA representative approached Genden's table to propose that Hoshino and his senior colleagues move to public seating. It was the NRA's preference that the meeting be presented as a scientific discussion, not a negotiation. After conferring with Iriya, Hoshino agreed. This meeting, early in the research process, was unlikely to become contentious.

A few minutes sitting in the public gallery dispelled Hoshino's assumptions about media indifference. The journalists expected a headline. Many had been sleeping in their newsrooms since parliament was dissolved three weeks ago, covering an election cycle scheduled to conclude in six days. Voters were expected to treat the election as a 3/11 referendum, and the NRA was the principal achievement of the DPJ's post-disaster reforms.

This atmosphere of anticipation was intensified by the unexpected presence of Chairman Tanaka. He was not a panel member, and no procedural matters requiring his approval were on the meeting's agenda.

Fujimoto, the geologist, and Tsutsumi Hiroyuki, the panelist who had been granted a waiver for his industry-funded research, wished Tanaka had not come. The meeting was meant to provide the public with an opportunity to audit the emerging dialogue between Genden and the panel. Tanaka wouldn't join that dialogue until the panel's findings had been submitted. His presence risked the inaccurate suggestion that today's discussion was conclusive enough to merit external vetting; a peer review process had already been instituted for this purpose and would not occur for several months.

Hoshino, Iriya, and their respective staff were surprised to see Tanaka at the NRA's table but presumed there had been other instances, unknown to them, of commissioners attending review meetings in an observational capacity, and further presumed that Shimazaki and the other panelists had known Tanaka would attend.

The panelists' presentations focused on a fault that projected toward Tsuruga 2. In the immediate vicinity of the reactor, the fault was surrounded by a zone of shattered rock where evidence of previous seismic activity was visible. The shatter zone was labeled D-1. The fault (which would receive its name shortly after the meeting) was G. Trenching had revealed a secondary fault on the north wall of the trench. It traveled in an irregular pattern, and the question of whether it reached the reactor remained unsettled. Iriya's team named the G fault for Genden—a gesture of acknowledging responsibility. The smaller fault was K, from the NRA's shorthand name in Japanese. Iriya surmised that its discovery pleased the agency.

The G fault was visually distinct and, if active, plainly a threat to the reactor. Genden believed enough of it could be exposed by trenching to assess its relationship with geologic markers indicating the beginning of the Late Pleistocene, 120,000–130,000 years ago, which the NRA had adopted as its threshold for an active fault finding. During their presentations, the panelists summarized the survey methods that produced the maps identifying G, and the scale of the damage that could occur if the fault activated. They provided general geomorphological context for the site and displayed images of trenching activities so the public could appreciate that the layering of geologic deposits beneath the plant was complicated and unclear.

Neither Tsutsumi's nor Fujimoto's presentation deviated significantly from Genden's expectations. Miyauchi's and Suzuki's were less comfortable. Suzuki emphasized the fundamentally deterministic nature of the standard the panel was required to apply: If there is evidence for the possibility that a fault has been active during the Late Pleistocene, and the evidence cannot be affirmatively refuted, the fault must be considered active. Miyauchi's presentation dwelled on the nearby presence of the Urasoko fault.

During Miyauchi's presentation, Hoshino began to regret his willingness to leave Genden's table. Taken together, Suzuki's emphasis on evidentiary certainty and Miyauchi's fearful characterization of the Urasoko fault could produce the inaccurate impression that the reactor was seismically guilty until proven innocent. He reminded himself that Miyauchi's behavior was consistent with the predictions Genden's consultants had made. Iriya's team was prepared to respond. They would have their chance tomorrow.

Iriya's sentiments were similar. He remained silent for most of the meeting. The scope of the panel's review was limited to active faults under safety-critical structures. Nothing the panelists said about the Urasoko fault would stick.

Following the panelists' presentations, control of the meeting returned to Shimazaki. Reporters began preparing to depart; they anticipated a perfunctory closing statement. Instead, Shimazaki solicited concluding opinions from the panelists. When none replied, he referred to specific passages in the meeting materials and said, "The available evidence points to the presence of an active fault." These were familiar words, but their position in relation to the rest of the meeting surprised Hoshino, and he noted that Shimazaki was avoiding eye contact with Genden's table.

Shimazaki kept talking, summarizing the concerns the panelists had raised. "Motor mouthing," one reporter remembers. "Like he was trying to run the clock." Members of the audience recall that he seemed suddenly uncomfortable with the public nature of the meeting, unprepared to modulate the mood of the room. Every sentence made the reactor sound more dangerous. Hoshino reached for his phone, thinking to text Iriya and instruct him to make an immediate rebuttal. His hands weren't steady enough to tap the screen, and he breathed slowly to control his adrenaline. He thought of shouting. Antinuclear activists interrupted regulatory meetings; they had done it at Ōi. Why couldn't he? He tried to lock eyes with Iriya, but if Iriya was looking for him, he realized, he wouldn't know where in the crowd to look.

Before closing the meeting, Shimazaki asked if Tanaka had anything to contribute, "just in case." Watching the video, it's apparent that Shimazaki doesn't expect a substantive response. The question is posed with a chuckle, between assertions that "I've never been much good at conclusions," and "This one's not much of a conclusion, but we'll live with it for now." Shimazaki expected—and his colleagues expected—what Hoshino had expected of Shimazaki: *Thanks, see you tomorrow.*

Tanaka said yes, he'd speak. When he did, his tone struck people in the room as a self-contradictory blend of offhandedness and gravitas. One member of the NRA staff compared it to giving someone a piece of shocking news in place of the words you'd typically use to sign off a phone conversation.

"In light of the evidence presented today," Tanaka said, "a definitive statement is necessary: The reactor must not be allowed to apply for restart."

25.

Camera strobes accompanied the remaining words of Tanaka's statement, as his gravitas vanished and he stumbled through an attempt to provide context for his decision. "If you're a nuclear company and you want to conduct more research, feel free to do more—to make sure you're satisfied with your own research," he said. "Actually, I would like that. Please go for it." In the press seating, reporters were edging past each other toward the door. Some dialed as they left. Others remained in their seats, texting the headline. They had witnessed history: the first nuclear plant decommissioned by Japanese regulators. Tsuruga 2, like Fukushima, had never been safe.

The meeting had not concluded. Betraying no surprise, Shimazaki conferred with staff about follow-up procedures, prompting Iriya to put his hand up. "The utility has raised its hand," Tanaka pointed out, taking care to wait until Shimazaki had turned off his microphone.

"They might like to add a word or two," Shimazaki observed.

"Yes, a word or two," Iriya said, adopting Shimazaki's phrase, then summarized the critical information the panel hadn't considered yet. "We'll conduct more research," he promised. This time the phrase was Tanaka's.

Shimazaki asked the staff to provide closing remarks. Staff noted that all relevant issues had been addressed. No meeting would occur tomorrow.

Hoshino's immediate judgment, that the meeting had been a frame-up, planned by Shimazaki with the panelists' consent, had a rich and recent history to draw from. Genden's consultants had warned that Shimazaki's blacklisting of veteran experts would produce an antinuclear regulatory environment. During Genden's preparatory meeting with the panelists, Suzuki had established the NRA's authority to conclude the review before Genden completed their research. If the intention had not always been an ambush, why had Tanaka issued a finding when essential trenching data had not been integrated? Why had NRA staff made liars of themselves on matters of administrative significance, including the broken promise of reconvening tomorrow so Genden could respond?

The NRA was determined, Hoshino concluded, to decommission Tsuruga 2 on the basis of the Urasoko fault. Because Urasoko was outside the scope of the panel's review, they were willing to evaluate smaller faults prejudicially. They had said G was active and would enforce their finding, even if they discovered it was wrong.

It was true the panelists shared a distaste for the siting of the plant near the Urasoko fault. None, however, had considered what he would do if trenching data revealed that G was harmless. Today's meeting, they had agreed, was preliminary. Each was shocked by Tanaka's remarks, Shimazaki included. Months of research remained, and would occur in the shadow of an egregious misunderstanding. The entire nation now mistook their hypothesis for their conclusion.

Shimazaki was grateful that tomorrow's meeting was canceled. The NRA's position would be compromised if a premature ruling was subjected to premature debate. He resolved not to question or criticize Tanaka, whose reputation was inseparable from the agency's ability to survive its vulnerable infancy. This decision was Shimazaki's to make; his authority was closest to Tanaka's, and his seniority among scientists was the greatest of the NRA's five commissioners. Without his intervention, the number of people who understood that Tanaka had profoundly misspoken would remain as small as it had been when the meeting concluded. Beyond Shimazaki and the panelists, hardly anyone would know.

Genden's executives were waiting for Hoshino when he returned. They had drafted a press release. The question, before publishing it, was whether Hoshino could explain what had happened. Because Genden was a smaller company with fewer political connections, Hoshino speculated, and because the media had sensationalized the relationship between the Urasoko fault and the smaller faults under review, the NRA had staked its credibility on finding an active fault at Tsuruga 2. Without knowing that Tanaka's appearance had surprised the panelists, Hoshino and his superiors could only presume the ruling was premeditated.

Staff at Tsuruga 2 were disposed toward cynicism. Murabe Yoshikazu, the site superintendent, suspected that Tanaka's ambush—six days before an election—was politically motivated. It would help Noda, the prime minister who had installed Tanaka by emergency fiat. It would also help Tanaka. The press had branded him pronuclear. Giving the industry a bloody nose would secure the trust of liberal and centrist media outlets, and the opportunity to secure it was rapidly departing. There would be no public theater at the industry's expense after the LDP returned to power.

From the moment of his nomination, which he initially declined, Tanaka had been bruised by the political demands of his chairmanship. When his name was leaked to the press, he was required to go before parliament without preparation. He assumed he would face questions about disaster recovery, his staff recalled, and would answer based on his experience leading the government's response to the JCO incident—at the time it occurred, the most severe radiation leak in Japanese history. Instead, he was forced to declare his intentions about restarting the reactors in Fukui, establishing a pattern of political pressure that reached its peak before the first meeting of the Tsuruga panel. "Regarding politics, the less we overheard, the happier we were," one member of Tanaka's staff reflected. "It's plausible Tanaka decided there was more to gain than lose by punishing Genden. It's equally plausible he misunderstood the purpose of the meeting."

Hoshino and Iriya had known Tanaka a long time; the industry had known him. They blamed Shimazaki, whose public statements as vice-chairman had resurrected Genden's misrepresentations of the Urasoko fault. Shimazaki had stacked the panel with antinuclear partisans. He had known Miyauchi would mention Urasoko, and Suzuki would attempt to shift the burden of proof to Genden. His rambling conclusion had been a deliberate attempt to deprive Genden of the opportunity to protest the cancellation of tomorrow's meeting.

When Genden's press release was ready, Hoshino accompanied the public relations staff to the press club at the NRA, where they hand delivered it. The experience did not replenish their faith in the fourth estate. *You're saying they're not geologists, they're geo-something-ists. This guy—Google says—works in the geology department. Why would the government appoint the wrong experts?* The reporters meant well, lacked context, and were out of time to acquire it. "I hear you there's no evidence, evidence that's needed," one said. "Tanaka talked. You know the headline."

The headline was "Tsuruga Decommissioned." Another article quoted an unnamed NRA source: "An undergraduate in geology could tell you G is active." Hoshino attributed the quotation to Suzuki, who had favored didactic statements in meetings with Genden. Suzuki thought the quotation was nonsense. He doubted the source was real.

A day of press coverage was enough for Iriya to accept that Genden had forfeited the opening maneuvers. He compared Tanaka's statement to Genden's misrepresentation of the Urasoko fault. In each instance, influential people had adopted a convenient deception. Reversing a mistake of that nature was almost impossible.

Industry-wide, the implications were stark: Six sites belonging to four utilities, hosting seven Kanden reactors and fifteen reactors in total—including every site where the future construction of additional reactors remained realistic—were scheduled for active fault review. The review's panelists, the utilities realized, were not ceremonial consultants like their predecessors. Japan's nuclear restart could not proceed without their permission.

At the NRA, the gravity and brevity of Tanaka's comments called for an assessment of their legal significance. Before the Tsuruga meeting, the NRA's lawyers hadn't given much thought to the active fault review. The panelists weren't NRA staff. They didn't make NRA decisions. They conducted specialized research and reported their findings in a peer-reviewed forum. No one had anticipated that the panels might be stripped of the administrative distinction that separated them from the agency.

The NRA had been launched without regulations in place, with the intention of operating under its predecessors' rules until improved standards were devised. Proposed regulations were constantly crossing the desks of the agency's legal staff, most of whom were not attorneys; they were civil servants who had studied law. The most significant rule under development provided "backfit" authority, allowing regulators to idle or decommission noncompliant reactors. American and European regulators maintained this authority by issuing operating licenses. In Japan, licenses were required to construct a reactor. Afterward, its fitness for operation was presumed.

During early attempts to remedy this omission, the NRA recognized the legal risk of abandoning construction licenses in favor of operating licenses; electricity companies had been granted construction licenses lawfully and would not relinquish them. Instead, it was decided the backfit rule would specify the NRA's authority to compel existing plants to comply with construction standards enacted in response to Fukushima. This was a complicated and contextual way of addressing the shortcomings of Japan's licensing system and depended on a corresponding measure to establish the NRA's post-Fukushima safety review; without this measure, there would be no occasion for verifying backfit compliance.

When Tanaka spoke at the Tsuruga meeting, the backfit rule was still a proposal. The lawyers weren't sure what authority could be invoked to render Tsuruga 2 ineligible for restart. Tanaka's comments also violated Japan's due process protections. Genden had not been provided timely notice of Tanaka's decision or an opportunity for rebuttal. Nor could it be said—in this transitional, rule-making phase—that the NRA policies affecting Genden had been made clear.

"Shimazaki never inquired about those vulnerabilities," a member of the legal staff recalled. "He knew what Tanaka had exposed him to. Reform was more important."

A professor from another active fault panel returned to his office after watching the Tsuruga meeting. He heard the phone ringing through the door while he unlocked it. It was a journalist, cold-calling panelists. The journalist had an angle: "Academia strikes back." He thought the geoscience professors on the panels had known the fault was active and welcomed Tanaka's decision. This professor was in a difficult position, wanting to suggest that all was not determined, without being misapprehended as a critic of his colleagues. The journalist, he could tell, was surprised by his hedging, and asked outright, "Is it active?"

"Maybe," the professor said.

"That's your honest answer?"

"No." The professor considered leaving it there, in the silence of expectation, but added, "I don't know. If you called the other panelists, if they were honest, they would say they don't know."

"What good are you to society?" the journalist said. "You clowns." And hung up.

There was a tap on the door, a graduate student who worked in the professor's lab. "Is your family all right?" the student asked, letting himself in. "It was ringing before you came, several times."

Initially it didn't make sense, a journalist trying the same number repeatedly, to reach someone who wasn't on the Tsuruga panel. On a brief delay, the professor registered his student's presumption of a family crisis; in Japan this was impolite, but the student wasn't Japanese. He had come to Japan without knowing the language or manners, to study with the best geoscientists in the world, to study fault systems that affected major population centers. There was a relationship between his student's attraction to Japan and the reaction of the journalist. They trusted the research community he represented. His admission of uncertainty transgressed that trust. The journalist was against a deadline and the professor had sent him to his editors empty-handed, but his question had been posed sincerely: "What good are you to society?" He had arrived at this question gradually, in a conversation that began with the assumption that geoscience had conquered the influence of vested interests. That would be comforting, the professor thought. It would give everybody what they wanted after Fukushima: a sense that science is a reckoning force, it does not conceal or mystify, and will not become the object of confusion and disagreement.

26.

Genden's executives heard from the company's major shareholders the following morning. They had read Genden's press release and wondered pointedly what was gained by antagonizing the regulators whose minds they needed to change. This question had been raised during the drafting process, but the mood in the room prevailed: No response would seem rational except a robust refutation. None of Genden's executives could recall a time when a Japanese nuclear company had alleged regulator negligence. Genden would be the first.

If a larger company had done this, or Genden had done it at a different time, it would have been a matter of significant intrigue. At present, it was minor in comparison with the possibility—raised by Tanaka's comments—that Genden would go bankrupt and drag Japan's regional electricity monopolies into an industry-wide recession. The monopolies had founded Genden to keep nuclear energy off their balance sheets until it was reliably profitable. They still owned the company and guaranteed its debts.

Floating on the finances of its shareholders, Genden had performed the specialized work that regional monopolies preferred to avoid. The breeder reactor workforce depended on Genden. When Japan exported nuclear technology, Genden was chosen to operate abroad. When the monopolies wanted to test the profitability of next-generation reactors, Genden prepared to build Tsuruga 3 and 4. These activities enhanced Genden's reputation, but the company's revenue depended on selling electricity from two existing reactors, Tokai 2 and Tsuruga 2. After Fukushima, restarting Tokai (the plant closest to Tokyo) was politically implausible, and Tsuruga 2 became the likely determinant of the company's solvency.

Tanaka's verbal decommissioning of Tsuruga 2 raised the question of whether Genden was creditworthy. If creditors stopped lending, the monopolies were obligated to maintain Genden's liquidity or pay its debts and shutter the company. With one temporary exception, all of Japan's nuclear plants were idle. None of the monopolies were in a position of financial strength. Their ability to absorb a potential Genden bankruptcy was severely diminished.

They sent Genden on an apology tour of the company's creditors, where the first question often implied criticism of Genden's leadership: You didn't see this coming? More than once, a creditor pointed out that it was difficult to imagine TEPCO or KEPCO in an ambush. Hoshino restrained himself from bringing up the price Japan had paid for TEPCO's ability to influence regulators.

When the Tsuruga panel reconvened, Hoshino and Iriya were scandalized by a slide Shimazaki presented to illustrate the potential consequences of fault displacement. A cartoon reactor building was pictured mid-explosion, flung into the air. No one at Genden was willing to consider the possibility that the slide was not deliberately prejudicial. During another panel meeting, Shimazaki rose from his seat and made several full-body gestures, with his own height as reference, to demonstrate the scale of potential displacement under the reactor. At Genden, this was mocked as "earthquake calisthenics" for its resemblance to morning exercises taught in elementary schools. The panel's written submissions used Genden's slides, and one had been altered in a manner that steepened the apparent angle of a slope the K fault passed through. The error was the result of an inattentive copy-and-paste job by NRA staff, but Hoshino was certain the panel had intended to exaggerate the likelihood of a landslide.

Genden's trench investigation, progressing rapidly, had not located displacement from the G fault in nearby layers of sediment that formed during the Late Pleistocene. According to NRA criteria, G was inactive. The panel's focus shifted to examination of the K fault. Hoshino and Iriya were astounded. K had not been shown—as G had—to project under the reactor.

The press covered the ongoing meetings, including the panel's adoption of K in place of G. Both faults appeared in the crush zone labeled D-1; without further context, what struck Genden as a tremendous change in the panel's trajectory appeared to the public as one exercise of scientific discretion in a process composed of hundreds. Certain editorial boards spoke in Genden's favor but failed to scrutinize the panel's procedural gymnastics. Instead, the idea that Shimazaki was ideologically opposed to nuclear energy began to appear, at first in publications funded by the industry, then in major newspapers sympathetic to corporate interests. His Fukushima tsunami warnings, previously the source of his credibility, were described as evidence of partisanship, or a seed of partisanship that germinated after Fukushima. His defense was taken up by critics of the nuclear industry. Both sides presumed—one happily, the other in discontent—that Shimazaki had deliberately set the stage for Tanaka's verbal decommissioning of Tsuruga 2.

Okumura Koji, Japan's leading paleoseismologist and the Nuclear Safety Commission's foremost geoscience expert before the NRA launched, was an early critic of the active fault review. His NSC research had revealed NISA's willingness to rubber-stamp the industry's risk evaluations, including Genden's mischaracterization of the Urasoko fault. The NRA was NISA with a new sign on the door, and Okumura had assumed that his history of holding regulators accountable would make him valuable to the agency's reform agenda. Instead, he had been blacklisted, at a moment when the public was looking for people to blame.

Tanaka's verbal decommissioning of Tsuruga 2 shocked Okumura, and the panel's pursuit of the K fault in place of G persuaded him the panelists were motivated by ideology. Suzuki, he was convinced, represented a group of scientists who wanted to join with progressive politicians to set a regulatory precedent for further decommissioning. "Of the people upset when Shimazaki announced the exclusion of former NSC experts from the active fault panels, Okumura was the most offended," a colleague recalled. "He was the best field researcher in Japan, and he was determined to involve himself."

After Tanaka's appearance at the first Tsuruga 2 review meeting, Fukui Prefecture advocated for Genden's right to present their rebuttal, and center-left columnists chose to portray Fukui's governor as a doctrinaire supporter of nuclear energy—the reactor is a middle finger to earthquake risk, but let's hear from the criminals who built it.

Governor Nishikawa Issei, entering his third term, regarded himself as the steward of Japan's best nuclear oversight mechanism: the requirement that nuclear plants obtain operating consent from host governments, which Fukui Prefecture had been the first to impose in 1994. His office frequently released sensitive information to the public after electricity companies chose to withhold it. Fukui also maintained the nation's only prefectural department of nuclear safety, which enabled Nishikawa to conduct investigations without relying on the questionable candor of political contacts in Tokyo.

Nishikawa wielded authority in a gruff, unpredictable way. His office frightened the electricity companies. By law, a licensed nuclear plant could operate in perpetuity. The rule of local consent, however, was unwritten. The host prefecture or city could introduce as many contingencies as it liked. Nishikawa and his staff felt they possessed rare expertise about the political demands of applying leverage to the restart process. By decommissioning Tsuruga 2 while an investigation was still pending, Nishikawa argued, the Tsuruga panel would force the nuclear industry to attack the active fault review's legitimacy. The NRA's authority would be jeopardized before a single reactor had been vetted under post-Fukushima standards.

Tsuruga's mayor, Kawase Kazuharu, also held local consent authority over Tsuruga 2, but municipal hosts of nuclear plants are fiscally captive. Kawase, a rotund, cheerfully belligerent former nightclub owner, told the press who flocked to Tsuruga after the first panel meeting that the city was careening toward an economic crisis.

The press clung to Kawase. He was folksy. He over-disclosed. Provocative questions made him loud, not humorless. Before granting interviews, he required the assurance that no mathematics would be involved, and this admission of dimness made him human. His linebacker build, baggy pinstripe suits ("can't boogie in a British silhouette"), and bubble-era boardroom hairstyle frequently appeared next to pictures of his antinuclear foil in city politics, a diminutive councilwoman who dressed for Fukui weather.

The postindustrial landscape in Reinan (Atomic Broadway's geographic name) made Kawase's warnings seem credible. Tsuruga, formerly connected to the Trans-Siberian Railway, was once expected to become one of Japan's great cities. Now the city and its surroundings are economically depressed, and appeared to me, on my first visit, as if someone had deposited the American rust belt in the world's best scenery: mountain forests and narrow valleys, lit by rays of sunlight slanting through a permanent canopy of clouds; spray-drenched sandstone cliffs overlooking the Japan Sea; and the restless, opalescent surface of Wakasa Bay, where fifteen nuclear reactors pumped coolant.

Fourteen were built to sell electricity, and all were eventually idled after Fukushima. But Japanese employment law proscribes most layoffs, and lifetime employment limits the availability of skilled workers—fire the workforce you trained and you'll never replace it. Not one full-time employee of the electricity companies in Fukui had been laid off as a result of the nuclear shutdown. The largest job losses were absorbed by subcontracting companies that imported the majority of their workers from outside the prefecture.

Local firms that specialized in operations support or radiation management went out of business. Only a small number of jobs were lost, but they belonged to Reinan residents. Former employees of these firms found jobs in the disaster area or settled for short-term contracts in the prefectures neighboring Fukui. I knew a few who signed thirty-five-year mortgages in Tsuruga less than a year before the disaster. Among the majority who chose to work in Fukushima, where family couldn't accompany them, hardly any figured out how to maintain their family lives remotely. Most eventually divorced.

Owners of restaurants in Tsuruga fretted about the economy. Local politicians habitually claimed that the city had the highest concentration of restaurants in Japan, and a similar density of hotels. Many were family owned. Their revenue constituted the primary relationship between imported labor and the income of Tsuruga's middle class. Biennial reactor refueling, which required a surge of visiting workers, was conducted at a leisurely pace for the benefit of these businesses. After Fukushima, the refueling bonanzas vanished. Blue-collar workers were spending as little as possible, and white-collar workers were subject to austerity: no boozy dinners on the utility companies' accounts.

I lived in Tsuruga during the shutdown, in Chuo-cho (Central

Area), the miracle mile surrounding the highway on-ramp. The tomato-red light of a pachinko parlor's electric sign—visible from the highway—flooded my apartment every night through my ill-fitting curtain and the slit of privacy glass it was meant to cover. If I couldn't sleep, I walked to the pachinko's parking lot and talked to the taxi drivers who idled there. Their best fares (six thousand yen each way to the closest nuclear plant) were gone, but most thought business would come back, or told me it would, to avoid sharing their anxieties with a stranger. "Operate it, decommission it, turn it into a theme park, somebody has to drive you there," one driver summarized, and a moment later a friend of mine—a maintenance technician at Tsuruga 2—crossed the parking lot, wearing his work uniform. "My son's in there," he told us, gesturing at the pachinko. "Gambles all night. Lost his job." When my friend went inside, the driver refined his argument: "One industry isn't the economy."

From the mayor's office, the economy and the nuclear industry appeared identical. Over the course of the industry's development, Tsuruga had plugged holes in its budget by expanding the nuclear tax burden. Targeted taxation of road users who transited Tsuruga to reach KEPCO plants was ineffective during shutdown. As was *kifu* (literally, "funding"), an arrangement that permitted utility companies to donate to the city's budget. On more than one occasion, having run out of reasons to impose new taxes, the city attempted to annex the town where KEPCO's nuclear division was located. To political rivals who suggested that Tsuruga diversify its economy in anticipation of Japan's post-nuclear future, Mayor Kawase said, "Bullshit. We need Japan to need nuclear."

The gambler's father had voted for Kawase, although grudgingly, without illusions about the relationship between the mayor's rhetoric and his son's job prospects. He was a neighbor of mine and worked under Hoshino's supervision at Genden. When the evenings turned warm, he loitered on the concrete banks of the river that flowed past my apartment building, watching the carp swim upstream toward a dead end, an outfall pipe they had come from and forgotten. Among the people I knew who were employed by the industry, his predictions about the restart process most often came true. The events of his life had attuned him to Atomic Broadway's patterns and omens.

Tomo Genshiryoku (*Friend of the Atom,* his pseudonym in my

notes) was born in the town where Genden built Japan's first nuclear reactor. His father worked for an engineering firm Genden contracted during construction. When the plant was finished, Genden gave the contractors lifetime employment. "This was a village when you were born," his father said during their first visit to the plant together, and showed Tomo a corridor where hundreds of uniforms were hanging. "A village always seems smaller than it is."

Genden hired Tomo in 1985, after he graduated from high school. In 1986, the Power Reactor and Nuclear Fuel Development Corporation (PNC), a quasi-governmental research entity, commenced construction of Monju Nuclear Station in Tsuruga. Monju was Japan's first commercial-scale fast breeder reactor. Genden would contribute the largest share of the plant's workforce. Tomo accepted an assignment to Monju with the feeling of having begun his life's work, the future of electricity, the same as his father.

Over the years that followed, Tomo watched the vacant site become a nuclear plant. There was hardly a corridor of the new buildings he didn't tread, a system he hadn't inspected, or a waking hour he spent away from work. When his father's former colleagues cycled through Monju, they had heard that Tomo was there, and their reverent way of describing his father made Tomo, among his colleagues, a protean figure, capable of whatever was required because he was the product of perfect timing: At the moment Japan's nuclear industry was born, he was in the phase of boyhood when the seriousness of his father's work would make the most solemn possible impression.

The Monju site overlooked the small harbor belonging to the fishing hamlet of Shiraki. Tomo never met anyone who lived there, but when he bought newspapers, their covers occasionally showed a picture of Ministry of Science officials visiting Shiraki, to maintain the blessing of village leaders. It amused Tomo to imagine the scene— eminent bureaucrats from Tokyo explaining technical drawings and bowing slowly, from the waist, to elderly Japan Sea fishermen. "A village always seems smaller than it is."

This cyclical sense, of Monju in the role of his birthplace and his life as a repetition of his father's achievements, acquired a monastic intensity. He gave up drinking, friendships, and hobbies. His son was born. His father was diagnosed with cancer and retired early to begin treatment, then surprised Tomo by asking if he would request a transfer to Tokai, so grandfather and grandson could know each other. Tomo chose to remain in Tsuruga.

By 1994, there was less plant to build and more to operate. Tomo cannot remember a period when he wasn't shocked by the indifference that the majority of PNC staff, who were research scientists, manifested toward the routine responsibilities of operating a commercial-scale nuclear facility. Important projects that Tomo had worked on would disappear from the schedule, and nobody could explain why. This experience, of effort expended without purpose, followed Tomo on his days off. He struggled to communicate with his son, whose first year of elementary school had revealed serious behavioral difficulties. His partner told him she planned to leave.

His father's cancer was no longer treatable. They spoke by phone, mostly about Monju. Tomo described a workplace rewarding enough to justify his absence during the years his father had wanted him home, and discovered that it was not a memory of his best period at Monju. It was fiction. His father died in December.

In January, the Great Hanshin earthquake was felt in Tsuruga. Tomo was between shifts, alone in his company apartment, where he experienced the obviously personal illusion that the tremors were reducing gravity's effect and he would soon float away.

He asked to be transferred. In June he was assigned to Tokai, where his father had recently died and his fond misrepresentations of his previous job awaited him among family and friends. Roughly a year after his father's death, Tomo heard a radio bulletin about Monju, where an incident of undetermined severity was ongoing. In the cafeteria the next day, staff who had worked at Monju gathered, silently at first, until someone said an accident had been inevitable, the way Monju was managed, and another volunteered that the accident was serious: A pipe had ruptured in the plant's secondary cooling system.

The welds on the secondary cooling system, unbeknownst to his colleagues, had been Tomo's responsibility.

The accident caused a months-long panic among Reinan residents, sustained by the gradual failure of an attempted cover-up. The eventual release of video footage from inside the plant revealed that the leak had been severe, and workers sent to assess it did not appear prepared for the chaotic environment of a sodium leak. Accustomed to deference regarding technical matters, PNC persisted in labeling the accident "inconsequential" after the press and public learned otherwise. Shortly after a news conference where he stipulated to his participation in the cover-up, a senior PNC manager killed himself.

When it occurred, the sodium leak was the first nuclear accident in Japan significant enough that the facility's name—Monju, named for the lion-taming left hand of the Buddha—was thereafter presumed to refer to an event, not a place.

The defect in the pipe that caused the accident was located where a thermometer had been installed during the manufacturing process, and the related inspections were the responsibility of the vendors who supplied the pipes. Had Tomo undertaken a voluntary, redundant inspection of manufacturer welds, he would have been the first safety inspector ever to make this gesture, but nonetheless found himself clinging to the thought that his individual effort at Monju, because it failed to prevent the accident, had been meager and misdirected.

He spent two years in Tokai and ten at Tsuruga 2, where the safety culture was strong and his response was ambivalent; he discovered that his colleagues who had worked in Tsuruga while he was at Monju were now years ahead of him in their effectiveness as nuclear safety professionals. In 2009, he was assigned to Genden headquarters in Tokyo and fell in love with a woman he met at a gospel concert. In February 2011 they learned she was pregnant, a happy reason to marry.

He was at work during the earthquake on 3/11, and his first thought, not knowing where the shaking originated, was for his son in Tsuruga, then his mother in Tokai. Both lived in seismically active regions, near several reactors. He slept in the office. At home the next day, he told his fiancée the next few weeks would require long hours. They canceled their honeymoon. There was never a thought given to calling off the wedding or waiting to start a family.

I met Tomo when the active fault review was in the headlines. He had a top-heavy, elbows-out gait, like a boy trying to mimic a man. His distractible manner and fast mind—from teaching himself a profession most educated people would struggle with—made him seem twenty. Not the heroic twenty he remembered from Monju. Still-growing-up twenty. He was forty-nine.

Genden shrank its payroll during the active fault review by transferring employees to other electricity companies. Tomo's responsibilities multiplied. When I encountered him in bars, he was drinking too much (and too quickly) to enjoy himself. I was with him one evening when he startled the owner of an *izakaya* by declaring, "I work for the nuclear industry. The nuclear industry caused Fukushima. Don't you agree?" The owner was standing in the heat of the open kitchen, sweating. There were voices coming from the party room, and an audience on TV burbled between points at a tennis match. The owner parted the curtain of utensils hanging where his gaze met Tomo's, and in the silence that followed, it seemed to me that Tomo was noticing a faint mark on the owner's forehead—a common mark in Tsuruga, from years of wearing the safety helmets distributed to nuclear plant employees.

The door slid and a friend of Tomo's appeared in the *izakaya*'s entryway. He was older, slightly built, and depended on a cane. The room was warm, but he kept his coat on and looked weightless in it, the way a bird looks in its feathers. Tomo brightened and made small talk while Feathers adjusted his hearing aid, but when the small talk lulled, he said, "They're making me work more." Feathers dug in his coat pockets and found a pack of gum, a brand that claims to whiten teeth. He ate boiled soybeans and drank whiskey with the gum in his mouth. Tomo wobbled silently in his chair. Eventually, Feathers spit the gum into a napkin and said, "Nobody but you makes you work."

This analysis appeared to relieve Tomo's gloom. He talked about the cow his late aunt had kept in her backyard in Tokai, an impractical last connection to the family's agrarian past. During his childhood, he thought it understood him when he talked. He never told an adult. He wondered what strange things his daughters privately believed. His oldest, born shortly after the Fukushima disaster, had been diagnosed with asthma, and his youngest, now five months, already suffered from allergies. His wife, after two difficult births, had developed chronic pain in her pelvis. His mother's Alzheimer's was progressing, he added, and although he was not an only child, he was the only child who visited.

The room filled with customers and the background noise was too much for Feathers. We left him finishing his drink and crossed the street to save three stools at a bar. Tomo showed the bartender a bottle I'd brought from Mihama and asked him to pour it for the regulars. "It has a mineral flavor," Tomo explained. "You get it where there's water and bedrock—perfect for brewing sake."

"Or running a nuclear plant," the bartender said.

Tomo fumbled the bottle and it slammed sideways on the bar. He stared at the bartender, who made a close-call gesture and started putting out cups. He hadn't meant to spar with Tomo, just show him their past conversations were remembered.

"I'm talking about the water table," Tomo said. "You need potable water in bedrock."

"And a nuclear plant uses salt water?"

"An ocean, a lake. A nuclear plant will operate anywhere you let it." They laughed at how that sounded, and Tomo assured the bartender that certain locations were off-limits.

27.

Six days after the first meeting of the Tsuruga panel, the LDP, out of power for three years, won a parliamentary majority under the leadership of Abe Shinzo, whose nuclear-dependent energy platform, however unpopular, had cost him few votes. Over the following months, the Ōi panel determined the F-6 fault wasn't active. Critics of the industry turned their attention to Tsuruga 2, where the NRA's independence—and by extension the solemn, self-mortifying promises of the National Accident Commission—would be preserved or eroded during the Abe era.

By April 2013, the Tsuruga panel had conducted four public meetings. Genden's investigation failed to match the pace of the review. Hoshino asked for more time and protested when he didn't get it. His repeated objections were ruled obstructive, and Genden was removed from the review process. In May, the panel confirmed its draft report, which found that the D-1 shatter zone, containing the G and K faults, was active.

Iriya and Hoshino were terrified by the likelihood that they had allowed, in a span of five months, the bankrupting of Japan's oldest nuclear company. Each knew from his experience of Genden's past deceptions that institutions in Japan—public or corporate—rarely retract flawed decisions, and the Tsuruga panel's decision had acquired the inertia of a formal report. Outside the door to their division, someone put a Dumbo puppet on the hand sanitizer pump. "The puppet store ran out of pigs?" Iriya joked. Pigs that would fly, he meant, when Tsuruga 2 was exonerated.

Genden's legal staff provided an alternate perspective: Tanaka's premature statements had become a premature ruling, and rulings must meet administrative standards established by law. Finish the trench investigation, they advised. Petition the NRA to review your results. If the fault isn't active and science fails to persuade, other remedies remain.

The investigation was a traveling affair, sometimes staged at the plant, in modular offices intended for the construction phase of Tsuruga 3 and 4, other times operating out of company apartments near city hall or hotel rooms adjacent to the train station. As the supervisor of this "delicate frenzy" (his deputy's description), Iriya passed through overlapping layers of exhaustion and intellectual intensity. I once watched him fall asleep against a window in a crowded bar, and his colleagues observed him in superstitious silence, as if he might not wake. His team completed their investigation in July 2013, two months after the Tsuruga panel's report was filed. They found no evidence to suggest that K passed beneath the reactor, and Iriya was confident that he could constrain the date of the fault's most recent movement. He prepared to petition the NRA to overturn the panel's findings.

It would be difficult to exaggerate Iriya's ambivalence about attacking the NRA's experts. Genden's mischaracterization of the Urasoko fault during the Tsuruga 3 and 4 application process had wounded his faith in the benevolence of his work. When Okumura and the Nuclear Safety Commission intervened, Iriya learned that he was not the sole advocate of difficult truths; the system was devised to discover them. Now, rather than pursuing the threat from the Urasoko fault, the NRA was preoccupied by a minor fault and the regulatory leverage it provided. The system was worse after Fukushima, Iriya concluded, and he could not escape the thought that he was responsible. If he had blown the whistle in 2004, before the NSC intervened, a precedent for transparency and cooperation might have been set. He resolved not to repeat this omission; he would provide the active fault review with the most comprehensive evidence regulators ever received from a nuclear company.

The excellence of Iriya's new research was apparent when it crossed Shimazaki's desk in July. Several months were required to study the entire submission, a delay that convinced Genden another ambush was being prepared. The panelists, however, indicated an immediate willingness to reopen the review. They had worked longer than planned, under circumstances of public controversy none had anticipated, but agreed that Genden's evidence merited a rigorous response.

Tsutsumi Hiroyuki, the Tsuruga panel's moderate, was relieved to embark on an audit of the initial findings but worried the political significance of the outcome would test the integrity of the geoscience community. He was aware of several large-scale disagreements between Japanese geomorphologists and geologists, beginning with the nation's first seismic hazard maps, published in the late 1970s. Geologists with ties to the nuclear industry had attempted to discredit the maps, jeopardizing the historic bond between geology and geomorphology in matters of risk assessment. Tsutsumi was also aware that antinuclear activists in Japan periodically attempted to influence fault classification, sometimes by provoking conflicts among government-appointed scientists.

Many panelists reported receiving strange emails, and strange calls at their homes and offices. Okada was targeted by activists who held him responsible for the Ōi panel's failure to act on Watanabe's initial warnings about the F-6 fault (Watanabe no longer considered F-6 active, but argued that the NRA's standards weren't strict enough). Reporters called panelists who had publicly expressed distaste for the media's handling of the active fault review, or panelists who weren't reviewing the reactor in question—a Monju or Mihama panelist for information about Tsuruga or Ōi. "They asked if you had any criticisms of what your colleagues said," one panelist recalled. "They thought the panels were like corporate board meetings, propelled by agendas and rivalries, and they could get the story if they figured out the resentments."

Okumura had begun denouncing any scholar whose work connected active fault mapping to public policy, including his predecessor as chair of Hiroshima University's geology department, widely admired for his mentorship of young scientists. Okumura's criticism attracted sympathizers because Tanaka's verbal decommissioning of Tsuruga 2 had established the panelists as gatekeepers for Japan's nuclear restart process, a responsibility none of them had been warned about or wanted. "Everyone from Shimazaki down to the staff was confused about the authority of the panels," Tsutsumi reflected. "Each of us reached his own conclusion, which makes a fertile environment for misunderstandings."

The active fault review's mechanism for taming dissent was borrowed from academia: peer review. After the investigation of Tsuruga 2 was reopened, Tsutsumi encouraged Shimazaki to invite the peer reviewers (every member of every panel) to tour the site. A matter between five scientists could be shared by many, and minority opinions integrated constructively. The peer reviewers' visit was scheduled for January 2014. Out of sixteen panelists, twelve indicated they would attend. "It was rubbernecking of the most genteel variety," recalled Fujimoto Koichiro, the Tsuruga panel's geologist. "There was a controversy—a spectacle—and everyone felt it was happening in their neighborhood."

At Genden, the opportunity to present the company's research to an expanded group of experts catalyzed a surge of preparatory activity, reminiscent of the unity of purpose Iriya had discovered when he was first assigned to Genden. "A little good news gets everyone pulling in unison," Hoshino reflected. "We had been coming to work every day wondering if Genden's creditors would call in the company's debts. Journalists would wisecrack me in the men's room at the NRA, *Why stress? Cash in your vacation time. This time next year, there won't be a Genden.*"

Winter in Tsuruga is oppressive. A day rarely passes without precipitation, and sleet falls more often than snow or rain. For nearly a week before the peer reviewers' visit, Iriya, with five of his staff and three employees of the geoscience firm Genden had hired, worked sunrise to sunset around the trench at Tsuruga 2, ensuring the weather did not compromise it.

Half of the reviewers visited on January 20 and 21, the rest on the twenty-third and twenty-fourth. In both instances, the program began with a presentation at Genden's public relations center, followed by a tour of the trench. To maintain the safe flow of visitors through the site, the reviewers were divided into small groups, each accompanied by a member of Iriya's team.

Hoshino floated between groups. Differences of technique and experience were soon apparent. Some panelists wanted to view specific elements of the trench from several vantages, or double back after examining a related element farther along. A few sketched their own logs. Occasionally, a reviewer would ask a question the geomorphologists on the Tsuruga panel were not prepared to answer. As the tour wore on and some experts completed their work before others, Iriya discovered that the reviewers were candid in private. Okada volunteered his assessment that he didn't see anything in the trench to indicate that the K fault was linked to the D-1 shatter zone (in other words, K could not meet the NRA's definition of an active fault). "I had wondered if my perceptions were warped by hired science and corporate self-interest," Iriya recalled. "When Okada offered his opinion, I wanted to hand him my megaphone and say, *As a favor to my sanity, please repeat yourself.*"

Watching the reviewers gravitate to Iriya, whose intellect and honesty epitomized the qualities of the Genden employees who had raised concerns about the Urasoko fault in 2004, Hoshino felt the long, serendipitous pattern of how time unfolds in an institution of Genden's size. A decade earlier, Iriya had identified with the regulator's position more than Genden's. Now he could contend with the NRA's experts by sharing their endeavor. There were few people in the nuclear industry, Hoshino realized, who maintained this purity of intention. Protecting Iriya, keeping him apart from infighting or wavering, insulating his sensibilities from the tedious tug-of-war that characterizes regulatory negotiation: Hoshino wanted this responsibility.

The second day of reviewer visits included an altercation. To demonstrate the independence of the active fault review, Suzuki had declined to tour the site in his assigned group and chose his own path around the trench. As he passed Awata Yasuo, an active fault reviewer from the government's industrial research institute, Awata made an inquiry about the visible traits of the K fault. Suzuki might have interpreted the question as an insult, or maybe his answer failed to satisfy Awata; like any quarrel whose witnesses were uniformly intrigued but later felt it beneath them to admit their interest, this one has been recounted in versions, the tersest belonging to Awata and Suzuki, whose words the wind obscured from their colleagues.

It is generally agreed that Awata invited Suzuki to climb into the trench and improve his familiarity with its features. This was an attack on Suzuki's ability, as a geomorphologist, to conduct field research in the environments he mapped remotely. The conversation continued until Awata raised his voice to be heard above the wind and said, "You have no idea what you're talking about." Suzuki walked away.

Among the other reviewers, Awata was known for his awkward candor. His outburst was nonetheless significant. "No one had wanted to be the first to say something which could be construed as an endorsement of the Tsuruga site," one reviewer recalled. "When Awata questioned Suzuki's judgment, he made it comparatively mild to say, *Have you considered an alternative explanation?*"

Hoshino wasn't aware that Suzuki's antagonist was eccentric, and remembers the pleasure of watching a person he scorned endure a difficult moment. When the names of the panelists had been announced a year earlier, Suzuki's had been one of a few familiar to Genden. As a technical adviser to activist groups, Suzuki had experience with the press, and his public pronouncements seemed authoritative. Hoshino had presumed that Suzuki's role in the earthquake research community was second only to Shimazaki's. This perception could not be reconciled with the experience of watching a colleague belittle him in the presence of their peers. Suzuki, Hoshino decided, commanded the respect and attention of the media, but among fellow scientists this willingness to engage the public could be made a liability, the way Watanabe's alarmism had attracted derision at Ōi.

The tour of the trench ran long, and several debriefings were due afterward. Exhausted when he returned to his hotel room but utterly awake, Hoshino attempted to settle his mind by remembering, in pleasurable detail, the expression of contempt Suzuki showed when he was invited to climb into the trench. *Keep it up,* Hoshino thought, addressing his image of Suzuki. *Keep thinking you know.*

28.

Under NRA criteria, a fault that passes beneath safety-critical equipment is active if it displaced during or after the Late Pleistocene, a threshold of 126,000 plus or minus 5,000 years. In movies and television, the dating of geologic features is achieved with conspicuous precision by measuring carbon isotopes. This method is applicable to sedimentary environments containing organic material, if that material does not exceed 50,000 years in age (NISA's active fault threshold until 2006). Older strata can be dated by measuring argon isotopes, but deposits of argon-datable material are unusual in Japan, and alternative dating methods rarely produce results precise enough to accommodate the NRA's regulations.

At Tsuruga 2, the feature in need of dating was a plane of unconformity (visible irregularity) crossing the sedimentary layers the K fault passed through—the area where K, beyond dispute, had demonstrated displacement.

Before individual layers can be dated, the surrounding stratigraphy (boundaries and relationships between layers) must be established, in Tsuruga's case by sedimentological research. Among the fundaments of sedimentology is the law of superposition: Unless a better explanation arises, older material—deposited first—is at the bottom. The "unless" in this law is particularly important in coastal environments like Tsuruga, which are subject to a variety of erosional and depositional processes.

Layers at Tsuruga are numbered from bottom to top. Layer 1 sits above the bedrock. The G fault, which had prompted the active fault review and was presumed to travel under the reactor, had not shown any evidence of displacement that reached layer 1, let alone the more recent layers above it. K, on the other hand, had demonstrated displacement in the boundary shared by layer 3 and layer 5 (the 4 designation was reserved for features resulting from the displacement). In order for any dating techniques to be considered reliable, the nature of the depositional relationship between 3 and 5 would have to be determined.

From Genden's perspective, their efforts to determine the date of K's displacement should not have been necessary. None of the trenching had produced evidence that K projected under the reactor. But the panelists raised the possibility that K was structurally related to other faults in the D-1 shatter zone; if K had moved during the Late Pleistocene, the entire zone might be active, including faults beneath the reactor (G, for instance). Genden claimed this was counterindicated by sedimentary data that distinguished K from G and D-1. The displacement sense of the K fault (the apparent direction of its past movements) was also distinct.

Genden's investigation nonetheless included a dating result for the displaced area between layer 3 and layer 5, obtained by means of tephrochronology, the science of analyzing tephra (volcanic ash) deposits—in this case, tephra extracted from layer 5. Tephrochronology can provide dating information only by correlating a tephra deposit to a geologic event that has already been dated. To arrive at their estimate of the age of the tephra in layer 5, Genden conducted chemical analysis to correlate it to ash deposits whose possible origins had previously been studied. The result was an estimated age of 127,600 years, which, in concert with paleoclimatic evidence, appeared to place it narrowly in the territory of harmlessness (as defined by the NRA's rules).

The panelists responded to Genden's findings on April 14, 2014, at the first public meeting since their investigation was reopened. Fujimoto presented first. As the panel's sole geologist and an experienced researcher in mineralogy, he was most qualified to review Genden's tephra evidence. He displayed Genden's charts of the chemical similarities linking the layer 5 tephra to a deposit in Mihama, which in turn was linked to tephra in Lake Biwa. The depositional history of the Biwa tephra was supported by marine isotope analysis that affirmed an estimated age of 127,600 years, originally suggested by researchers who studied the Biwa tephra in 2004. In Fujimoto's opinion, chemical variations between the layer 5 tephra and the comparison tephras exceeded the standard deviation. This was a meaningful discrepancy because the chemical profiles of tephras associated with the volcano in question (Mount Daisen) were tightly clustered, increasing the likelihood that ash deposits of differing ages could be mistaken for each other. He also noted that the distribution of the layer 5 deposit, as well as the ratio of certain tephra components to others, suggested the tephra had been moved and redeposited. The tephra itself could not be more than 127,600 years old, but it might have arrived at the site later, indicating that layer 5 formed more recently. On the basis of these uncertainties, Fujimoto concluded that Genden's evidence did not merit a revision of the panel's 2013 findings.

For Miyauchi, the question was not whether Genden's data was reliable but how precisely this data had been calibrated. Using Genden's slides, he identified contradictory mapping of the trench's stratigraphy and the tephra distribution in layer 5, raising doubts about the significance of the boundary between the upper and the lower regions of the layer, a critical element of Genden's analysis. Without precise stratigraphy, it was reasonable to conclude that K had not displaced within the last 100,000–115,000 years, but he wasn't confident the displacement did not occur during the interglacial stage that marked the beginning of the Late Pleistocene.

If this had been the conclusion of his presentation, it might have supported Genden's findings, by affirming that the displacement was borderline and could be moved past the border with additional research. Miyauchi's summation, however, was damaging. He theorized—outside the bounds of regulatory consideration—that K had demonstrated multiple displacements in the older layers of the trench. In the case of a fault that had apparently triggered multiple times in short succession before the Late Pleistocene, and whose present characteristics were not fully known, he felt the question of whether its most recent displacement had occurred 119,000 or 131,000 years ago shouldn't distract the panel from its fundamentally hazardous nature. To Genden's consternation, he reiterated his concerns about the Urasoko fault and vowed to apply a threshold of 400,000 years to the K fault because the two could be related.

Suzuki followed by reminding the panel that the tephra Genden had collected was extremely sparse. Tephra density and distribution are fundamental to the reliability of tephra dating. Tephras that cannot be identified, extracted, and analyzed without specialized methods and equipment, including the tephra in layer 5, are cryptotephras, and a dedicated branch of tephrochronology has arisen to facilitate their study. In analyses of cryptotephras that are particularly sparse and distal (geographically remote from the original eruption), the difficulty of arriving at reliable conclusions increases. The layer 5 deposit was difficult.

Suzuki also elaborated on concerns that Fujimoto and Miyauchi had raised about characteristics of the K fault that remained unknown. Initially, K had appeared to be a strike-slip fault (opposing sides of the fault plane pushing across each other in a mostly lateral motion) but now appeared to be a reverse fault (one side pushes upward). Determining displacement sense is usually the first step in characterizing a fault. Characterizing K, which showed varying displacement sense at different points, had proven challenging. In one location it changed direction entirely. Genden had mapped K from the north wall of the trench to a pit southwest of the trench, claiming the displacement rapidly attenuated and disappeared as the fault traveled in the direction of the reactor. The total length of the fault, as mapped, was roughly forty-five meters.

The claimed attenuation was drastic. More than a meter of displacement had reduced to a few centimeters, and Suzuki pointed out that faults don't shed that much displacement over the course of tens of meters. Some element of the fault's behavior therefore remained unexplained, and he wasn't prepared to accept that its southern terminus had been located. At stake was the regulatory significance of the fault. In its forty-five-meter form, it didn't threaten Tsuruga 2. If the fault's attenuated displacement indicated, for instance, segmenting rather than termination, the possibility remained that it continued far enough to reach the reactor or connect with a fault that did.

To mitigate these concerns, the panelists desired a wide variety of information. K should be followed down to the bedrock and pursued beyond the bounds of present trenching. Nearby faults that showed similar strike characteristics should be identified and studied. The layer 3–layer 5 boundary should be examined in multiple locations.

Data from older displacements of the K fault was requested, as well as information about other shatter zones that might affect K indirectly.

From Genden's perspective, this was an outrageously diffuse and expensive array of suggestions. Suzuki, who had delivered his critique verbally in lieu of a full presentation, infuriated Iriya's team by following his requests with an offhand speculation: "It has occurred to me that the shatter zone was formed by multiple strike-slip movements occurring together. Maybe K was formed as a kind of tangent to this process, a reverse fault that sits on top of it. The structure of a fault like that would be extremely difficult to ascertain." In short, we won't be satisfied until you dig up every inch of the site, and when you do, it's unlikely the resulting data will produce a conclusion.

Iriya considered Suzuki's demands the equivalent of asking his team to disprove the existence of extraterrestrials. In his discouragement, he failed to register the significance of Tsutsumi's presentation, the last of the four and the most responsive to Genden's evidentiary concerns. Among NRA staff, active fault reviewers, and the broader geoscience community, Tsutsumi's comments are well remembered.

The K fault, he hypothesized, had been produced by compressive force exerted on its surroundings during the formation of the Urasoko fault system. Much of its displacement would therefore be cumulative, the result of many movements too small to discern. This would make characterization of K according to its displacement sense nearly impossible. Because the displacements in question constituted a temporary phenomenon, present only during the Urasoko fault's formative period, studying this phenomenon would not improve the panel's comprehension of the site's present seismic hazards.

He reminded his colleagues that their inquiry had been launched to investigate the G fault, which, if it were to displace, would plainly pose a threat. None of them had foreseen the possibility that G would be inactive, but a small fault at the periphery of the trench, discovered accidentally, would show evidence of recent displacement.

"What should be done next?" he asked. "I haven't a clue. I wonder whether it's appropriate for us to conduct this research. Its object is a phenomenon beyond the present capacity of human comprehension."

He summarized Genden's position: Because the K fault was neither seismogenic nor reached the reactor, and therefore could not be identified as the potential origin of displacement that might affect a safety-critical component of the plant, its inclusion in the review took too many liberties with the intended meaning of the words "earthquake fault." The NRA, he countered, was operating within its warrant. Nuclear companies would no longer decide how the regulator applied its rules. He nonetheless considered it dubious, in light of the K fault's unconventional qualities, to apply the 120,000–130,000-year rule uncritically. The rule had been drafted with seismogenic faults and primary displacement in mind, and presupposed a relationship between past and future displacement, but K—regardless of when it last moved—could move tomorrow, or never move again.

His colleagues had concluded their presentations by expressing the opinion that the panel's original findings should stand. Tsutsumi omitted this. Instead, he encouraged the consideration of displacement in "earthquake-resistant designs." In one reading, Tsutsumi's final comment was a way of telling Genden to be careful what they wished for; if they were allowed to proceed with the restart process, the challenge of making the reactor safe would be tremendous. In another reading, Tsutsumi was communicating his dissatisfaction with the scattered, piecemeal research the panel had produced, by inviting the intervention of engineering expertise. If this reading was accurate, Tsutsumi had ventured the most provocative statement an active fault reviewer could make, however select the audience for a message so narrowly coded.

29.

According to Tsutsumi, lingering questions about the K fault were "beyond the present capacity of human comprehension." If a panel of active fault experts can't determine whether a fault is active, what can they tell you?

A great deal, if the knowledge sought corresponds to the cumulative pattern of scholarly learning, and you seek it from the appropriate expert. In the natural sciences most of all, knowledge is produced—even when it constitutes a breakthrough—by refining the discoveries of colleagues and predecessors.

Tephrochronology provides an instructive example. Its ultimate object is the isochron, an event in the geologic record whose date has been narrowly constrained in multiple evidentiary contexts. The volcanic eruptions that produce tephra leave several markers. At the lowest level of methodological complexity, a tephrochronologist might study tephra associated with a major eruption that occurred during recorded history. The date of the eruption would connect the tephra to a known isochron. Research published by tephrochronologists who had located similar tephras in separate locations would help to form hypotheses about the climatic mechanisms responsible for spreading it beyond the eruption site (primary deposit from an ash cloud, distribution in rivers or lakes, rafting on ice floes). The age of the geologic features that contain the tephra could be constrained to a narrow range on the basis of this information.

Some prehistoric eruptions are well studied. Many are not. The more obscure the origin of the tephra in question, the less suitable it is for research. If a doctoral student supervised by one of the handful of specialists in tephrochronology—or the few among them primarily interested in cryptotephra—proposed to study the tephra deposit that Genden's evidence relied on, the response would be unfavorable. It is neither close to its hypothesized origin nor linked to well-studied eruptions.

The Tsuruga panelists represented the collected eminence of relevant thought in a nation widely admired for its geoscience research. But they were not tephrochronologists. Tephrochronology's body of expertise is owed at least as much to paleoclimatology as it is to structural geology, and paleoclimatologists are not represented in the academic societies whose members were invited to join the active fault review. The utilities were fond of complaining that Shimazaki had hobbled the investigative process by over-relying on geomorphologists. In Tsuruga's case (and in the cases of the other three panels where an active fault finding either was reached or would have been reasonable), adding more geologists would not have addressed the absence of critical research specializations, several of which did not typically participate in active fault investigations or seismic risk assessments. Even the Ōi panel, which produced a swift, unequivocal finding, had deemed it necessary to seek an off-the-record opinion from an external expert on landslides.

Of the specializations critical to the active fault review, sedimentology appears most frequently in the panelists' backgrounds. This familiarity is one way of understanding the panelists' response to their uncertainty about Genden's tephra evidence: They wanted to develop a sedimentological profile of the surrounding area, every stratum accounted for and its origins understood. Genden suspected an attempt to convict the reactor by uncovering additional faults, but the panelists were only resorting to their tools.

If our imaginary tephrochronologist and her doctoral candidate were sedimentologists instead, she would still discourage him from studying Tsuruga, where the panelists had been asked to develop findings about a very small area, within a narrow chronology the NRA imposed, in spite of the destruction of depositional continuity by the construction of the plant. Separated from its contextual methods of characterizing the environment, sedimentology loses its explanatory power.

Paleoseismology, the applied science of studying prehistoric earthquakes (Okumura Koji's specialization), draws from sedimentology and maintains a heterodox relationship with dating methods, the better to incorporate radiocarbon dating, tephrochronology, electron spin resonance—whatever helps. It is the method of first resort for most geotechnical siting concerns. Its expertise spans academic and professional science.

Published paleoseismology research includes analysis of distributed faulting (the triggering of secondary, non-seismogenic faults during an earthquake), but its chief focus is primary fault displacement, occurring along the fault that caused the earthquake. If a researcher must study a non-seismogenic fault to determine the risk of surface rupture at a critical location, it helps to have more of the fault to study. The panelists had asked for more, and Genden insisted the entire fault had already been uncovered, but there was no disagreement about the fault's relative size: It was small. In the peer-reviewed paleoseismology literature, there is no methodological precedent for evaluating a non-seismogenic fault of K's size.

The Tsuruga panelists had been asked to assess the distributed faulting hazards associated with a tiny, non-seismogenic fault whose critical dating mechanism was a sparse, remobilized cryptotephra, in a sedimentary environment that had been artificially limited in space and time—a fool's errand in geoscience, irrespective of its urgent and admirable intent. Even the degree of foolishness was unclear. Some experts, Okumura foremost, were better qualified than others, which enabled them to point out shortcomings in their colleagues' research. It was impossible to say whether their qualifications enabled them to judge the K fault accurately.

Every panelist I met agreed that the Tsuruga research moved the active fault review into methodological territory they would have been reluctant to volunteer for, had they been forewarned. With only one exception, all said they would have overcome their reluctance and honored their commitment to issuing a finding. The reasons they gave were uniform: The active fault review had been established after a lack of regulator expertise and independence had been identified as the root cause of Fukushima. To abandon the original intentions of their service because the questions that arose were too methodologically fraught would have constituted an unforgivable act of timidity.

When the panels had issued their findings and the press turned their attention elsewhere, most panelists agreed that four out of six sites could have gone either way, cleared or condemned, without controverting the evidence. The Tsuruga panel had epitomized the quandaries confronted by its successors. "I flatter myself by thinking a hundred panels could review Tsuruga 2, with a different combination of qualified experts each time, and sixty would rule the way we did," one of the Tsuruga panelists recalled. "In my sober moments I realize it could be sixty ruling the opposite way."

The April 14 meeting also introduced the concerns raised by the peer reviewers during their inspection of the trench in January. These would be treated as supplementary information, NRA staff determined—the administrative equivalent of footnotes. The panelists would decide which concerns, if any, would be pursued during the remainder of the reexamination process. Hoshino and Iriya were certain they had been ambushed again. What had been the point of conducting an intricate program of visitation if the resulting assessments were extracurricular?

In the following days, Hoshino asked the NRA to explain the peer reviewers' role, and the answers he received were partial or unclear. There weren't any policies governing the administrative conduct of the active fault review, he discovered. The system had been Shimazaki's to devise and was his to operate.

At the institutional level, it was unclear how many NRA policies had been inherited from the NSC and NISA, and how many were undergoing revision or replacement. Since returning to power in 2012, Abe and the LDP had declined to emphasize institutional reform within the NRA, advocating instead for the immediate application of revised safety standards to the nation's nuclear facilities, to reassure the public and restart the reactors. In July 2013, when Genden completed their trench investigation, the NRA advised utilities that applications for conformity review—an exhaustive safety review resulting in renewed operating permission—could be submitted, initiating the restart process. Four utilities submitted immediately, having successfully forecast what would be required.

The process was patterned on the initial licensing of a new reactor, and conformity reviews were equally labor-intensive. NRA staff, less than a year in their new roles, were confronted with a crush of review and inspection responsibilities that had no equal in the history of nuclear energy. These reviews were hobbled by an absence of attention to the unresolved details of administrative reform. Some processes in the conformity review could be carried out with the benefit of an updated review manual. Other processes had last been systemized decades earlier, or not at all.

Genden had known for decades that the regulators didn't have a work management system to match the utilities, but Hoshino never gave the issue much thought. The NRA's predecessors rarely audited the safety data gathered at nuclear plants. The utilities made their submissions as mandated, omitting anything extraneous or inconvenient, and NISA reviewed what they already knew—by virtue of its inclusion—required attention.

The nuclear companies had benefited. Because this arrangement was biased to leave the status quo undisturbed, regulatory lapses posed no consequences for the utilities. If regulators were not satisfied with a utility's submission, the mechanism for resolving the substantive issues was informal: The compliance staff at the utilities called their counterparts at NISA and negotiated a solution. In the vast majority of cases no formal infraction was recorded.

The poor administrative practices of the active fault review, Hoshino realized, were nothing new. Genden was the victim of a broken system they once helped build, and the authority of the panelists—who were scientists, not administrators—was absolute.

This authority, Hoshino decided, could be challenged only by the panelists' colleagues, academic geoscientists with decades of research experience. The panel's next meeting was scheduled for June 21. Hoshino invited two prominent geoscience professors to speak on Genden's behalf.

One was Okumura Koji, formerly of the Nuclear Safety Commission. Hoshino and Iriya came to understand that Okumura considered the active fault review "a mortal hazard to geoscience" (his words), particularly the Tsuruga findings. Speaking to a reporter, Okumura had recently described one panelist, who played a critical role on two panels, as "worse than an empty chair." Increasingly, his willingness to question the panelists' motives—to accuse them of serving an antinuclear agenda—had emboldened other scientists excluded from the panels to publicly condemn the NRA.

When he excluded NISA and NSC experts from the active fault review, Shimazaki had expected the utilities would hire several of them, including Okumura. Shimazaki hoped that an adversarial exchange between Japan's leading geoscientists would help fulfill his promise "to neither restrict nor promote" the commercial use of nuclear energy. It had not occurred to him that Okumura would incite their colleagues to attack the credibility of the system.

Toda Shinji, of Tohoku University, also accepted Hoshino's invitation to attend the June meeting. Like Okumura, Toda had been an adviser to the NRA's predecessor agencies and a critic of Genden's insufficient attention to faulting hazards at Tsuruga 2. Before Fukushima, he contributed research to projects funded by the electricity industry, but after the disaster avoided forming financial relationships that would embroil his work in controversies of nuclear patronage. When Okumura took a consulting position with Genden, Toda declined to do the same. Privately, Toda felt that a legal mechanism should exist that would permit regulators to suspend the operation of plants where the initial licensing process had been egregiously flawed, and he felt Tsuruga fit this description. But there was no such mechanism, and convicting the K fault presumptively struck him as contrary to the regulations, contrary to the neutral posture of scientific inquiry, and a distraction from the conversation about the Urasoko fault the site genuinely needed.

Iriya wasn't certain the use of independent experts would improve the reexamination process. Both sides agreed the Urasoko fault was a threat. Both agreed the K fault could be characterized chiefly by what was unknown, and in many respects unknowable. Experts sympathetic to Genden could give the finest presentations of their careers, and the panelists wouldn't hear anything they didn't already believe. The zero-risk standard the panel had imposed would be unaffected, and the burden of proof would remain with Genden. "We already agreed with the NRA about everything," Iriya recalled. "Except what should be done."

On June 9, the NRA informed Hoshino that the official record of the peer reviewers' site visit had been finalized. They wanted to ratify the document during the June 21 meeting and asked Genden to have their response ready within a week. Hoshino balked. If the NRA's version was anything like the April meeting had intimated, he could imagine Genden's response would run a hundred pages. The NRA asked him to send what he could by the sixteenth and offered to replace the submission with an updated draft before the twenty-first.

Genden submitted on the sixteenth and on the eighteenth notified the NRA that an updated draft was forthcoming. Iriya offered to visit the NRA in order to explain Genden's revisions. The NRA declined his help, on the grounds that they had stipulated to accepting an updated draft and were prepared to digest it.

After the sixteenth, Hoshino and Iriya rarely went home. The final stages of completing their revised draft could be performed only by employees familiar with the panel meetings. The completed submission was sent to the NRA in the late afternoon on the twentieth. Iriya's staff hoped for a brief reprieve, to recompose themselves before the meeting on the twenty-first.

Within the hour, Hoshino received a call from Kobayashi Masaru, head of the NRA's earthquake and tsunami division, who embarked on a lengthy harangue. At issue was a disagreement that had arisen earlier in the day, when Genden provided their list of attendees. The NRA had objected to the inclusion of Toda and Okumura. Hoshino had taken a taxi to the NRA to state Genden's case for their use of independent expertise. In the absence of any response from the NRA after he returned to his office, Hoshino had assumed the NRA understood that Genden was free to select its advocates. Kobayashi was calling to dispel that assumption, and to inform Hoshino that, irrespective of the NRA's assurances that Genden's updated draft would be used, it had changed too substantively and would not be entered into tomorrow's meeting. According to an NRA rule that Hoshino was certain Kobayashi had just invented, a week would be required to read it.

By 8:00 p.m., Hoshino had returned to the NRA to present his objections. From the ambush of the first panel meeting to the mishandling of the peer reviewers' comments in April, he told Kobayashi, the NRA's administrative conduct had been consistently inexcusable. Opportunities for rebuttal had been offered and withdrawn without notice. Genden's own documents had been mischaracterized during panel meetings. Too few NRA staff had been present at important stages of the review, or the staff present lacked relevant expertise. In the beginning, Hoshino and his colleagues had presumed they were being targeted for ideological, opportunistic enforcement; now they wondered how much of the unfairness they encountered was the result of ineptitude. "I don't know what prevented the NRA from understanding that the shortcomings of the process could be exploited by the industry and its political allies later," Hoshino reflected. "We were frustrated because we wanted to protect our interests, but we also wanted to admire the NRA, and we had discovered it was weak."

In the argument that followed, Hoshino experienced a form of indignation he associated with adolescence, of enduring mistreatment from an authority figure because the terms of their relationship had been unjustly arranged. The meeting had begun to drag, and Kobayashi continued to insist on the exclusion of Genden's submission and their independent experts. To break the deadlock, Hoshino threatened to provide the media with proof of the NRA's misconduct. It occurred to him that his threat had substance. He could prove that Genden had been promised a second day of meetings to rebut the panel's initial claims, and never received it. He could prove that the panelists had ignored concerns raised by their own peer reviewers. He could prove that the NRA had agreed to read Genden's updated submission and the week-in-advance rule Kobayashi now cited was fictitious. It would be the first time a nuclear utility had resorted to radical transparency; it would make a good story.

Kobayashi left the room to seek flexibility for Genden's submission and experts. When he came back, he said none was available. In Hoshino's understanding, only Shimazaki could overrule Kobayashi. The remainder of Hoshino's original, sympathetic image of Shimazaki—the avuncular scholar making a diligent examination of the trench in foul weather, with a towel around his neck to keep warm—was replaced by a notion of what had just occurred, of Kobayashi relating Genden's litany, and Shimazaki, prejudiced against Genden from the beginning, shrugging at the enumeration of his many serious errors.

After returning to Genden, Hoshino and Iriya attended an emergency meeting with leadership. Iriya proposed a boycott of tomorrow's meeting. If the NRA had invented a rule to give themselves a week before accepting Genden's revised submission, Genden should make it known that they would be happy to attend in a week, when the panel was prepared. Without their experts, without their evidence, what was Genden's role, except to fill the seats at their table?

Genden's executives overruled the idea. The date of the meeting had appeared on the NRA's public schedule two weeks ago. Reporters glancing at the circumstances would conclude Genden was stalling. The NRA could advance this narrative by omitting their role in the exchange of vital documents and pointing out that Genden's final submission had been sent less than twenty-four hours in advance.

Toda and Okumura would attend, whether or not their presentations were permitted. Both had already traveled to Tokyo. On the morning of the twenty-first, Okumura made it a point to sit with Genden's staff. When he was asked to move, he relocated to the press seating, causing a stir among the reporters, whose ignorance about yesterday's disagreements he was happy to relieve.

The meeting conformed to Genden's low expectations. The NRA's self-serving version of the peer reviewers' comments was entered into the record without Genden's input. Addressing the exclusion of Okumura and Toda, a member of the NRA staff explained to the audience that panel meetings were "not a place for the utilities to execute their public relations strategies," a comment that even the panelists found offensive; Toda and Okumura were dissenting scientists, not hucksters. When Hoshino began to raise concerns about the administrative process that preceded the meeting, Shimazaki interrupted him to explain that bureaucratic matters were a waste of the panel's time. As Iriya had warned, the meeting would transpire without providing a reason for Genden's attendance.

After the meeting, the feeling at Genden recalled the disenchantment and sublimated fury caused by Tanaka's verbal decommissioning of Tsuruga 2 a year and a half earlier. Management had clung to the possibility that the reexamination of the site would be characterized by due process. The events of the past week had made their hope seem foolish. "The last faith in the institution had gone out of the room," Hoshino remembered.

Now the question was whether to publish the information that

Hoshino and Iriya had tried to submit to the meeting, a detailed analysis of the concerns shared by the peer reviewers and Genden. The panel's actions were plainly prejudicial, Iriya's superiors decided. He was instructed to adapt Genden's submission for public disclosure.

Hoshino was anxious about the increasingly pugilistic nature of his role, but Genden's legal staff reassured him. On multiple occasions, he had used the appropriate channels to pursue his concerns. After several careless responses, it was reasonable to become upset. His record of timely objections would be helpful when Genden's attorneys pressured their counterparts at the NRA.

"Legal wanted to be certain the NRA said no." Iriya recalled. "After June, they felt *no* was the only plausible interpretation."

30.

As Genden's complaints increased in number and intensity, the NRA rotated its employees. Genden was presented with a new point of contact, Morita Shin. Morita had worked for the International Atomic Energy Agency (IAEA) until 2012 and was currently being groomed to replace Koyabashi Masaru as director of the NRA's earthquake and tsunami division. His colleagues considered him the agency's most gifted nuclear safety expert and recalled his frustration upon arriving in Tokyo, where escalating hostility between the NRA and the nuclear industry could not be solved by applying his technical skills.

When Genden attended closed-door meetings at the NRA, Morita was present. Previously, Genden would send Hoshino and Iriya. Over time, Iriya attended less often. Hoshino was consistently present, in partnership with someone the NRA had rarely seen—the head of Genden's legal department.

Hoshino talked and Genden's legal staff took notes, with the goal of substantiating Genden's allegations of misconduct. In an attempt to provoke Morita to clarify NRA policies, Hoshino often compared them unfavorably with their IAEA equivalents. "That cut him," Hoshino recalled. "I'm sure he was disappointed, coming from a superb institution, to see the NRA poorly managed."

Morita told Hoshino that his behavior was "worse than the antinuclear activists." He told the media, off the record, that the NRA was "prepared to deal firmly with Genden's conformity review application, dare they file it." He warned Hoshino to be wary of making enemies, because many of the same people, himself included, would be involved in Tokai's conformity review. The first two comments revealed potential bias; the last—about Tokai—was illegal. Japanese law prohibits regulators from suggesting that the discussion of one administrative process could prejudice the outcome of another.

"I treated his outbursts as progress," Hoshino said. "The more lines he crossed, the more leverage we gained."

After each visit to the NRA, Genden published meeting minutes before the NRA could finish theirs. Hoshino began submitting information disclosure requests (Japan's equivalent of Freedom of Information Act inquiries) that targeted Shimazaki's staff, and the results were posted to Genden's website. As more was made known, Genden's legal department believed, the NRA's negotiating position would suffer.

The willingness of Okumura and Toda to speak on Genden's behalf made it easier for other experts to accept the possibility that Genden was more credible than the Tsuruga panel. A growing number wanted academic geoscience societies to terminate their relationship with the NRA. Occasionally, members of these societies provided Genden with evidence that the active fault review had been conceived under questionable circumstances. One expert claimed that his society's procedure for board approval had been skipped when its list of panel nominees was compiled, because board members would have raised concerns about the political biases of certain experts. Genden also obtained messages Shimazaki had sent asking for Fujimoto's participation by name, and developed the hypothesis that Fujimoto was Shimazaki's personal asset—the only geologist he trusted to accommodate the geomorphologists' hatchet job.

The academic societies hadn't expected this schism. At one extreme were a small number of geoscientists (perhaps five or six), including Okumura, who judged the active fault review a politically compromised failure. At the other extreme was another small group, including Watanabe and Suzuki, who were astonished by their colleagues' reluctance to accept the policy implications of their research. Many scientists apparently believed they could continue mapping faults, publishing in journals, and attending conferences, and engineers could continue building above those faults, and no tension or disagreement need ever arise. According to Okumura, Shimazaki tacitly endorsed the antinuclear agenda of Suzuki's group.

Between the poles of this discourse were positions associated with particular experts. Okada represented one, closer to Okumura than Suzuki, but moderate. Okada surmised that a small number of panelists were more interested in punishing the nuclear industry than practicing science, but he stopped short of claiming the panels had been doomed by design. He didn't blame Shimazaki or dismiss the notion that scientists could help reform Japan's nuclear regulator. As disagreements between his colleagues intensified, he reserved his disapproval for the problem of logic that resulted from applying narrow standards of risk (it seemed clear to Okada that the panels were being asked to apply a zero-risk standard) to a scientific inquiry that concluded with an aye or nay vote—active or inactive. Only bad science, he felt, could come from placing all the subtleties of the panels' research into this bureaucratic double bind.

Okada's critique was popular among his colleagues. In 2014, the Active Fault Society elected him president. It helped that he had been the face of intransigence at Ōi. All the panelists had worked for the active fault review much longer than initially planned; against the backdrop of their collective exhaustion, Okada's insistence on the highest possible standards of fact—at the cost of delaying his own retirement—struck many of them as heroic.

Amid this plurality of opinions, Japan's community of earthquake experts suffered a striking loss of research productivity. "It's more than the drain on the panelists' time," explained Tsuruga panelist Tsutsumi Hiroyuki. "At any given moment, there are thirty or forty Japanese scholars active in a particular research specialty related to earthquakes. If you polarize or factionalize the community, you've gone from a pool of thirty potential collaborators to fifteen or fewer. The effect is immediate."

When estranged experts encountered each other at conferences, they found their differences were reconcilable, with two exceptions: First, if Tsuruga 2 was, by overwhelming consensus, the riskiest reactor under consideration for restart, would allowing it to enter conformity review amount to an acknowledgment that seismic risk standards had not become stricter since Fukushima? Second, if it was widely agreed that every nuclear site in Japan contained fault traces, did this mean that using small, non-seismogenic faults to evaluate geologic safety was a politically inflammatory distraction? Or did it mean that Japan was not a wise place to operate nuclear plants? No proclaimed aversion to politics would be sufficient to insulate Japan's earthquake experts from difficult questions that could be answered only by inclination, intuition, and perspective.

The collapse of solidarity in the geoscience societies put Shimazaki at a disadvantage. Responding to concerns from prominent experts was time-consuming. The review was already behind schedule. Genden's disclosure requests had created a paperwork backlog. In an episode the NRA staff remember as the worst of this chaotic period, Genden representatives visited the NRA to present a list of complaints, and the staff who received it, startled by the intensity of the browbeating they received, bowed in apology to their Genden counterparts. The indignity of the incident was compounded afterward, when Shimazaki summoned the staff involved and criticized their timidity with uncharacteristic vehemence.

Emboldened by Shimazaki's vulnerability, Genden's legal staff urged Iriya to prepare for the Tsuruga panel's draft report (anticipated in November 2014) by developing a comprehensive critique of the active fault review's administrative and scientific shortcomings. Iriya reviewed transcripts of panel meetings and produced visual representations of their omissions and inaccuracies. During this period, his wife told him he spoke to the panelists in his sleep, addressing them by name, disputing their characterization of the K fault. "Running simulations," he told her. ("Please call them thoughts," she replied.)

Iriya accompanied Hoshino to a meeting where Hoshino inadvertently stopped using a title or honorific at the end of Shimazaki's name. Iriya concealed his surprise and let Hoshino work. "I hadn't realized how much the conflict was hardening him," Iriya said. He mentioned it to Hoshino after the meeting. Hoshino was shocked by his own behavior; no matter his dislike for his NRA counterparts, he had always addressed them respectfully. "Even Morita. I was paid to provoke him, and I called him by his title. The difference was, I looked at him and thought, *I'm going to make you upset.* I looked at Shimazaki and thought, *I'm going to make you lose.*"

Allowing Genden to threaten whatever they liked was initially the consensus recommendation of the NRA's lawyers. A nuclear utility suing the new regulator during the aftermath of the Fukushima disaster would be suicide for the utility in the press, and no judge would be willing to hinder the NRA, especially at a site where profound seismic risks were undeniably present. Shimazaki was nonetheless solemn about the implications of Genden's litigious posture. "They believed we were trying to destroy them," he recalled. "If they couldn't run Tsuruga 2, mutual destruction was second best."

Shimazaki's term was due to expire in September 2014. The other departing commissioner, diplomat Oshima Kenzo, was admired by the antinuclear movement. He had survived the atomic bombing of Hiroshima, investigated Chernobyl for the UN, and served on the National Accident Commission after Fukushima. Neither Shimazaki nor Oshima sought a second term. The press made little of Oshima's decision; his technical expertise was too limited to justify a long-term commitment to the NRA. Shimazaki's decision appeared in the headlines as a resignation. Speaking to a wire service reporter, "an official at a utility who asked not to be identified because of the sensitivity of the topic" impugned Shimazaki's motives while simultaneously diminishing the significance of his accomplishments, calling his departure a "small victory" for the nuclear industry.

Stakeholders concerned about the erosion of the NRA's independence under the Abe administration were alarmed by the surprise nomination of Tanaka Satoru, a nuclear engineer with lifelong industry ties, to fill one of the vacancies. The other nominee, Ishiwatari Akira, was an academic geologist. Although Shimazaki and Oshima were replaced almost simultaneously, the order of events led the media, out of sensationalism or lack of context, to describe Tanaka Satoru as Shimazaki's replacement, advancing a narrative where an industry antagonist had been replaced by an industry partisan, even though Ishiwatari, the geologist, was expected to assume the majority of Shimazaki's duties.

More than Shimazaki's lame-duck status during his final months, the polarized nature of the commentary surrounding his departure made the work of the panels vulnerable. "You're trying to remain objective while the chair of the panel sees his name in the newspaper every day, and some people love him, some hate him, but they all say he's partisan," recalled one of the NRA staff detailed to the active fault review. "Of course you feel some anxiety about your credibility, and what the incoming boss is making of this, what happens when he takes over."

From the outset Ishiwatari faced two distinct pressures, his staff recalled. The first was political: An election was forthcoming, and pro-nuclear Abe was poised for reelection by a landslide. With Shimazaki and the active fault review lately in the papers as the antinuclear movement's favorite elements of the regulatory system, the political capital necessary to protect the panels was more than Ishiwatari, a career academic, possessed at the beginning of his tenure. Second, the geoscience societies were eager to acquit themselves of their obligations to nuclear safety. Ishiwatari, a specialist in the formation of metamorphic rocks, was an odd choice to help lead the NRA; senior members of earthquake-related academic societies had been offered the nomination he eventually received, and moved swiftly to exclude themselves, not wishing to be associated with the political entanglements that had wearied their colleagues.

The Tsuruga panel submitted its draft report in November. Before the report was ratified, the NRA's legal staff attempted to assess whether the panel's first finding—the only active fault finding issued before the backfit rule took effect—posed legal complications for the new report, which reached the same conclusion. Because the first report was linked to Tanaka's verbal decommissioning of Tsuruga 2, the significance of Tanaka's comments would need to be determined. Did the panels possess the authority to prohibit a reactor from continuing the restart process?

When this question arose in 2012, Shimazaki's decision to disregard the possible consequences of Tanaka's pronouncement had enabled the NRA to continue its work and earned him the admiration of the agency's legal staff. After Shimazaki's departure, the prospect of publicly diminishing the authority of the active fault review seemed too much a betrayal of the system's commitment to sound science and original research, and a betrayal of the quiet, firm objectivity that enabled Shimazaki to distinguish the NRA from its predecessors. "It was also unavoidable," a member of the legal staff remembered, "that our ability to take ourselves seriously as the world's strictest regulatory regime had come to depend on the idea that Tsuruga 2 couldn't be permitted to operate."

If Genden raised the backfit issue in litigation, the NRA's legal staff were hopeful the court would accept an alternate source of regulatory authority. It could be argued that the NRA's pre-backfit directives were de facto cabinet orders. Genden would be forced to claim that the NRA had no authority whatsoever, a distasteful argument after Fukushima. Genden's specific allegations of administrative negligence could be litigated at the level of detail. For instance, the law guaranteed that Genden would receive timely and substantive responses from the NRA, but nowhere was it guaranteed that these responses would come from the Tsuruga panel.

Genden could nonetheless present a strong case. Tanaka's comments had been sufficiently premature to raise the question of whether the NRA had acted in a biased manner, and judicial review of this question would expose a period of administrative confusion at the NRA to the cold light of cross-examination. The Tsuruga panel's first report, although it must have seemed a long time coming to the panelists, had been issued three months too soon. "We knew the backfit rule was a cornerstone of the proposed regulations," one member of the NRA's legal staff remembered. "We behaved like we had it. We should have waited."

Surveying the initial group of post-Fukushima lawsuits, the NRA's legal staff concluded that some district court judges felt the time had arrived for judicial review to play a larger role in nuclear safety. In May 2014, the chief judge of the Fukui District Court had ruled against further operation of Ōi 3 and 4. The NRA's legal staff were not opposed to the involvement of the courts; as a maturing institution, the NRA could benefit from additional oversight. But the Ōi decision put the risks of litigation into context. A lawsuit in Tsuruga would draw the same judge, who had expressed grave reservations about the restart process, and a ruling in Genden's favor on the backfit issue could nullify the NRA's preliminary risk assessments at Fukui's other reactors, effectively delaying all restarts.

Internal discussions of this risk were limited to NRA leadership, who feared the consequences of its disclosure. They were aware of legal doctrines that could be invoked to connect the agency's early regulatory gestures, which predated the backfit rule, to the entirety of its subsequent work. Succession of illegality, for instance, determines whether one illegal act—regulating without statutory authority—renders the acts that depend on it equally illegal. A court loss to Genden could expose the NRA to a wave of backfit-related litigation from a multitude of utilities and antinuclear groups, threatening the first year of the agency's work and by extension the present system. "It was a narrow risk, but it was existential," a member of the legal staff recalled.

The authority ascribed to the panels by Tanaka's comments would be rescinded. Having arrived at their decision, the legal staff felt it sensible: The NRA would retain complete discretion over Tsuruga 2's restart prospects. The agency's seismology staff could raise the panel's concerns again during conformity review, under the legal protection of the backfit rule.

On December 3, the NRA issued a "clarification" of the administrative significance of active fault findings: They were reference materials to be included in future regulatory discussions. Where active fault findings had been issued, the reactors in question would be permitted to continue the restart process.

The brevity of this clarification produced immediate confusion—at Genden, at other utilities that owned plants where active fault findings remained possible, in the media, and among local governments affected by active fault review. "We had been calling the NRA and sending letters for two years, asking what authority the panels could exercise," one municipal official remembered. "Now we're told the panels are extraneous. We're not going to accept that until it's explained." Tanaka had overseen the clarification's scope and wording. He elected not to acknowledge that he acted prematurely in 2012. The NRA's legal staff accommodated this omission because they didn't believe Tanaka could admit his mistake without becoming petulant and lashing out at the press. Instead, the Tsuruga panelists were instructed to maintain secrecy about the improvised nature of Tanaka's decommissioning order.

To contextualize this policy adjustment without exposing Tanaka, the legal staff resorted to a ploy: They would pretend nothing had happened. Tanaka's comments in 2012 would be described as his personal paraphrase of the panelists' contributions to the first Tsuruga review meeting. Attention would be called to his subsequent remarks, which invited the utilities to conduct further research.

Stakeholders were asked to believe that the NRA's chief executive could speak the precise words of a decommissioning order, leave those words in the air for two years, and intend nothing by them. Hundreds of related inquiries had been strategically ignored in the meantime. The Tsuruga panel's first report included an explicit description of the panel's regulatory obligations: assessing whether the reactor "conforms to post-Fukushima standards." None of this would matter in the legal scenario the clarification was meant to foreclose. If Tsuruga 2 were allowed to apply for conformity review, there would be no administrative determination—at least none before backfit—for Genden to dispute.

When the NRA issued its clarification, Hoshino was in his office with Genden's lawyers. "Shock at first," he recalled. "That's it? We win? We were starved for context. Weeks went by and the NRA would say, 'Nothing changed. We don't know what you're referring to.' I wondered how people were living this down at the NRA, witnessing this disingenuous way of describing the biggest loss of face in the history of their agency."

The panelists were immediately in a difficult position. "The utilities resented the rigor of the active fault review," a member of the NRA's seismic safety staff recalled. "But they participated wholeheartedly out of respect for the authority they believed the panelists held." After the panels' limited role was made clear, more utilities were prepared to imitate Genden's belligerence. They consulted experts who had been vocal about the shortcomings of the active fault review (including Okumura Koji) and published these experts' opinions as "independent reviews" of the panels, exacerbating the schism in the geoscience community.

Ishiwatari was eager to retire the active fault review and urged the remaining panels to reach swift conclusions. Shika panelist Shigematsu Norio—who had demonstrated that the suspected fault movement at Ōi was most likely a landslide—was incredulous, as was Fujimoto Koichiro, also on the Shika panel. "Tanaka's motives at Ōi and Ishiwatari's motives at Shika were different, but the same bad advice was repeated," Fujimoto recalled. Shigematsu, who believed that three distinct faults at Shika could be active, joined his colleagues in providing Ishiwatari with the troubled history of the advice he had given and disabusing him of any expectation it would be followed.

The panelists' memories of the period following the NRA's clarification are dominated by the intensifying impatience of the agency's new leadership and the utilities alike, and by the sense that none of the stakeholders understood why the panels' research was important, irrespective of its administrative significance. "When we began, our relationship with the restart process was an implied relationship, which provided the appropriate amount of responsibility," explained Miyauchi Takahiro, the member of the Tsuruga panel whose concern about the Urasoko fault had been vocal and unwavering. "Tanaka's comments in 2012 made us exclusively responsible, which subjected our research to political interference. After the agency clarified its position in 2014, we had no responsibility at all, which made us vulnerable."

In board meetings, the other utilities acknowledged they had been wrong to criticize Genden's tactics. The NRA's clarification was proof: You could fight the regulator and win. Iriya in particular, as the triumphant leader of an underfunded research team at Japan's smallest utility, became a cult figure for staff at other nuclear companies, who had labored under the burden of low public esteem since Fukushima. Iriya's former superiors at TEPCO, who called him an embarrassment when he left for Genden, took credit for him now and ascribed the quality of his work to his TEPCO training. Annual financial reports were forthcoming, and Genden, the only utility whose entire generating capacity was suspended after the disaster, was also the only utility that reversed its post-Fukushima losses in 2014, an outcome attributed to careful management of the service fees its customers (the other utilities) were obligated to continue paying. Iriya's achievements and the company's surprise profitability acquired a mystique; the nuclear industry had never been able to think of itself as the protagonist of an underdog story.

Five days after the NRA issued its clarification, Genden received a proposal from their risk consultants, recommending a probabilistic assessment of faulting hazards at Tsuruga 2. The proposal also recommended investigating the possible relationship between the K fault, the D-1 shatter zone, and the Urasoko fault. This would require a significant expansion of Genden's research, amounting to the development of a detailed stratigraphic profile of the entire site, down to the bedrock, as the panelists had suggested. The rationale of these recommendations was plain: Recent Urasoko movements might not have displaced K, but the next one could, a concern outside the scope of regulations but critical to the reactor's safety. In the words of their own consultants, Genden "had not developed a compelling case."

The proposal had been created in response to materials provided by Genden in August. It was not clear whether the consultants grasped the significance of the clarification the NRA issued on December 3, or considered its effect on the likelihood that Genden would commit to a costly expansion of research.

When the proposal arrived at Genden, there wasn't time to discuss it. The peer review meeting for Tsuruga 2 was scheduled to occur in two days. It would be the first time the peer reviewers who raised concerns during their visit to Tsuruga could express them without the intervention of the NRA. Hoshino hoped the reviewers felt the panel had censored them and would use the meeting to retaliate.

Expectations rested on Okada Atsumasa, the senior expert in the active fault review since Shimazaki's departure. Okada had been responsible for slowing the Ōi panel to a judicious crawl when the press, the public, and the government wanted a quick conclusion. During the peer reviewers' tour of Tsuruga 2, his criticisms of the panel's conclusions had been detailed and explicit.

The meeting was chaired reluctantly by Kato Teruyuki, president of the Seismological Society of Japan (SSJ). More than other academic societies, the SSJ had been implicated in the Fukushima disaster; the earthquake and tsunami had exceeded seismologists' estimates and occurred in a region classified low risk. Three years earlier, Italian seismologists had faced criminal prosecution for similar failures of prediction. Kato was aware of the schism among the panelists and feared that any mediating stance he adopted at the peer review would aggravate tensions in his own organization.

Before leaving the NRA, Shimazaki had asked Kato to chair the peer review. Kato agreed as a gesture of respect. It was difficult to find an earthquake expert in the SSJ whose research was not indebted to Shimazaki, and Shimazaki's 2004 attempt to warn the government about tsunami risk at the Fukushima reactors provided a notable exception to the SSJ's nuclear safety failures. When the NRA circulated their initial solicitation for panelists, which nearly eliminated Okada from candidacy on the basis of his age, every academic society had objected and listed Okada anyway—except the Seismological Society, chaired by Kato, whose list satisfied Shimazaki's criteria to the letter.

As Kato had predicted, the meeting was contentious. Experts whose criticisms had been mild when they visited the trench were more explicit now. Watching online, Hoshino detected the possibility that the meeting would become a conduit for the broader discontent affecting the geoscience community.

Okada participated with vigor, raising issues that implied harsh judgment of the Tsuruga panel's performance and the NRA's evidentiary standards. In one tense exchange with Fujimoto, the panel's geologist, Okada demanded an explanation of several assumptions the panel had made about the site's stratigraphy.

"You've raised some important questions," Fujimoto acknowledged.

"And you've recorded displacement in locations where other phenomena might be more plausible," Okada said. "It's a self-serving interpretation."

As the tone of these exchanges grew increasingly strident, Kato was presented with two undesirable scenarios: He could surrender the meeting to grievances or veto them. Either decision risked an argument. If the meeting became contentious enough, it would wind up on the front page of most newspapers.

I waited a long time to correspond with Kato about his role in the peer review. His colleagues said he hated to discuss it. "I wanted to forget about this meeting as soon as [it] ended," he recalled. The initial request to chair the meeting had surprised him. He was a geodesist, specializing in measurements of Earth's shape. His active fault expertise was negligible, and his familiarity with issues of nuclear safety even less. At the point in the meeting when the dilemma of intervention arose, it was plain to him that the need for moderation—his express responsibility—was acute, but he couldn't determine which of the reviewers' criticisms, if any, should be preserved on the basis of their scientific merit. "I had no idea how I should behave or comment," he reflected.

This crisis of confidence was not apparent in the meeting. The peer review was not the place, Kato insisted, to question the standards the panel had used or identify missed opportunities for additional research. Reviewers should limit themselves to the contents of the panel's report and ensure that their comments, even when critical, presented opportunities to revise the report constructively.

If you had watched the meeting, you might have presumed (as many did, as I did) that Kato had misjudged. Okada was likely to consider this narrow construction of the peer review's scope prejudicial and unscientific. The room was full of people who admired Okada and shared his concerns. If he objected, they would object too, and Kato's authority would evaporate.

Okada did object. His colleagues echoed his concerns about the panel's interpretation of trenching data, but accepted the limits that Kato imposed on further discussion. Only Awata Yasuo, who had engaged Suzuki in a verbal confrontation during the peer reviewers' trip to Tsuruga, insisted on pursuing concerns that Kato attempted to rule out. The meeting concluded without serious conflict.

To understand the effectiveness of Kato's intervention, I debriefed panelists who considered themselves Okada's supporters. These were not people who agreed with Okumura that the entire system was compromised. They were, like Okada, sincere in their belief that scientists should collaborate with regulators and disappointed by the shortcomings of the research this collaboration had produced. "Analytically, I agreed with everything Okada said at the peer review," one remembered. "Regarding what should be done, I agreed with Kato. The panel's report was flawed, but I wanted it upheld."

Most peer reviewers recalled similar thoughts: The Tsuruga panelists, whatever one felt about their motives and methods, were enforcing an urgent, necessary decision. "Tsuruga 2 was seismically compromised," remembered a peer reviewer who had been critical of Suzuki after the first panel meeting. "The regulations were too narrow. I'm grateful for scholars like Suzuki, if they appear when the time and place are right."

Tsutsumi Hiroyuki, the Tsuruga panelist who had insisted a defensible conclusion could never be reached, also suspended his ambivalence. When the time had arrived to draft the panel's second report, he had not hesitated to support an active fault finding. "The evidence remained unclear," he recalled. "Did the proximity of the Urasoko fault—which was outside the scope of our review—lead me to believe the site was dangerous? It did, and I acted on that belief." He no longer accepted the argument that the panel should be blamed for the controversy surrounding its work. "The reactor never would have been built if the threat from the Urasoko fault hadn't been hidden. Regulators agreed to hide it, and the scientists who reviewed the regulators' work played along. If blame is important, it's possible to obtain records of those meetings, including the names of the scientists who failed us."

The circumstances of the active fault review, several of the peer reviewers also pointed out, had evolved since Okada first raised his concerns. All six plants selected for review were undergoing assessment, and Tsuruga was no longer the only site where evidence was disputed. At Higashidori, the water-retaining properties of the site's soil introduced ambiguities. The panel there, led by Awata Yasuo, anticipated an active fault finding of their own, bitterly contested by the utility. At Shika, controversy had arisen over the question of whether subsurface faults presented a risk of surface displacement. The Shika panel included members who had been critical of the Tsuruga panelists, as well as the geologist responsible for turning the Ōi panel away from an active fault hypothesis. At Mihama, where the evidence was inconclusive, panel members were apparently prepared to forgo an active fault finding, reversing the burden of proof the Tsuruga panelists had employed and leading several critics of the Tsuruga panel to reconsider which of these two imperfect outcomes was more troubling. Only the Monju panel, reviewing a noncommercial reactor, was proceeding without controversy.

The NRA's withdrawal of the active fault review's regulatory authority occurred a week before the peer review. "Before, we had to ask ourselves whether it was appropriate to allow Suzuki and Shimazaki to decommission a reactor," one peer reviewer recalled. "After, the question was whether we could allow them to issue a strong, nonbinding indictment. The flaws in their report were minor compared to the flaws in Tsuruga 2's original siting process. It wasn't a difficult decision."

This decision, another panelist recalled, was not without its risks. "We had to decide what was worse: allowing Tsuruga 2 to resume operation near the Urasoko fault or abusing the regulations to avoid this. We accepted the Tsuruga panel's decision to address the immediate threat, and permitted the creation of a place in the process where abuse can occur. We won't always like what happens there."

For Tsutsumi and many of the peer reviewers, another consideration, further outside the panel's regulatory mandate, had to be weighed: Tsuruga 2 was closer to the Urasoko fault than the proposed site of Tsuruga 3 and 4. If Genden were allowed to certify the seismic safety of Tsuruga 2, it would be difficult to block construction of additional reactors. Between a possible extension of Tsuruga 2's license and the construction of Tsuruga 3 and 4, followed by their own license extensions, a nuclear plant might operate near the Urasoko fault for another century. The estimated periodicity of the Urasoko fault was five thousand years, plus or minus two thousand, and its most recent movement had occurred approximately four thousand years ago; a crude estimate on the basis of these numbers would put the likelihood of an Urasoko-generated earthquake striking the new Tsuruga reactors at roughly one in seven. The panelists called it "the hundred-year fear." Classifying the K fault active could eliminate this risk, and the resulting opinion was nearly unanimous: K was active.

31.

Irrespective of Kato's intervention, Hoshino and Iriya felt the events of the peer review vindicated Genden's request to delay ratification of the panel's report. Kato's comments were in one respect helpful: If the concerns Okada and others had raised were admittedly serious, and the peer review was not the place to raise them, they would need to be addressed elsewhere before the Tsuruga review concluded.

Hoshino urged Genden's leadership to adopt a new strategy for pursuing the peer reviewers' concerns: Attach them to Iriya's critique of the panel meetings and publish the resulting document. The NRA couldn't expurgate Genden's complaint if it was already available to the public.

This revised critique, a list of sixty-three objections, was published on January 6, 2015. Among Genden staff, the months that followed were characterized by a sense that the press was beginning to acknowledge contradictions in the panel's assessments. Hoshino forwarded the YouTube video of the peer review meeting to friends, family, and his colleagues at other utilities. Their response was much like the response of reporters: They didn't understand the substance of the debate but understood that several scientists appeared convinced that the Tsuruga panel had been careless.

In communications and meetings with the NRA, Hoshino pursued the resolution of priority items from Iriya's list. As late as Friday, March 20, NRA staff assured him their superiors were receiving Genden's inquiries and a response was forthcoming.

Since early March, the NRA's website had shown the commissioners would meet on Wednesday, March 25. On the twenty-third, Hoshino learned that the commissioners' agenda included Tsuruga 2; they would ratify the Tsuruga panel's findings, to the effect that Genden's reactor was located above an active fault. For the fourth time since the active fault review began, Hoshino realized, the NRA had promised to address Genden's objections and willfully betrayed this promise.

Hoshino wasn't able to reach the NRA official responsible for negotiations with Genden until the following day. This official had recently left Japan to visit the IAEA in Vienna, a circumstance he raised in his own defense, proclaiming—to Hoshino's disbelief—that he hadn't known the panelists' report was complete, nor that it would be confirmed before Genden's concerns were addressed.

On the twenty-fifth, Iriya watched the meeting on his phone in a doctor's office, waiting for an appointment. "My colleagues wondered if the commissioners would address our complaints before ratifying the findings," he remembered. "I knew they wouldn't." The only words in the meeting that might have been meant to placate Genden came in Tanaka's reiteration of the role the panel's report would play: important reference material, nonbinding in nature, to be considered during conformity review. Hoshino had tried to learn what "important reference material" meant. The NRA told him, *A future commission meeting will establish rules for the use of reference materials by conformity reviewers*, or, *The conformity review staff will determine how to use relevant reference materials.* Neither reply was an answer.

The panel's final report was worse than Iriya expected. The draft version had specified D-1 as the shatter zone of concern, pinning the active fault finding on the hypothesized relationship between D-1 and K. The final version referred to "all shatter zones potentially affecting safety-critical equipment." This would enable the conformity review to continue scrutinizing possible faulting hazards even if Genden could improve the evidentiary basis of their argument that D-1 and K were unrelated.

During the weeks that followed, the NRA's customary assurances evaporated; they hardly took Genden's calls and offered no reply to inquiries about the panel—none beyond the insistence that the panel could not revisit its work, because the panel no longer existed.

The NRA's legal staff had instituted this policy of calculated non-response. The panel's new report was protected by the backfit rule, the reopened investigation had satisfied Genden's right to receive a thorough review of their evidence, and the withdrawal of the panel's authority protected its work from legal liability. No further improvement to the NRA's position could be gained from continued dialogue.

"We were proud of ourselves," recalled one of the lawyers who designed the NRA's legal strategy. "The entire first year of regulatory activity had been jeopardized. In the space of a few months we eliminated our exposure."

Genden filed an information disclosure request with the NRA, seeking all communications involving the Tsuruga panelists that occurred between the peer review meeting and ratification of the panel's final report. The NRA claimed there was nothing to disclose. When he consulted colleagues at other utilities, Hoshino learned that Japan's public information laws primarily apply to documents "retained" by the government, and many agencies claim not to retain internal correspondence.

Determined to preserve their right to dispute the panel's findings, Genden addressed legal threats to the panelists and sent them to the NRA. In a feat of unexpected stupidity (the consensus description among NRA staff), these threats were forwarded to the panelists at their homes and offices, turning a futile repetition of Genden's complaints into an embarrassment for the agency, whose reputation in the geoscience community continued to decline. The forwarding of Genden's threats also drove a wedge between NRA leadership and the panelists. "I was numb to whatever Genden's lawyers wrote," Tsutsumi remembered. "But the NRA told us that our authority had been rescinded in order to protect us. Apparently we were never protected. I admire the courage of the panelists who delivered active fault findings after we received those letters, knowing they would be exposed to similar threats."

Irrespective of the NRA's sense that the gravest risks to their authority had been avoided, the relative success of Genden's confrontational posture continued to inspire other utilities to meet force with force. In Kyushu, where taxpayers had petitioned the district court for an injunction to prevent Japan's first indefinite nuclear restart, the utility company publicly threatened to countersue individual plaintiffs for lost revenue, a tactic that would have seemed politically improvident a year earlier.

"Genden moved the needle," one of the NRA's legal staff later reflected. "We were so excited to save the NRA from Genden's lawyers that we failed to notice we were already living in their idea of our world."

After reducing legal vulnerability at Tsuruga 2, the NRA believed its role in pivotal restart litigation would be the role of an interested observer. The cases that mattered now had been filed by antinuclear activists against utility companies. The Kyushu case was among them. The utility involved was not one of Japan's largest, but the significance of its restart plan had grown exponentially after the Fukui District Court ruled that Ōi could no longer operate.

Tanaka held the NRA at a distance from the Ōi ruling. "I won't say anything about it," he told reporters. "Our conformity review at Ōi will not be affected." From the NRA's perspective, the invalidation of Ōi's exigent restart could be useful, if it helped to establish conformity review as the sole legitimate source of operating permission after Fukushima.

Any ruling against a nuclear utility in Japan is newsworthy, but not without precedent. At the time of the Ōi ruling, utilities had occasionally lost in district courts, and on one occasion one reactor (Monju) lost in Nagoya High Court. A nuclear plant had never lost in the Supreme Court, which often establishes precedents that benefit plaintiffs without ruling in their favor. Two cases filed in the 1970s and decided by the Supreme Court in the 1990s generated the case law necessary to ensure the long-term viability of antinuclear litigation. Together, they established the administrative standards that regulators must meet before their obligation to public safety has been satisfied, and moved the burden of proof to the utilities, who necessarily hold the critical evidence in lawsuits alleging safety failures at nuclear facilities.

Many of KEPCO's arguments in the Ōi trial attempted to dissuade the court from exercising jurisdiction. "Control of discretion—the idea that courts shouldn't decide technical matters—was the argument Kanden needed to make," a lawyer involved in filing the suit reflected, "but it's a delicate argument, telling judges what they're not allowed to do, and Kanden's lawyers were indelicate." Control of discretion principles notwithstanding, the court had an overriding obligation to protect the rights of citizens, and in light of KEPCO's nonchalance, the plaintiffs' lawyers were not as surprised as they might have been when the ruling went in their favor.

KEPCO declared that the Ōi reactors would be restarted irrespective of the ruling. It might not have been wise, in a political climate leery of the nuclear industry's motives, to advertise the company's indifference to a court ruling that raised grave concerns about public safety, but KEPCO's posture was perfectly legal. While the case was undergoing appeal, the effects of the judgment were held in abeyance.

The plaintiffs in Kyushu filed on May 30, 2014, nine days after the Ōi ruling, with the assistance of many of the same lawyers. They sought a victory without precedent: an injunction that would prohibit restart, effective immediately. The plaintiffs' chances struck the utilities as too remote for concern. The completion of the NRA's conformity review in Kyushu was imminent. An injunction would pass within a hairbreadth of indicting the NRA's new regulatory regime, including the firm authority of the backfit rule—not a fight any judge would want.

A separate injunction request was subsequently filed at Takahama, and a ruling was expected after the Kyushu ruling. Takahama was Kanden's least controversial facility, in part because it held the strongest safety record in the company's fleet (the nation's largest, nine reactors with restart potential). If an injunction was improbable in Kyushu, an injunction at Takahama would be the equivalent of a meteor striking the crater made by a previous meteor.

On April 14, 2015, twenty days after the Tsuruga active fault findings were confirmed, the Fukui District Court issued its Takahama ruling, nearly a year before it was expected. For the first time in Japanese history, a court had granted an injunction against a nuclear plant. Effective instantly, Takahama 3 and 4 were forbidden to operate. The industry's restart agenda, delayed for several years by the active fault review, would be suspended indefinitely while Japan's leading electricity company battled a sitting judge.

The legal reasoning in the Takahama ruling, authored by Chief

Judge Higuchi Hideaki (who previously presided over the Ōi law-suit), was shocking enough to overshadow its effect. Higuchi targeted the NRA, which was not a party to the case. Even at sites where conformity review was complete, his ruling stated, the NRA's administrative shortcomings had put public safety at risk.

32

Higuchi's ruling was based on the NRA's handling of ground motion calculations. More than the risk of fault displacement, the potential consequences of ground motion have shaped the past few decades of seismic safety regulation at nuclear facilities. After the 2007 earthquake near Kashiwazaki-Kariwa produced shaking that exceeded the design specifications of the site's seven reactors, the reforms that followed—including the multiplication of ground motion requirements for every reactor in Japan by 1.5—revealed the dangerous inadequacy of previous thresholds.

In his April 2015 ruling, Higuchi determined that the ground motion threshold at Takahama (700 gal) was equally inadequate. The strongest earthquake-induced ground motion recorded in Japan was 4,022 gal, produced by the Iwate–Miyagi Nairiku earthquake in 2008. The thresholds for ground motion mitigation used by two of Japan's largest residential construction firms bracketed this value. Mitsui Home rated its structures for 5,115 gal. Sumitomo units were rated for 3,406. In comparison, the 2011 earthquake recorded ground motion as high as 2,699 gal. The 2007 ground motion recorded near the Kashiwazaki-Kariwa plant, which led to the reassessment of ground motion thresholds at nuclear reactors, was 1,699 gal, more than triple the amount seismologists expected in the vicinity of stable bedrock formations, where reactors are sited because bedrock is less vulnerable to shaking.

In consultation with the NRA after the Fukushima disaster, Kanden had raised the ground motion threshold at Takahama from 550 to 700. The same value was applied at Ōi. This did not mean that

Kanden expected the plants to fail if they were affected by ground motion exceeding 700 gal, but beyond this threshold emergency measures would be required.

Higuchi wasn't satisfied. None of these ratings approached the ground motion recorded at Kashiwazaki-Kariwa. Construction firms were plainly capable of building structures with additional seismic resilience, which revealed Kanden's reliance on the bedrock beneath the reactor to guarantee seismic safety. "Seven hundred was a strange number, and in some respects conspicuously low," recalled a Fukui District Court employee familiar with Higuchi's consideration of the case. "If the threshold had exceeded 1,000, representing the number that Kanden and the NRA had discussed but increased by a factor of 1.5, as regulators demanded after Kashiwazaki-Kariwa, I think [Higuchi] would have considered dismissing the injunction request. He knows it's not his responsibility to judge the standards. It's his responsibility to judge the administrative prudence that produced the standards."

Journalists struggled to obtain context for Higuchi's surprise ruling. Japan's courts are open by constitutional decree, but the Supreme Court has interpreted this article of the constitution narrowly, and individual courts have developed idiosyncratic standards for the release of records in ongoing civil litigation. Many won't release anything until the case concludes, then limit their release to the final ruling. In nuclear industry litigation, lawyers for the plaintiffs usually release filings and evidence, but without a complete record to put these documents into perspective, newspapers are reluctant to cite them. On the nuclear beat, it's more fruitful to hang around the press club in the building that houses Fukui's prefectural legislature. Kanden's office is across the street.

After the Takahama injunction, Kanden's spokespeople were in no mood to give long press conferences, never mind visiting the legislature's cafeteria to tell journalists what they knew about Judge Higuchi's pets, children, and hobbies. Consequently, coverage of the ruling contained several presumptions about the man who had ruled and his motives.

Japanese wire services ran some prose to the effect that Higuchi's ruling was unexpected in light of the Japanese judiciary's conservatism. This boilerplate was taken up by global news outlets. He was a "maverick," they declared. This statement often constituted an article's sole contextualization of his career and appeared in configurations that reduced it to a truism about his two recent rulings against Kanden, without attribution to the type of source a journalist consults for truisms (a professor, a spokesperson, a politician). One newspaper wrote, "The presiding judge, Hideaki Higuchi, is considered a maverick in Japan's traditionally conservative judiciary, having issued a similar ruling against separate reactor restarts in Fukui Prefecture last May."

Another newspaper consulted a law professor who repeated a popular hypothesis: Higuchi had been embittered by a series of undistinguished assignments, and he issued the injunction to punish the judiciary. The judges I consulted found this laughable; Higuchi's appointment to the chief judgeship in Fukui already disproved it. Higuchi had been appointed *after* Fukushima, they explained, when high-profile nuclear litigation affecting the nation's most important electricity company was certain to be filed in Fukui. If a career in local courts tended to produce an independent streak, Higuchi wouldn't have been sent where a steady hand was needed.

According to former classmates and colleagues I contacted, Higuchi had been a private, cautiously spoken student whose early blooming rhetorical abilities were distinguished by their adult plainness and economy. The son of a civil servant in the capital city of a medium-sized prefecture, he attended one of the city's best high schools and ranked well, then enrolled at Kyoto University, where he studied law without apparent zeal. He didn't intend to enter the legal profession until he passed the bar exam. Many classmates who studied harder and more purposefully did not; the pass rate in 1978 was less than 2 percent.

His first judgeship was in Fukuoka, where his early cases included an attempted murder, committed by a poor man with a difficult life and an exhausting job, who stabbed his wife after she cheated on him. "A case like that during your formative years will change your view of the court," explained a criminal attorney from Higuchi's law school class. "The institution no longer seems omnipotent; what comes in broken goes out broken."

After two years Higuchi was assigned to the district court in Shizuoka, under the chief judgeship of a widely admired legal writer whose commitment to the subtle, cumulative effect of fastidious rulings made him a natural mentor for a young judge suspicious of the judiciary's grandiosity. "I would say [Higuchi's] attitude toward law was blue-collar rational," a former colleague reflected. "He had an indifference to the cultural and social elements of jurisprudence, and an attachment to the inner logic of his profession that I associate with tradesmanship." People who subsequently worked with Higuchi or litigated in his court recalled him as a moderate jurist whose interest in analytical simplicity introduced an element of conservatism to his rulings.

Lawyers working the Nagoya High Court recalled seeing Higuchi on his way to the chambers of the chief justice before the earthquake struck on March 11, 2011. "After Higuchi issued the injunction, I realized what that moment suggested," one of the lawyers said. "Higuchi and the chief justice were together during the disaster. They would have discussed it. I don't believe Higuchi would have been assigned to Fukui if he had hinted at any bitterness toward the nuclear industry." Another lawyer who spoke to Higuchi shortly before his departure for Fukui recalled that Higuchi was certain his new assignment would require him to preside over aggressive litigation filed by anti-nuclear activists, but wasn't particularly worried, because it was his impression that the plants in Fukui were already being subjected to extremely conservative safety measures.

Lawyers, judges, and legal scholars familiar with Higuchi's career suggested an alternative explanation for his decision: Higuchi was the kind of judge who wouldn't be willing to rule solely on precedent in a significant case. "The precedent is control of discretion," explained one law professor, referring to the principle that assigns legal authority for technical decisions to regulatory agencies. "Before applying it, judges should examine the regulator's conduct for rationality. Most judges grant rationality presumptively, since it doesn't need to be superb, just adequate. A judge who disdains presumptions might rule against a government agency unexpectedly."

Pro-industry commentators insisted it had been Higuchi's intention to weaponize the injunction process so that other judges who had responded to Fukushima ideologically could pattern injunctions after his. In a variety of analyses, which included the mainstream political voices of prominent law school professors and the editorial board of Japan's largest newspaper, a "zero risk" threshold was attributed to Higuchi's ruling. Masuda Jun, a Tokyo High Court judge who had retired into academia, gave the critique reproduced most widely: "It seems [Higuchi] set out with the idea of imposing a zero-risk standard. Judges are not experts on nuclear power plants, so it is imperative they humbly heed the counsel of science. I doubt [Higuchi] took that into consideration."

The NRA had been established to eliminate the possibility of a future accident on the scale of Fukushima. Higuchi's standard was explicitly identical. The risk of an accident was acceptable. The risk of an accident that resembled Fukushima, affecting millions of people and requiring decades of decontamination, was not. "That's a mainstream interpretation of acceptable risk," said a senior editor of a newspaper that criticized Higuchi's ruling. "The editorial board doesn't always read court documents; they request executive summaries. A few board members understand risk communication. The rest think 'low risk' and 'zero risk' are rhetorical terms."

After the injunction, a contact of mine at Kanden told me his colleagues had discovered a controversial ruling Higuchi made in a traffic case. Two vehicles collided, and one driver admitted to being in the wrong lane, but Higuchi held both parties liable. This proved Higuchi was a certain kind of judge, my contact said. He was willing to ignore overwhelming evidence if it enabled him to draw attention to the cleverness of his ruling.

I reviewed local coverage of the accident. At first glance it seemed what my contact claimed: One driver was responsible, both were held liable. When I visited the district court to request the file, I learned the ruling wasn't Higuchi's. I assumed Kanden's people would discover their mistake and find other rumors to circulate. Instead, I started hearing from journalists in the Fukui press club: *We know what kind of guy Higuchi is; there's this traffic case.* I never heard it from the reporter for Japan's leading center-left newspaper, whom Kanden wouldn't have trusted. At lunch with a reporter who mentioned the case (and attributed it to Higuchi), I asked if he had read the ruling. The reporter said no, someone who covered the courts had alerted him to it. He changed the subject and asked if I had seen *The Big Bee,* a thriller in theaters then, which dramatizes Japan's nuclear energy debate by introducing a robot helicopter under the control of an antinuclear terrorist. I hadn't. "It's very realistic," he said. "The sets—not the plot," he added, sensing my review of his credibility.

There was credibility to review. He was the reporter of record for prefectural politics.

When I lived in Tsuruga, I knew the retired director of the public relations center and kiddie museum located at the entrance to Kanden's three-reactor plant in the neighboring town of Mihama, where the head office of the company's nuclear division is sited. Tadao had retired into the PR center from a previous career as a reactor operator. He began this career at Kanden's hydroelectric plants and volunteered to train into nuclear when the Mihama plant—Kanden's first, and Japan's first pressurized water reactor—was announced. The first group of operators trained on-site while Westinghouse Electric Corporation supervised the reactor's construction. Tadao and his colleagues were amazed by the expertise and competence of Westinghouse's chief engineer and equally amazed by the clueless, mean-spirited incompetence of his omnipresent deputy. "We wondered whether the technology we were getting came from Americans like the chief engineer, or his pet moron," Tadao recalled. "Now I know it was both."

When I met Tadao in 2014, he was seventy-five. All he wanted from retirement (which had recently become complete after a series of plum part-time assignments) was to play recreational sports with fellow retirees at the local civic center. "It's nice to have a body," he explained, elaborating that in childhood his body had belonged to the privations of wartime, and in adulthood to Kanden. He had adopted nervous habits to forestall the decline of his focus and dexterity—for example, tearing restaurant napkins into strips and tying the strips in Chinese knots. After coffee and a cigarette he wasn't steady enough and watched his hands lose the trick. "The disappearing man," he mused, without bitterness.

His father had worked at power stations, too, a job that exempted him from conscription during the war, and the family moved depending on his assignment. Tadao was one of six children. At school, teachers knew which families were food-poor, and he was allowed to skip gym. "I sat with the malnourished children. We had big heads compared to our bodies. I hate to see a balloon on a string. It reminds me."

The electricity work he started after graduating from vocational school was like his father's and kept him in remote parts of the country, where poverty lingered longest after the war. Nuclear work was a revelation. After training he stayed awake with his colleagues deciphering English blueprints and writing Japanese operating manuals. This new vocation, compared with working at a dam, seemed to emanate the nation's future.

Mihama 1 was finished in 1970, in time for the World's Fair in Osaka. To advertise the domestication of American nuclear technology, Kanden arranged a press junket in the control room. When the switch was thrown, the electricity would power a display at the fair. Tadao's team worked the day before the junket. They brought the reactor to criticality and conducted a shakedown of the control room. The calm afterward, when the reactor was generating electricity and required no intervention, gave Tadao an unfamiliar sensation of serenity. What you notice when history finds you, he later reflected, does not include your old hunger.

The next day, for effect, Kanden told the press that the electricity sent to the World's Fair during the junket was the first the plant had generated. Tadao saw news footage from the fair and thought, *I did that,* then realized he hadn't. "It's good when a lesson happens early," he told me. "My work was important, but someone else owned it." Reactor operators in Fukui are familiar with this anecdote and use it to describe Kanden's corporate demeanor—never too awed by history to use it.

I was friends with another former plant employee Tadao's age, named Seiji. Seiji's father had been a farmer, drafted during the war and killed in the Pacific. The war ended, the system of myths and morals that made it possible for families to endure the loss of fathers and husbands ended, and Seiji entered the workforce with an animated sense of the cost of going along. He worked for the contractors who perform ground-level maintenance at Japan's nuclear plants. He tried to unionize them.

Highly skilled workers at Japan's nuclear plants are unionized. If the objective of a strong union is good pay and job stability, these unions are strong. They rarely fight the company. They don't need to. Electricity companies train their own and don't like to lose them.

When they need flexibility from their workforce, they get it further down the labor pyramid. Seiji recognized the urgency of unionizing the industry's manual labor supply. As his successes began to accumulate, there were days when his car was followed from meeting to meeting by a convoy of yakuza vehicles and secret police, the latter to monitor both sides: communist elements helping the unionizers, and mobsters hired to intimidate them. Seiji had the shit kicked out of him—bones broken, organs lacerated—several times. Bricks were thrown at his windows so frequently he stopped replacing them. Later, industry partisans decided he was an antinuclear activist and ridiculed these stories, as if they were fabricated. In Fukui, I met people who were there when he was beaten, or saw the immediate aftermath. Most had no opinion about the nuclear industry but couldn't see why an old man should be slandered for enduring the violent gangsterism that followed construction money everywhere in Japan during the bubble era.

Seiji resisted the notion that he was antinuclear. He moved in those circles but felt the plants should run if the workers could be paid fairly and their safety protected. Of the nuclear energy policies under political consideration, he favored the widespread restarts the LDP had proposed. He couldn't see what was antinuclear about saying the plants shouldn't rely on workers who had accepted radiation-significant jobs because their loan sharks had coerced them. Or hadn't been trained to tie a secure knot around the pipe they were repositioning. He had seen contractors force workers to surrender their personal seals (equivalent to a signature in Japan) so their radiation records could be forged. Criminal exploitation of the industry's workforce was profoundly antinuclear, he reasoned.

Like many of his colleagues who had endured a lifetime of manual labor and dosimeter tampering, he ailed physically and immunologically. His appearance and mannerisms were afflicted, his eyes always watering, for instance, and his eyelids fluttering to expel the fluid, until a thin foam gathered on his cheeks. When he explained how many times an electricity company would ask contractors to repair a deteriorating pipe instead of replacing it, told you how much radioactive waste this type of misjudgment could save the company from accumulating, how they figured legally they were safe if the pipe burst because they could fault the contractor, you were certain he had been there. His anger seemed the source of his cognitive health, which defied age and frailty.

It was easy to think of Tadao and Seiji in a pair: the boy who worked, in Seiji's words, as a human load-bearing component in the nuclear plants after the war took his father, and the boy who worked in the control room after the electricity business exempted his father from conscription. Tadao's story about the World's Fair fit with a story Seiji told about union drives at the Kanden plants. At other utilities, the ballot boxes were often empty because the workers feared retaliation. At Kanden, when the first box was opened, a loaf of cigarette butts hit the table and disintegrated, releasing a plume of ash into the faces of several managers. "The contractors had been jamming their butts in there and ashing their cigarettes into it," Seiji said. "Of all the utilities who resorted to intimidation, Kanden found a special way to piss off their people."

Tadao's and Seiji's anecdotes showed the two traits that industry professionals outside Kanden (and a few inside Kanden who were detached from the culture of their workplace) mentioned when they were unhappy with the company: a prejudicial commitment to myths and symbols, and the unwillingness to acknowledge any merit in arguments put forth by their critics.

In 1976, Kanden had been the first nuclear utility to cover up a significant equipment failure at one of its plants and consequently became the first utility subjected to a high-profile leak, pursued by a prominent investigative reporter. The national government sent representatives to local governments in Fukui to apologize for their failure of oversight, setting a precedent for the current practice of obtaining local permission to construct and operate nuclear reactors.

After additional accidents at the Mihama site, Kanden founded the Institute for Nuclear Safety System (INSS), a think tank that produces much of the industry-side research published in academic journals. Critics have accused Kanden of launching INSS when a policy response to accidents at Kanden plants became inevitable, in order to lend their agenda the imprimatur of academic science. Mihama's safety record hasn't improved since the institute was established; in 2004, four workers were killed when a steam pipe ruptured in the turbine building of Mihama 3. A fifth later died from his injuries.

Most of Kanden's accidents have resulted from work management problems inherent to multilayer contracting, a model that
Kanden and its competitors borrow from the nation's construction
firms, whose network of affiliated companies exists primarily to cheat
Japan's labor laws. Besides encouraging mid-level supervisors to pass
risk judgments downward, the nuclear industry's dependence on contractors creates logistical hazards. A complicated pipe-fitting task in
a primary radiation area of a plant might require twenty workers.
Finding twenty who aren't dosed from other jobs could realistically
require the hiring of five contractors. Problems of coordination will
inevitably arise between them, and none will assume responsibility
for mistakes or incidents.

A well-connected nuclear professional once walked me into Kanden's offices and explained to one of his contacts there (the person
responsible for managing Kanden's contractors in Fukui) that my
project was supported by a legitimate publisher and would probably
find a wider audience in English than Kanden's activities typically
attracted. He requested that the company, for the sake of the world's
impression of nuclear restarts in Japan, not subject me to their characteristic aloofness toward the press. I was invited to speak with the
Mihama site's leadership team a few days later.

The star of the meeting was Okamoto Takuo, single-handedly
responsible for overhauling the aging Mihama 3 reactor. He didn't
like the contracting system, and said so. He drew a diagram of the
labor force at Kanden's plants on the whiteboard and showed where
critical issues of communication arose between Kanden employees,
contractors, subcontractors, sub-subcontractors, and so on. Some
of the measures taken at the operations level to mitigate these
issues were box-ticking exercises, or fancy names for pleading with
contractors to reduce the pace of their work (because there's no
such thing as a minor injury in the media-scrutinized setting of a
nuclear plant), but I couldn't dispute the gist of Okamoto's arg
ment: Kanden did their best within the existing labor environment.
I knew it was true that Kanden's contracting system was simpler than
TEPCO's, and I was certain the corporate culture of TEPCO did
not permit a comparable degree of candor about the consequences of
segregating management from labor. Before Fukushima, some Kanden employees have told me, Kanden's nuclear division was defined

by the sense that TEPCO had done too little to ensure contractors' work was performed safely, but Kanden had unjustly acquired, after a few media frenzies, a reputation as the cavalier exploiter of the contracting system.

I found that Kanden employees often described themselves in relation (or opposition) to TEPCO. Kanden's electricity was essential to heavy industry in western Japan; TEPCO served the populous residential and service hubs of Kanto. Kanden was nuclear dependent and ran pressurized water reactors exclusively; TEPCO had built its reactors in fits and starts, choosing boiling water reactors because they were cheaper and simpler to operate. TEPCO, Kanden's employees felt, was a political institution, too interested in its relationship with consumers, too eager to acquire generating capacity, too attached to the idea of being the nation's largest utility.

Kanden's willingness to wield history, witnessed by Tadao when Mihama's first reactor-critical shift was credited to the team who repeated it for the cameras, shaped the company's interpretation of the Fukushima disaster. TEPCO's safety culture had failed. Kanden would determine the fate of Japan's nuclear industry. From the perspective of the company's executives, this was overdue. They didn't run shoddy plants. They provided nuclear electricity where the nation needed it most. They would meet the new regulatory requirements and carry on.

My meeting with Okamoto occurred eleven months before Kanden announced they would seek a license extension for Mihama 3, to run it until 2036. Construction already occurring at the plant made it clear to anyone with a connection to the industry that an application to extend the license was forthcoming. Reporters and editors knew, but none published. Instead, they worked their sources for advance knowledge of the date when Kanden would make an announcement, so they could scoop their competitors by a day or two on a story that would barely differ from Kanden's subsequent press release. Kanden kept their decision unconfirmed because the politics of seeking the first license extension after Fukushima were complicated, and the longer they waited, the less aggressive their decision would seem.

As a rule, Japanese newspapers don't brawl with the government or major corporations in order to reveal official decisions before they're official, even if the decision has been withheld in an attempt to manipulate public opinion. In their routine treatment of Kanden's non-announcement, the press missed a better story—not the question of when Kanden would disclose their intentions, but the conspicuously immediate nature of a complicated financial decision.

A separate category of post-Fukushima regulations governs the work that must be completed before a reactor can apply for a license extension. The additional expense of overhauling older reactors is significant enough to disqualify many of these reactors from long-term profitability. Mihama 3, the only Mihama reactor eligible for restart, was no exception. Of the three reactors in Kanden's fleet that required costly upgrades to qualify for license extensions, Mihama 3 was the oldest, and its generating capacity was the smallest by 33 percent.

The planned restart of Mihama 3 was a fait accompli, irrespective of its financial implications, because the plant had established the company's claim on history. The town of Mihama (population 9,643) also hosted the headquarters of Kanden's nuclear division. The current municipal building was paid for by Kanden, which knocked down the old building and put its headquarters on the site. If Kanden made the cold, rational decision to decommission Mihama, the town's popular, long-serving mayor (a staunch Kanden ally) would have been horrified. Mayor Yamaguchi, who gives the best interview in Fukui, once told me that the tiny Polynesian nation of Tuvalu was depending on Mihama to do its part, by restarting Mihama 3, to reduce Japan's greenhouse gas emissions and save Tuvaluans from rising seas. Loyalty this grandiose is not purchased cheap or swiftly, and Kanden would not discard it, not in the town most convenient to Reinan's nuclear hub, but—at one-seventh the population of that hub—much easier to influence, through a far smaller tax burden.

When I raised the problem of Mihama 3's profitability to Kanden employees, they were familiar with it but took for granted the soundness of Kanden's motives: The plant's historic value and the company's relationship with its best client town were compelling reasons to seek an extension. I developed the impression that most people at Kanden felt it was the company's prerogative to make business decisions for cultural reasons, and were glad of the Mihama decision, which affirmed, in a dark time for the industry, a lofty version of its history and intentions.

Kanden employees who had worked on the active fault reviews

were especially superstitious about Mihama 3, which was more likely the location of an active fault (as defined by post-Fukushima regulations) than Tsuruga 2, but escaped an adverse finding. The initial disagreement among panelists at Ōi, where the evidence was forcefully in Kanden's favor, had distracted the media from the Mihama panel, which was hardly scrutinized. Some Kanden managers were aware that the Ōi panelists had privately disagreed with the Mihama findings. Were the panels switched, an active fault finding at Mihama would have been likely.

This company, improbably unscathed by the active fault review, believing itself a paragon of nuclear safety and economic vitality, aloof from the press and public opinion, convinced that Fukushima was not their affair except as a vindication of their superior investment in technology and expertise, was the company enjoined from operating their safest plant by the Fukui District Court in April 2015, because Judge Higuchi concluded that the plant, if run long enough, would produce a disaster to rival Fukushima. Like the contractor accidents that Kanden felt they didn't deserve because they treated their contractors better than most, like the overwhelming attention of activists and reporters to one panelist's premature pronouncements during the Ōi active fault review, Higuchi's decision struck Kanden as the minor, temporary annoyance unjustly occupying the center of their universe. Higuchi would complete his term, and his successor would overturn the injunction. Kanden was indignant to be left waiting and simultaneously indignant that Higuchi would be transferred before he could deny their appeal. A reversal in high court would have established a precedent, and other utilities could have used it to protect themselves (utility lawyers worried that one in three judges had been driven to ideology by the Fukushima disaster). In the district court, the reversal would count only in Fukui.

"The injunction occupied our thoughts, compromised our finances, and misrepresented our commitment to safety," recalled a retired Kanden executive. "It never tested our resolve. We believe in the soul of the company."

33.

Chairman Tanaka addressed the media the day after the injunction. "I haven't read the ruling carefully," he said. "It's full of errors and misapprehensions." He cited one example: "The whole world knows we have the strictest regulations, but this judge doesn't." The NRA would not reconsider its ground motion thresholds.

For the active fault panelists and the NRA's legal and seismology staff, who had worked for three years to establish the agency's safety culture and protect it from litigation, Tanaka's comments were bewildering. "You didn't read it, but it's stupid, and it's stupid because everyone knows we're wonderful?" was one panelist's summary. At internal meetings, Tanaka refused to acknowledge the right of the courts to participate in nuclear safety reform. The legal staff blamed themselves for his aloofness. Beginning in 2012, when Tanaka made his surprise appearance at the first Tsuruga meeting, they had insulated him from the consequences of his careless proclamations.

Key members of the legal staff left the NRA after Tanaka's response to Higuchi's injunction. Senior members of the seismology staff left, too, or transferred to research positions far from Tokyo. Among them was Morita Shin, the former IAEA official who returned to Japan to help establish the NRA, then found himself across the table from Genden's attorneys at the height of the Tsuruga conflict. His departure crippled the earthquake and tsunami division, where he was next in line to lead. The Ministry of Economy, Trade, and Industry immediately granted him a clandestine exemption from the NRA's "no return" rule, which prohibited METI from hiring senior NRA officials. The rule was the cornerstone of the NRA's safety culture, but Morita was willing to abrogate it, according to former colleagues, because the agency no longer commanded his loyalty; its mishandling of the Kanden litigation and the active fault review had forced Morita to engage in information warfare with the press and the utilities when his team should have been focusing on the implementation of post-Fukushima safety standards. He chose to suspend his civil service career, and rejoined the IAEA.

When I reached Judge Higuchi by phone, he declined to comment on the upheavals his ruling caused, except to say, "I hope the NRA will approach my ruling analytically." I asked if he was aware of quiet support for the injunction within the NRA's seismology division. "You're the first person to suggest that," he said.

When Higuchi ruled at Takahama, Shimazaki had been retired for seven months, not too soon for his reputation to enjoy a renaissance among staff disappointed by Tanaka's response. "Tanaka is an excellent scientist and a person of integrity, but he created more public misunderstandings than he avoided or repaired," said one conformity reviewer. "Shimazaki knew how to let science say difficult things."

I debriefed Shimazaki two years after he left the NRA. The remaining active fault panels, continuing their work under Commissioner Ishiwatari, were poised to issue active fault findings at Shika and Higashidori. The affected utilities didn't plan to demand—as Genden had—any formal reconsideration. They preferred to take their chances with conformity review, where new evidence would be reviewed by NRA staff, not academic scientists. Ishiwatari wasn't inclined to extend the panels' mandate. They had been working longer than initially anticipated, they operated outside the NRA's administrative authority, and their findings generated controversy.

"When I conceived the active fault review, I hoped its allegiance would be to science first and the NRA a distant second," Shimazaki said. "Both sides—the NRA and the utilities—seem glad to be rid of it. That suggests the panels were appropriately independent."

Hoshino, Genden's designated pugilist in the active fault dispute, had used words that meant "avuncular" or "grandfatherly" to describe his first impression of Shimazaki. I found Shimazaki boyish but felt our impressions were nonetheless the same. He was not a room-filling presence. He stayed out of the media after leaving the NRA, but when we met he had begun interpreting data from a recent earthquake in Kumamoto, which led him to discover that the NRA had underestimated ground motion produced by vertical faults. He alerted the NRA and submitted his findings to the high court considering Kanden's appeal of Higuchi's Ōi ruling. The NRA checked its Ōi calculations and came up with a value (644 gal) below the threshold that Kanden and the NRA had recently established (856). Shimazaki learned that the rechecked calculations had been manipulated to avoid embarrassing the agency, and lodged a formal protest. The NRA admitted its errors, but Chairman Tanaka declined to stipulate to Shimazaki's calculations, which gave results as high as 1,500 gal.

Center-left newspapers were alarmed that Tanaka had insisted for more than a month on the reliability of the NRA's calculations, only to reveal they were bunk. The month had passed while the NRA quietly took their calculations to the University of Tokyo, where experts couldn't reach a consensus, then to the government's industrial research institute, where experts also split, teaching the NRA an undergraduate lesson in seismology: There isn't enough earthquake data in the world (yet) to establish a reliable method for predicting ground motion.

In the press, Shimazaki's claims and Tanaka's responses placed the two in direct opposition, but Shimazaki refrained from criticizing Tanaka's leadership. I was reminded of Tanaka's verbal decommissioning of Tsuruga 2 at the first panel meeting, which Shimazaki never asked Tanaka to withdraw. When I asked why he hadn't, he insisted there was no particular reason, and insisted he could have, an answer that suggested he had weighed Tanaka's failings against the man, and kept the man.

I asked if he regretted giving Okada the same latitude—to disdain his colleague on the Ōi panel who initially believed the fault there was active, to criticize the Tsuruga findings, to reject the deterministic nature of the NRA's regulations.

"I never figured him out," Shimazaki said. "I asked around. The best explanation I got was, *He's older now. He has strong opinions.*"

"You hadn't considered the possibility he wouldn't go along? He never collaborated with the people in the Active Fault Society who were interested in public policy."

"You can say the schism that developed during the active fault review was the result of existing differences, but that misrepresents the Active Fault Society. We liked each other and worked together. It never occurred to me that Okada would object to the nature of the process. Some scientists declined to be nominated. I assumed anyone who agreed to participate thought the system was well conceived."

These differences of opinion between the panelists, I pointed out, had led the utilities to suspect Shimazaki of manipulating panel assignments to increase the likelihood of active fault findings. Shimazaki confirmed that wherever possible, he had deliberately given two panel assignments to each of the panelists who had a history of criticizing the nuclear industry. If a panelist had criticized the safety of a specific site, he tried to assign them to it, provided this did not violate criteria related to the distribution of expertise, the panelists' logistical constraints, and potential conflicts of interest—especially if panelists had been recipients of research funding from the electricity industry.

I was confused about Okada's assignment to Ōi, I said. Was it necessary to assign him to one of the panels that commenced immediately (Ōi and Tsuruga) as an acknowledgment of his seniority?

Shimazaki said that had been a consideration.

I had seen a bibliography that showed Okada, a long time ago, conducted research at the Ōi site. Someone told me, I said, that Kanden had contributed funding to the project.

Shimazaki was quiet. Moving only his eyes, he examined my notes. "I missed it," he said, and made a sign that I should put down my pen. "At the beginning, too many things needed doing. Later I saw the paper you're referring to. If I had known, I wouldn't have assigned him to Ōi."

I had assumed that other considerations had overridden the question of Okada's old research. Revisiting my assumption, I found it rested on the criticism Okada had heaped on Shimazaki and the active fault review. Had someone been so unkind to me, I might have retaliated. If Shimazaki had disclosed Okada's conflict of interest, the press would have pursued it, and Okada's criticisms of the active fault review would have lost their credibility.

"You kept this to yourself," I said. "And let Okada trash the system."

Shimazaki didn't like my logic. His expression showed it. "At Ōi,

Okada was right. He took control of the process when the media was trying to run with an unfinished story. I can't say, about someone the panel needed, *He shouldn't have been there.*"

Okada's accidental assignment to Ōi made it five out of six—the number of active fault panels where the results fit one or more of the following categories: active fault, findings inconsistent with regulations, findings dependent on administrative idiosyncrasies or errors. If Okada had gone to Tsuruga instead, he would not have endorsed the panel's findings. That was his position when we met. Under Shimazaki's criteria for panel assignments, the experts who might have gone to Ōi included Suzuki Yasuhiro. Suzuki was combative toward the electricity companies and frequently collaborated with Ōi panelist Watanabe Mitsuhisa, who believed the F-6 fault was active.

"I wasn't ready for Okada to disdain the process so deeply," Shimazaki added, "but there's nothing negative to say about him, except I don't know why he agreed to take part."

"Because of you," I said. "Because you asked, and you are who you are." Shimazaki flinched. I might've missed it, but the table's legs weren't even and our water glasses shook. "I asked him," I said. "We only met once. He went out of his way to do it. He put himself on the record saying he would have done most things the way you did, including the exclusion of experts who previously worked for the regulator. He said the criticisms from the excluded experts went too far, and he didn't want those people to admire him."

"They're my colleagues, even Genden, and I heard their concerns," Shimazaki said, and nodded at the truth of it: He had heeded his critics. "But the sum of their counteroffensive exceeded me." The public disclosure requests, the legal threats, the steady pressure Genden applied to any discrepancy between the panel's stance and the opinions of the peer reviewers—each imposed a burden of time on the overwhelmed seismology staff. "The work was too important for caveats. Then you learn you have limits," he said. He raised a hand to show the limits were high, but you reach them. The gesture reminded me of his physical demonstration of the K fault's potential displacement, which Genden had found unbecoming of a scientist. "People who understand each other are supposed to agree," he said. "Mutual understanding also makes disagreements more violent. I didn't expect the utilities to applaud an active fault finding, but I told myself a conflict wouldn't change what everyone felt after Fukushima. I believed I had consensus." His expression paused on this thought. His hands were restless on the table, trying to reduce consensus to a gesture. He said, "I wasn't wrong. You can know everybody's intentions, and what they do will still confuse you."

I last interviewed Shimazaki in 2017. The active fault review was finished. He had resumed spending time with his colleagues in other contexts. He didn't reiterate his regrets, and tried to retract some. Any question that went beyond confirming my notes ended in a memory he had lost or changed. I suggested that his long absence from the media after he left the NRA demonstrated the ambivalence I remembered from our first meeting, and he told me he never would have left the public discussion if his files hadn't been shuffled when they were moved from his office to his home. He also said his wife was increasingly dependent on his care. He said it in a way that suggested the active fault review had occupied him during her last good years—but how else could you say it? In eight years of retirement, he had spent only one at home.

The old disappointments weren't entirely effaced. I mentioned the active fault component of conformity review at Tomari Nuclear Power Plant. The utility had submitted tephra evidence, as Genden had when the Tsuruga panel reopened its investigation. The NRA's new chairman, Fuketa Toyoshi, responded by demanding additional research. Five years earlier, Fuketa had sat next to Shimazaki on the bus after touring the ruined Fukushima reactors. Shimazaki had conceived the active fault review on that ride, while he described to Fuketa the destroyed elementary school he previously visited, where seventy-four students died in the tsunami because flawed emergency procedures prohibited them from fleeing. I said maybe Fuketa's vigilance at Tomari proved it had all come right: The standards the Tsuruga panel established had been adopted by the conformity review.

"It's one sample," he said.

"One sample of what?"

"They found the tephra by sampling the uppermost area of a single outcrop, I believe."

I was stunned. Claiming that a tephra sampled this way could establish the date of a fault's most recent displacement was like saying that your house must have been constructed in 1952 because a penny you found in your couch was minted that year. It made the strong suggestion that the utility felt the restart process was a box-ticking exercise; you couldn't omit required materials, but the reviewers wouldn't—or couldn't—analyze what you sent.

A submission this cynical vindicated the active fault review. The utility wouldn't have dared to send it to a panel of professors. Fuketa had been assisted in identifying the deficiency by Ono Yugo, a professor emeritus of geography and frequent critic of the nuclear industry. Ono had become suspicious when the utility claimed the fault in question hadn't moved for 200,000 years, a dubious number in local context. He presented his concerns to the Active Fault Society, which brought them to Fuketa's attention, prompting Fuketa to scrutinize the utility's restart application.

"Ono Yugo is Yoko Ono's cousin," Shimazaki said. "Not that it matters." But it seemed to; it betokened the absurdity of one anti-nuclear scientist putting his thumb in the dike on the NRA's behalf. The NRA's staffing problems were worse every year, and Fuketa was supposedly the chairman who would fix that.

I couldn't move Shimazaki off script again. I left feeling lucky I had debriefed him a year earlier, when more was on his mind: his supposed motives circulating in the press (none accurate), the opportunity to imbue the NRA's work with the benefits of scholarly inquiry, the schisms and stalemates that made this opportunity distant and insubstantial before its time. He had insisted he wasn't wrong to believe everyone felt the same after Fukushima. "I would have let them restart Tsuruga," he told me, meaning to prove his side of it. "Tsuruga 2. Not 3. Not 4." He knew this statement would please no one, but didn't soften or complicate it. I asked about the Urasoko fault. The other panelists had told me the K fault was borderline compliant, and they nudged it toward noncompliance because the Urasoko fault was two hundred meters away. They talked about the hundred-year fear—that Genden would use the Tsuruga 2 approval to justify licenses for Tsuruga 3 and 4, and a nuclear reactor would operate near one of Japan's longest faults into the twenty-second century.

Shimazaki said the risk of displacement under Tsuruga 2 in the reactor's remaining years, especially if Genden declined to seek a license extension, was small. A chastened approach from Genden, in acknowledgment of their past misrepresentation of the Urasoko fault, could have convinced him they were committed to operating the site briefly, to stay solvent. He might have reciprocated Genden's good faith by applying the regulations narrowly. I knew the panel would have followed him. Genden thought Fujimoto, the panel's geologist, was an ideologue, and bruised him in the press, but Fujimoto once told me he would have considered voting with Shimazaki and Tsutsumi in Genden's favor if he felt Genden was prepared to retire the site.

During my final interview with Shimazaki in 2017, I related part of a recent interview with Hoshino, who was admired by people in the industry for doing more than any other nuclear executive to undermine the active fault review. "I told Hoshino you gave two panels to the experts who didn't like the industry," I said. "I told him you assigned people to sites they had already called unsafe."

"And Genden didn't issue a press release?"

"I told him none of this was any secret."

"He didn't believe you." Shimazaki didn't know what Hoshino believed; he didn't look as if he knew.

"He said he would have done the same. He said there was only one way to feel after Fukushima, and you did what you felt."

Shimazaki passed a moment in thought. His hands were steepled over his mouth—a gesture for consensus, if unintentional. "I was trying for different and hoping for better," he said. "Better is beyond one person." He glanced around the loud, crowded place where we had sat to talk. Observing him while he observed other people, it occurred to me that I hadn't known Shimazaki in the part of his life that ended when he met Hoshino, or the brief part of his life when he knew the active fault review would exist without knowing what would happen to it. "It was a great two years," he said, as much to the room as to me, in a voice too soft for the noise.

34.

In 2015, I observed a meeting between a regulator and a Kanden executive. As I was leaving, I overheard the regulator say (referring to me), "He can't write about Genden. They've got no future." Both laughed. "Don't pick on poor Genny," the Kanden executive said. He mimicked the tone of a schoolmarm and attached the diminutive used for little girls.

I told this story to a room of Genden executives. "I don't know if you were right and the NRA was wrong," I concluded. "But I'd be pleased if my book made levity difficult." Hoshino was in that meeting. Afterward, he made it a point to shake my hand and show he spoke English. When Genden was incredulous about the time and access required to vet the information their people passed me, he instructed his colleagues—occasionally his superiors—to let me work. He embraced the risks associated with my project because the events I was investigating would define his career.

Hoshino had a son whom Chubu Electric had tried to recruit, not for his grades or his second-generation engineering pedigree, but because he rowed competitively and the company had a team. He declined because the job was a plum, intended to enable him to spend more time rowing than working, and everyone in the industry would know it. During our final debrief, I told Hoshino his son's decision reminded me of his swimming anecdote from high school, when he repressed his fear of the ocean in order to avoid wearing the cap given to weak swimmers. "You don't always have to win," Hoshino said. "You have to demand an outcome that tells the truth about you." His son was preparing to begin graduate study at Nagoya

University, where Suzuki Yasuhiro, the Tsuruga panelist least sympathetic to Genden's research, was a professor. "My son would recognize Suzuki, but Suzuki wouldn't recognize my son," Hoshino said. "I wish he would. He hoped Genden would collapse. Let him think of me."

Tsuruga 2 was in the preliminary stages of conformity review, where Tanaka once said it couldn't go. "We still don't know what 'important reference material' means, about the active fault finding," Hoshino said. "But we're getting our research into the meetings. It feels functional."

Of the two Genden reactors (Tsuruga 2 and Tokai 2) where conformity reviews were forthcoming, Tsuruga was drawing fewer manhours. Tokai was less lucrative, and wasn't on the restart list (too close to Tokyo, too sensitive to announce), but would require a license extension, which imposed imminent deadlines. Politicians in Fukui would have been scandalized if they learned that their support during the active fault review hadn't prevented the diversion of resources to Tokai, where the present mayor was antinuclear. Fortunately, they presumed the NRA was responsible for the pace of the review; Hoshino had declined to correct this presumption.

"If I'm remembered as the first Japanese nuclear executive to antagonize the regulator, I can accept that," he said. "When regulation is driven by ideology, corporate belligerence is important." He and I had shared a social environment and its psychological effects during the active fault review. His new confidence suggested he had outgrown that period, outgrown his regret that either side could triumph or fail, but the NRA would not soon become an agency worthy of its post-disaster mandate.

"What if you obtained a favorable finding during the active fault review?" I said. "The panel determines K isn't active, but you know they neglected to review your submissions. They rubber-stamped you."

"Our data, I wouldn't mind. We don't submit unreliable data." The Japanese regulatory system was driven by utility data, he pointed out. "If the regulator takes shortcuts, the result should favor the utility. Nuclear is essential to Japan's energy policy. The government has a responsibility to err on the side of promoting growth." In this version, the establishment of the NRA was a symbolic shuffling of institutional relationships. Regulators no longer worked for the part of the government responsible for growth, but regulated under the continuing presumption that their work served an economic agenda.

"You would accept a favorable finding, knowing the NRA had been negligent, without ambivalence?"

"I do hope the NRA stops making mistakes."

"Some ambivalence."

"I care about process. We were victims of the process."

We were in a room on the floor where Hoshino worked, separated from the open work space (Japanese offices look like American newsrooms) by privacy glass and walls that muffled words but transmitted sound. The shapes through the glass repeated: white on top, black on bottom, the pattern of professional attire.

"What if Shimazaki had let Tsuruga 2 run in exchange for a promise that you wouldn't try to build Tsuruga 3 and 4?"

"That's difficult to imagine. Shimazaki expended so much energy enforcing the deterministic nature of the fault review: It's either active or it isn't. If you're comparing a few years of Tsuruga 2 to a few dozen years of new reactors, you're considering the periodicity of past fault movements. That's probabilistic. Shimazaki avoided probability."

"If it happened, would you push to operate for another hundred years as hard as you've pushed to operate for twenty?"

"Prime Minister Abe is in Paris for a climate conference," Hoshino said. "He can't meet emissions targets without nuclear restarts. Running one reactor for a hundred years is the same, from a probability perspective, as running a hundred reactors for one year. China might build a hundred reactors. We're running safer plants than China." A black-and-white shape passed the privacy glass; if I worked for Hoshino, I thought, I would recognize his voice through the wall. The familiar rhythm of his speech might cause me to mistake these political statements for the shrewd observations he often made about Japan's regulatory culture, including his admission, which seemed distant now, that he would have constituted the active fault panels precisely as Shimazaki had, assigning critics of the industry to sites they considered dangerous.

"Tsuruga 3 and 4 would be farther from the Urasoko fault than Tsuruga 2," he continued. "They're more technologically robust. If Tsuruga 2 is approved for restart, I would use that precedent to advocate for 3 and 4." These were the intentions Shimazaki had ascribed to Genden. The peer reviewers had confirmed the Tsuruga panel's findings because they feared Shimazaki was right. "If you can't finish an APWR project that's halfway done," Hoshino concluded, "you're saying there shouldn't be new reactors. Japan can't afford that statement."

"Would the Hoshino of four years ago make the same arguments?" I asked.

Understanding the question before I finished, he overlapped it to say, "He would." He added another syllable, then hesitated. Earlier in the conversation, we had revisited his recollection of 3/11. Hoshino had stayed at the office and skipped sleep, spending the night monitoring Tokai, where he worked until 2010. The plant lost grid power and one of three generators when the tsunami struck. Climbing the stairs between the floors of Genden's office building to keep his superiors informed, Hoshino was immediately exhausted, and increasingly stupefied by the plant's good fortune. A new sea wall had been completed at the site two days earlier. It was built at the behest of the prefectural government and the plant's ground-level employees; when Genden drew up the plans, the company's executives were mocked by the other utilities for their costly allegiance to a democratic style of management. Had the project been scaled down or suffered delays, Hoshino realized, the plant would have been defenseless against the tsunami.

When the plant was stable the following afternoon, he walked ten kilometers beside the train tracks before finding a station where the trains were not too full to board. His leather shoes were the object of this memory—the discomfort of walking, the handicap they would have imparted during an emergency at Tokai. I recognized one of Hoshino's leitmotifs: uniforms and appropriate dress. The red swim cap. Iriya's wetted hair and natty clothes, which the press gossiped about and Hoshino was offended by their shallowness. The leather shoes that incorrectly implied he had never worked in a nuclear plant.

Describing his memory of meeting Shimazaki, the man's retiring presence, the calm in it, the frank generosity of his expertise, Hoshino had chosen the moment before they toured the trench together, when Shimazaki realized he left his jacket on the bus. In the high wind, in the cold, intermittent rain of a Japan Sea winter, Shimazaki wrapped a towel around his neck and refused to wear a Genden jacket. Hoshino received nothing from Shimazaki during this tour except an impression, but it had been enough for Hoshino to recall that he believed Shimazaki would remedy the failures of his predecessors.

"I would have made the same arguments four years ago," Hoshino said, letting his preemptive response to my question stand repeated. "If it was necessary."

I knew a regulator who wore sneakers with his business clothes for the reasons that occurred to Hoshino on 3/11. Hoshino kept wearing leather shoes to match the way he filled a suit. Walking with him to the elevator after our debrief, I could picture him as he pictured himself: sprinting in a disaster-stricken nuclear plant, the suit and shoes no hindrance, while men in practical clothes struggled to know where they were needed. The Dumbo puppet the staff had adopted as their mascot during the active fault review was still hanging near the elevator, and I said, "He doesn't fly until the plant runs?" Hoshino didn't know the movie, and I explained about the magic feather, the trick of perception that lets you believe in unlikely achievements. "If you can fly," he said, "you're not an elephant." He asked if I had kept abreast of Shimazaki's expert testimony in litigation filed by anti-nuclear activists. I said Shimazaki's testimony seemed to follow the issues, not the litigants, but Hoshino didn't think there was much difference. "Eventually," he said, "you figure out what you are."

35.

I had a favorite bar in common with Tomo Genshiryoku, the nuclear safety technician who helped to build Monju and demanded a transfer when he saw how the plant would be managed. The bar was a multigenerational family business in Tsuruga's old downtown, inherited by the present owner from his late father. When the owner had been a boy, the house across the street belonged to the family of his closest childhood friend, future mayor and nuclear stalwart Kawase Kazuharu. The bar owner, an outspoken hippie whose politics had never affected his friendship with Kawase, married the city councillor who became Kawase's left-wing antagonist. Before the disaster, the bar was neutral territory for industry employees and antinuclear activists, many of whom—if they were Reinan natives—liked each other more than they liked their allies in Tokyo. After Fukushima more journalists visited, and fewer employees from the plants. The journalists asked for the owner's wife, and the owner surrendered his role as the city's conduit of industry gossip.

Tomo and I visited the bar when the active fault review concluded, to celebrate having nothing sensitive to discuss. Talking with Tomo, whose stories from Monju were part of the city's nuclear folklore, put the owner in a nostalgic mood, and except for a courtesy visit from a tabloid reporter no other customers appeared.

Since Tomo had known it, the bar never changed. The shelves held clay liquor vessels from previous generations of the business, and the wall plaster was colored by decades of cooking steam. "Two daughters with my wife now," Tomo said when the reporter had left. The owner recalled that Tomo and his wife had brought their older daughter to the bar in swaddling clothes, during the first week of the nationwide nuclear shutdown. The other patrons reacted as if they hadn't known women were still giving birth. "Fear, joy, resentment, awe," the owner said. "You could tell what kind of person each of them was by the way they looked at that baby."

Tomo said his adult son recently had a diabetes diagnosis and hospitalization. "He was in social withdrawal for a few years, but after the disaster he couldn't tolerate loneliness. He works part-time in a day care. He's twenty-eight, living with us since eighteen. College in Nagoya, withdrew, went back, withdrew. That's his life. I like him doing what he does. I don't ask myself what's enough."

"I met him," the owner said. "The image I have is an image of the two of you together." The owner was in his mid-sixties. Maybe because his adulthood occurred in the home of his youth, his voice—his way of speaking—belonged to a younger man. It was a voice, Tomo once remarked, that remembered who you were before, whether or not you wanted that.

"When I started, I liked work," Tomo said. "Monju, I stopped liking it. Tsuruga 2, I learned to like it, taken with other things. Then Fukushima. We're assigned to help Tsuruga meet new rules. But we're fighting the people making the rules. Never mind Tsuruga—do Tokai, quietly. Now we're giving Ibaraki Prefecture the rationale to disseminate when Tokai joins the restart list. The government's rationale should be its own. There's law to that effect."

"Your man stopped by," the owner said, and mentioned the name of a Genden manager active in local politics. "He pumps the journalists. He recognized the wire service reporter. It was like watching a bug crawl someone before deciding where to nest."

The conversation went on about people they didn't like. Self-interested nuclear types and reporters. People who wanted assurances that nuclear power was always safe. People willing to exploit that desire. They talked about the new train station, built in anticipation of the bullet train. Downtown had the boarded-up look of the depopulating countryside. The only money Tsuruga could attract was already captive to its industries. They agreed the bullet train fit the same stale category of bubble-era opportunism—build it, then hold it hostage.

"Have you danced with your wife recently?" Tomo interjected. He was sentimental when he drank.

"She's busy," the owner said. He put a stopper in a liquor jar, tamping it with the heel of his fist. It had been open all night and now bothered him. "The journalists don't help." He cleared my plate, which I hadn't finished. "You call ahead before you bring sources here," he said. "You think I'm antinuclear." Before I could explain that my calls were a mark of trust, he elaborated his contempt for the press, whose patronage led the city government to keep a list of civil servants seen visiting his bar.

Tomo was impassive during this exchange, not turning from it, not involved. When it finished, he resumed the subject of family: His older daughter had enrolled in gymnastics and discovered a knack for the balance beam. He told the owner they were both lucky in marriage, and the owner relinquished his resentment, which I had never known him to attach to his wife. We left in a few minutes, and Tomo lingered in the snowfall outside, watching the air in the bar turn to vapor where it escaped through a window frame. "He wanted his father's life," he said. "He got his father's job."

While we looked for a taxi, Tomo took a call from his wife. She had a fever and earlier that night found their younger daughter teething on a loose battery. The call sobered Tomo. He stopped in the street to confess he'd been skipping work. Not absenteeism, simply claiming the time he could. "Enough I've been warned," he said. He looked at the trail of our footprints in the snow, down the center of a street that was busy all night when the reactors surrounding the city were under construction. "I told them: People warned you for decades, and look what that accomplished."

My friend Makoto called the next morning. He owned a tailoring shop in Tsuruga and did most of his work altering the uniforms of plant employees. Before Vietnam unilaterally withdrew from a nuclear development pact with Japan, he had visited Ho Chi Minh City with a friend who manufactured radiation-proof doors, sniffing for contracts. He was also a proud son of Obama, Reinan's second city, where residents had voted out the nuclear industry. "I'm the bubble, not the tide," he explained. "My choice? No nukes."

A shoe store had caught fire overnight, and two blocks of downtown were evacuated. The Dream Factory (Makoto's shop) smelled like charred rubber. Did I still own a car and want a driving tour of Reinan?

Makoto was in his late fifties, still affecting a thick, neat mustache from his early-adulthood years in Berkeley. His hair wasn't all gray, and he kept it long to complete the appearance of a lapsed bohemian. He had gone to California when being Japanese and being poor were the same, and once asked me to locate people whose houses he had lived in. One had owned a salon, Hair by Grace (a pun, she confirmed). When I told her I was calling on Makoto's behalf, she said, "He would have liked to be an Eagle Scout. An overactive boy." Another family were poor white from Oregon and, like the poor white branch of my family tree in the Pacific Northwest, had been integrated by the soil. No listed number, nothing in public records.

We ran his errands. Letters of complaint needed delivering to judicial scriveners and city bureaucrats. He had demolished the house in Obama his parents left him, but couldn't sell the plot until the owner of the neighboring house joined him at city hall to verify the property line. The neighbors were dead, and their son lived in Los Angeles. The property line could be confirmed remotely, but no one seemed to know how. Most houses in the neighborhood were empty or inhabited by the elderly, but remained deceptively handsome, maintained by the unsolicited altruism of a handful of younger residents.

His concern today was the tree. A crape myrtle grew on the property, near the unverified boundary. It would need to be transplanted before spring. Makoto had managed the demolition so the tree survived the process and stood now in its nude winter condition, chest-high, on a plot of bare soil surrounded by smashed concrete and wood scrap. Until the house was gone, Makoto hadn't realized it sheltered the tree from the wind and salt of Wakasa Bay. He watched from the car while the wind flailed the tree, then walked onto his property and mothered it with his eyes, letting the wind break across his back and inflate his untucked shirt. "Is it fragile, or is it flexible?" he asked when he came back. "For trees, is there any difference?"

We didn't stay long at the Wakasa Museum. It put Makoto in a low mood after we each paid 300 yen ($2.70) to get in. It was 100 before. ("We've remodeled," the manager explained, plainly surprised to find himself justifying the price in the presence of a visitor from another hemisphere.) The permanent exhibition was called *A Culture of Comers and Goers,* about the Horse Mackerel Highway in Obama, the Trans-Siberian Railway in Tsuruga, and the dominance of Reinan's ports during the seafaring era. The new objects of this way station culture (waste incinerators and nuclear plants) were omitted.

The Horse Mackerel Highway Museum, on the *shotengai* where the eighth-century trade route began, was a room where a TV played on a loop. A mackerel had been mounted to the wall, and the rest of the space was filled with reproductions of primitive fishing tackle. Makoto browsed the museum's tourist brochures and seemed disappointed to discover they were current. "If you visited this place, would you remember it?" he said. "That's a koan. Just now, this museum inspired it."

One of his childhood friends owned a nearby grocery shop in its fourth generation on the *shotengai.* When we stopped to visit, the storefront's steel shutters were halfway drawn. The door in the middle was open and the lights were on inside. Beneath an incense burner laden with ash, a television was playing a game show. Makoto called his friend's name and moved farther into the shop, crossing the threshold into the attached apartment, the kitchen, the living area. "They went out," he said. "Or evaporated."

At Myotsu-ji, a temple of international renown, the abbot was leaving when we arrived. I knew Nakajima Tetsuen from his role in the coalition of antinuclear activists, labor unions, and fishing cooperatives that forced two successive mayors to rescind their acceptance of national funding for nuclear reactors in Obama. Makoto knew Tetsuen's brother, Kaoru, who acted as the abbot at Myotsu-ji (under his Buddhist name, Kun-yu) whenever Tetsuen was busy organizing. When they were children, Makoto and Kaoru walked home from school together along the banks of the river that flowed past the temple. The oldest building on the grounds was a designated national treasure, built in A.D. 1258. "We played tag in there," Makoto said. "We'd have smashed a sliding wall and not known it mattered. I wonder how we never did." The trees around the temple were old-growth cedar, never logged, never firebombed, their trunks broad enough to inspire Makoto and Kaoru to daydream about building a dwelling inside one. They would sit in the copse and watch the hawks dive the carp pond, usually without extracting a fish because the carp were too heavy to lift. "When I saw new-growth forests for the first time in suburban Japan, and hungry carp in urban ponds, I understood that living things change their size to fit the world."

It was dark when we returned to Tsuruga. The windows were lit at the factory producing radiation-proof doors, visible from the expressway. Japan's general contractors were stockpiling the doors, Makoto said. "Good Fukushima money."

A year earlier, I met the Tsuruga Chamber of Commerce representative who managed nuclear initiatives. When the plants offered work, he made sure the greatest possible proportion was awarded to local firms, a responsibility at the center of a sensitive political question in Fukui: Why were the majority of nuclear laborers imported from other prefectures? "A lot of local firms that want the work aren't qualified," he said. "If they are qualified, they'll look outside the prefecture for subcontractors who'll take it cheaper, so they can load up contracts without removing their crews from other jobs." Using veteran crews on the short shifts necessitated by radiation rules was a waste of skill and money.

After Fukushima, the chamber of commerce was made responsible for maintaining the readiness of the local nuclear workforce, and they cheated the statistics to create the impression that contractors in Tsuruga were doing nuclear work. If a job went to a Tsuruga firm and it involved products or skills that could be used at a future date on a nuclear contract, the chamber counted it. Leaving out the us-versus-them bias, he said, Tsuruga needed the outside laborers more than the nuclear contracts. There was only one modern, commercial reactor inside the city, but hundreds of restaurants and a glut of hotels. And the good jobs at the utilities were never closed to locals. "If you graduate high school here and you look for an industry job, your likelihood of employment is nearly 100 percent. Maybe 30 percent of graduates stay in Tsuruga. Not all of them want industry work, and among those who do, many enter the industry and discover they would rather do something else."

Fukui was in the process of branding itself a hub prefecture for nuclear expertise, hoping local firms would be hired for decommissioning projects after Fukushima. As we parted, the chamber representative described a trip he took to a decommissioning center in Tokai. His Tokai counterparts sat him on a bench and told him it was made of metal from a decommissioned plant. "I thought: These people have no idea how the industry is seen. No one wants to sit on a piece of a nuclear plant." The chamber of commerce had since determined that decommissioning contracts were barely profitable compared with selling electricity.

Another time, I was riding the train into Fukui, holding a sign that said *OBAMA,* hoping someone would offer a ride. Across the aisle were two dark-suited men preparing a presentation on behalf of a company that sells components to nuclear plants. One said to the other, "Where's Obama?" "Fukui." "You don't say." Working their territory, it wasn't possible to get between Tsuruga or Mihama on one side and Ōi or Takahama on the other without passing through Obama. Eight out of eleven Kanden reactors were closer to Obama than any other city in the prefecture.

Money, my acquaintance in the chamber of commerce had learned, lays a narrow trail. The suited men were on it.

In November 2015, eight months after the Tsuruga panel issued its final report, the NRA scheduled meetings to begin reviewing Tsuruga 2 for restart. Iriya had two weeks to prepare. Of the eighty pages of filings he assembled, roughly forty addressed faulting hazards. Other utilities under review had submitted, on average, twenty to thirty pages total.

The NRA asked Genden to trim their submission. How to trim was unclear. The active fault panelists had been capable of contextualizing Genden's data. The regulators who would conduct the conformity review were drawn from backgrounds typical of the agency: civil service, engineering, and occasionally the electricity industry. None were seismologists or geoscientists, and few were veterans of seismic investigations. They needed Genden's eighty-page crash course.

Early in the initial meeting, NRA staff said they hadn't read the active fault report. This information was communicated cheerfully, Iriya recalled. The NRA apparently thought Genden would be glad to see a hated obstacle removed. "I often hid my disappointment with the NRA," Iriya reflected. "But that was a novel instance—hiding it in response to collegiality." He had long blamed the dysfunction of the active fault review on the NRA's inability to digest complicated data. The staff's indifference to the panel's report suggested the conformity review would display similar shortcomings.

He discarded the items in his agenda that referenced the report, and presented Genden's compliance strategy in the narrow context of the regulations. Many safety standards had been inherited from the NSC and NISA, he explained, and Genden would submit their evidence in the expectation that substantive changes, except where specified by the appropriate NRA departments, were not anticipated. The meeting soured. "They made it vehemently clear they wouldn't tolerate being associated with the NSC or NISA," Iriya recalled. Many of the NRA staff present—and most of the NRA—were former employees of these agencies. Iriya remembered Okumura's critique of Shimazaki's decision to exclude NISA and NSC experts from the active fault panels: The idea of making a clean break from the NRA's predecessors was a dangerous fiction.

Before Iriya went back to Genden, a senior NRA official pulled him aside to express surprise that he had brought "all different people this time." Iriya's team was the same. His NRA counterpart had forgotten every face. "I realized the reviewers were too oversubscribed to absorb anything," Iriya reflected. "Not the findings of the active fault panel, not the color of the walls in the conference room." He wondered if this was the inevitable result of Japan's decision to force every reactor into simultaneous shutdown after Fukushima: a precaution without precedent, leading to an insurmountable burden on regulators.

Hokuriku Electric and Tohoku Electric, it occurred to Iriya, would soon receive active fault findings. They would decline to object, in favor of moving directly to conformity review. Would NRA staff discard the Higashidori and Shika reports? Each of those panels included experts who had proven their objectivity by supporting Genden's objections at Tsuruga. Neither panel had been conducted under the judgment-distorting pall of a nearby threat on the scale of the Urasoko fault. As he considered the possibility that the limited competence of the NRA staff would return responsibility for seismic safety to the utilities, Iriya began to feel implicated and concerned. "I had learned from fighting the Tsuruga panel how little I previously understood about geoscience," he said.

Fujimoto Koichiro, the Tsuruga panel's geologist, was a member of the panel poised to issue an active fault finding at Shika 2, the newest reactor under review. None of the other Shika panelists were among the scientists whose motives the utilities impugned, and the panel included two structural geologists, perhaps the two whose research was best suited to paleoseismic investigations. Hokuriku Electric asked independent geoscientists—including Okumura Koji—to review the panel's findings, but their criticisms failed to generate controversy. Aware the NRA staff had omitted the Tsuruga findings from conformity review preparations, the Shika panelists structured their report to mandate continuous follow-up. Finally, they returned to their academic careers, and the active fault review, formerly the critical component of Japan's nuclear reform agenda, entered history.

I met with Fujimoto after the NRA ratified the Shika findings. He said the Geological Society of Japan had recovered from the contentiousness of the panels but the Active Fault Society hadn't. A severe earthquake had recently struck Kumamoto, and experts from both sides of the Tsuruga conflict were visiting the area, gathering data to vindicate themselves. Fujimoto wouldn't participate. His health had suffered during the active fault review, his academic research had stalled, and his reputation had been damaged by industry innuendo. "It could have been different if Tanaka hadn't chosen the least appropriate moment to declare Tsuruga 2 ineligible for restart," he said. Genden's research had elevated the standards of industry transparency. Had it been conducted in a collaborative spirit, it might have become a model for future seismic safety reviews, not a model of corporate belligerence. "Iriya was almost a scientist," he said, embroidering his esteem of Iriya with his contempt for the industry's expedient faith in its engineers. "And the NRA was almost the agency everyone wanted after Fukushima."

The judge who replaced Higuchi Hideaki on the Fukui District Court reversed the Takahama injunction on Christmas Eve 2015, eight months after it was issued. Commenting on the ruling, the governor of Kyoto Prefecture pointed out that Kyoto residents constituted the largest demographic in the plant's emergency evacuation area, but Fukui was the only prefecture permitted to participate in the restart process. The nonpartisan Union of Kansai Governments published a letter demanding to know how governments in Kyoto's position would be accommodated. No interjurisdictional arrangements existed to facilitate evacuations, and real-time nuclear accident data was available only to prefectures that hosted nuclear plants.

Takahama 3 restarted on January 29, returning Kanden to the electricity-selling business for the first time under post-Fukushima rules. Takahama 4 restarted on February 26. Three days later a transformer failure caused an automatic shutdown at Takahama 4, an unremarkable setback for a reactor idled several years, but severe enough to attract media scrutiny and prompt several days of repairs.

Takahama 3 was running and 4 was offline on March 9, when the Otsu District Court (in Shiga Prefecture) issued a new injunction, suspending the operation of both reactors indefinitely. When Higuchi ruled in Fukui, Takahama 3 and 4 weren't operating. By enjoining the plant while Takahama 3 was operational, Yamamoto Yoshihiko became the first judge in Japanese history to order the immediate shutdown of a nuclear reactor. He also became the first judge to acknowledge the legal standing of plaintiffs outside a nuclear plant's host prefecture, laying aside forty years of unwritten convention, previously upheld by the courts. Higuchi's decision had been significant for its attention to the NRA's administrative deficiencies. Yamamoto's acknowledged the gravity of the dilemma that had been pressed on prefectures excluded from the restart process, but subject to evacuation in the event of an accident.

According to Kanden, Judge Yamamoto's ruling would encourage a wave of injunction requests, and the consequences would be dire. The Otsu court had enjoined the larger of two NRA-approved nuclear plants in the country, reversing the industry's economic momentum less than a month before planned deregulation of Japan's electricity market. Utilities owning nuclear plants would be placed at a competitive disadvantage, and prices—perhaps even the grid—would be subject to historic volatility. The effect of the ruling would be industry-wide, Kanden argued, because theirs weren't the only plants where evacuation zones had expanded across prefectural borders and judicial jurisdictions. They cited media reports of a "Fukushima factor" in the judiciary, a notion helpful to the industry's contention that the rulings were expressions of conscience, not law.

As deregulation approached and Kanden's despairing predictions continued, investment firms issued advice. Hardly any advised giving up Kanden positions. The low price of fossil fuels made Kanden's importation activities lucrative, they pointed out, and deregulation would allow large utilities to consolidate their share of the electricity market. From two sides of the same mouth, pro-business groups decried the calamity of the Otsu ruling and congratulated electricity companies for their prospects in a supply-side system. "The real loser in court was the NRA, again convicted of failing to protect the public," a member of the Fukui governor's staff told me. "Some people—people who like to say ideology is a bad habit—were probably glad about that, for ideological reasons."

36.

I made it a priority to overhear the innuendo that circulated whenever Kanden's leadership visited the Fukui Prefectural Assembly, presuming while they petitioned their allies that the corridors of the building were consecrated to discretion. Shortly before their defeat in Otsu District Court, they sought help deflecting the concerns of an NRA safety reviewer who had been reassigned to Takahama from Sendai Generating Station (in Kyushu), the first nuclear plant to resume operation after meeting backfit standards. His involvement, Kanden felt, had needlessly reduced the pace of Takahama's restart process.

Kanden had quietly played an outsized role in the restart of Sendai 1 and 2. Work to conform the reactors to post-Fukushima regulations had been carried out by contractors affiliated with Kanden, under generous financial arrangements extended to Kyushu Electric. Like Takahama 3 and 4, Sendai 1 and 2 were three-loop pressurized water reactors. When they restarted, a precedent was set. Setting this precedent had been easier in Kyushu, farther from Tokyo (and the national media) than the Fukui reactors, farther from the liberal urban and political centers of Kansai. Kyushu Electric intended to restart four reactors on two sites, an ambition that attracted less attention than the eleven reactors Kanden owned on three sites near Takahama, nine of which were candidates for restart.

When the NRA's new reviewer arrived at Takahama, his mandate was regulatory continuity: Ensure the Takahama reviewers would benefit from the lessons their colleagues had learned at Sendai. One can imagine the intensity of Kanden's frustration when Flunky Frank (a sanitized translation of his nickname at Kanden's nuclear division) began to undermine the expensive precedent that Sendai's restart had set. Kanden invited the media to attend a demonstration of Takahama's regulatory compliance—under standards the present review would supersede—and Frank had the audacity to introduce himself to his Kanden counterparts by asking, off the record, whether they had misunderstood the review process or the event was a publicity stunt.

Disagreements followed. The siting of the Takahama reactors in relation to the ocean differed from the arrangement at Sendai, Frank pointed out, and Kanden's claim that the area was sheltered from tsunami risk only made sense if you didn't go back a few decades examining tsunami propagation patterns. These considerations were particularly important at Takahama, where Kanden intended to operate four reactors (Sendai hosted two, and both were newer units). Frank also probed the logic of Kanden's accident mitigation priorities, for instance their attention to potential pipe ruptures at large-diameter pipes—why not smaller pipes, where ruptures took longer to locate? Eventually, Frank rescinded his predecessors' approval of Kanden's ground motion calculations, which used older guidelines even though updated versions were available.

Absent the capacity for primary research the active fault review had provided, this capacious, vigilant posture—applying the regulations as a means of expanding safety culture, not limiting it—was the NRA's best answer to the demands of the National Accident Commission, and its best fulfillment of the promises Chairman Tanaka made during his confirmation hearings. The handful of conformity reviewers with technical backgrounds consequently developed a sense of themselves as the vanguard for public safety. This group was predominated by staff who had worked in the industry before joining the NRA, a career path the utilities disparaged under the presumption that no one capable of succeeding in the private sector would leave it to work at a second-rate regulator. In Kanden's estimation, Frank, a former industry employee, was necessarily a flunky; they weren't aware that he was among the slimmest demographic of conformity reviewers: The government had drafted him into his job.

When politicians in Fukui inevitably contacted the NRA to complain on Kanden's behalf, Frank's review team was appalled. "You hear that Fukui is corrupt, but in my experience the governor's office is clean, and the prefectural department of nuclear safety is excellent," one reviewer recalled. "It's the local politicians and low-ranking bureaucrats who compete to be the industry's handmaidens. Some of these people, when we ran into them, would relate Kanden's angst entirely intact."

Previous conformity reviewers had assessed the same systems at Takahama, according to the same regulations, but Kanden's engineers were astonished by Frank's rigor. Knowing only what he had demanded, not what his predecessors had failed to demand, I could extrapolate by Kanden's aggrievement that the difference was significant, and when the standards had been lower before, they were much lower. In the context of nuclear safety, this is grave. A gap like this can't last in a place like the NRA. Its continued exploitation leads to a failure too glaring to ignore, or someone exposes it.

Someone exposed it.

A few months earlier—roughly a year before the problems Frank encountered were exposed—the IAEA sent a team to Kashiwazaki-Kariwa, the world's largest nuclear plant, seven TEPCO reactors on Japan's back coast. TEPCO was in the early phases of developing its restart strategy for reactors 6 and 7. The company was aware the site was politically vulnerable. At the broadest level, Kashiwazaki-Kariwa 6 and 7 resembled the destroyed reactors at Fukushima, and the site had experienced earthquakes of unexpected intensity, including the 2007 event that prompted a reassessment of NISA's seismic safety procedures. Voters in the plant's host prefecture opposed restart three to one.

TEPCO claimed they couldn't afford Fukushima cleanup without operating Kashiwazaki-Kariwa and promised to adopt safety standards that exceeded the NRA's post-Fukushima requirements. Their request for a two-week mission from the IAEA fit their intention of volunteering the plant for additional regulatory scrutiny. After completing their visit in July 2015, the IAEA's team produced a flattering report that gently suggested a few potential improvements.

Some of the IAEA's suggestions surprised NRA staff who had formerly worked abroad. They revealed shortcomings that belied Japan's commitment to world-leading safety culture. The IAEA noted that TEPCO didn't regularly check, as nuclear utilities in other countries do, that employees were fit for duty—sober, psychologically sound, physically prepared for demanding conditions. Continuous training of operations staff omitted a pass-fail mechanism to take struggling staff out of rotation. The use of skeleton crews in parts of the plant that needed fewer staff during shutdown had created fire vulnerabilities. In the operations department, the IAEA was disturbed to learn that no one below the shift supervisor—not even the employees who operated the reactors—had been given a job description, obfuscating the chain of authority.

A separate IAEA mission was scheduled to visit Japan in January 2016, representing the Integrated Regulatory Review Service (IRRS), which provides comprehensive recommendations to regulatory agencies of member states once per decade. The report issued by the IRRS after its previous mission to Japan in 2007 had explained, while avoiding pointed language, that NISA's administrative habits often prevented the agency from independently verifying the truthfulness of the utilities' risk assessments. After Fukushima, the 2007 IRRS findings were used by reformers in the NRA to remind their colleagues to remain vigilant; promises about altering the spirit of regulatory activity had been made before. Until the disaster, NISA leadership had insisted these promises were nearly fulfilled.

The 2016 IRRS mission issued its report in late April, six weeks after the Otsu District Court ordered Takahama into shutdown. In the dry language of international atomic expertise, preceded by the standard IRRS disclaimer that the number of issues raised did not indicate whether this nation's nuclear safety practices were inferior to others, the report "obliterated the notion that the NRA was capable of developing the world's best regulatory system" (the words of a regulator who specialized in emergency response). Many of the issues raised by the report resembled issues that fueled the conflict between Kanden and the NRA's newest reviewer at Takahama, the sole commercial plant the IRRS team visited.

Chapter 4 ("Management System") included the following findings:

The NRA's management system did not include descriptions of regulatory processes that would allow them to be repeated in a consistent manner.

The system lacked mechanisms that would permit one regulatory process to trigger or inform a related regulatory process.

The NRA's operational manuals failed to explain the role of regulatory activity in relation to utilities' safety programs.

The NRA did not reliably assign grades to pending work on the basis of risk significance, resulting in failures to appropriately track and document important tasks.

The NRA had failed to identify which elements of the agency were responsible for critical categories of regulatory activity. Urgent matters often languished at the divisional level because division managers didn't know where these matters should be escalated, and wanted to avoid consulting superiors who didn't wish to be responsible.

This chapter included an assessment of the suggestion box available to NRA employees. During the lifetime of the agency, it had received ten "inputs." A number this modest implied an absence of openness and deliberation, the report speculated. NISA veterans, Fukui government employees, and Kanden staff chuckled at this item, which recalled the ballot boxes stuffed with cigarette butts during Kanden union drives.

Chapter 6, Safety Review:

The NRA had failed to establish basic procedures to govern the safety review process (conformity review, in the NRA's language), including procedures for selecting review personnel and assigning them to teams. Reviewers were not subjected to any documented selection criteria, and therefore did not necessarily possess core competencies needed for their work.

No procedures or mechanisms existed for inter-team communication (a critical activity, according to conformity reviewers, because one team often possessed expertise the others lacked). In the absence of these procedures, important information had to travel up the management chain before trickling down the other side of an inter-team divide, often in altered or less useful form.

No policy existed to specify how the review teams should report on their regular progress—to whom, at what intervals, and whether they had any interim reporting obligations whatsoever.

No policy existed to specify how safety issues discerned during the review would be opened, graded, documented, tracked, and eventually resolved.

Reviewers—and the NRA in general—lacked a mechanism for identifying safety issues arising as a result of human factors (the critical category identified by the National Accident Commission report).

Review teams rarely visited the sites they were reviewing, and visits were only confirmatory in nature, not exploratory or inspection-based.

Problems described in this chapter accounted for Iriya's disappointing interactions with conformity reviewers. NRA staff hadn't read the active fault findings and weren't conversant with the relevant regulations because there weren't any requirements for them to possess this knowledge, nor adequate opportunities to acquire it. Throughout the IRRS report, descriptions of the NRA's administrative conduct resembled concerns Genden had raised during the active fault review, vindicating Hoshino's insistence that these shortcomings could be manipulated by larger utility companies at the expense of public safety.

The problem of staffing safety reviews with knowledgeable individuals predated the NRA. NISA, operating under the Ministry of Economy, Trade, and Industry, had been staffed by career civil servants, few of whom came to their jobs with technical backgrounds. METI's dual role as a regulator of Japan's nuclear industry and promoter of its economic interests was not only, as the National Accident Commission report pointed out, a moral hazard to safety, but also an obstacle to the cultivation of technical expertise. METI's job rotation frequently moved technical staff into management roles and vice versa, disincentivizing specialization among mid-career regulators.

Following the establishment of the NRA under the Ministry of Environment, attempts to hire expert staff were hindered by problems of pay and prestige. The private sector offered more of both. Among nuclear professionals considering government work, METI, notwithstanding its role in the Fukushima disaster, remained the desirable posting. In the United States, highly skilled utility employees retire at fifty or fifty-five (the earliest retirement age with a pension) into a second career at the Nuclear Regulatory Commission, in pursuit of the stability and additional benefits provided by a federal job. In Japan, utilities offer skilled workers unsurpassed stability and benefits.

When the IRRS report was published, the average member of the NRA's regulatory staff was significantly older than his or her counterparts at comparable agencies, and nearly half of the employees responsible for managing the restart process were in the age bracket nearing retirement. Staffing pressures were exacerbated by a post-Fukushima exodus from JNES, the research agency that merged with NISA and the NSC to form the NRA. JNES experts provided on-demand technical advice to the NRA's review teams, the fail-safe mechanism for the scattershot composition of the teams themselves. JNES had been the target of pointed criticism after Fukushima as documents inevitably emerged that showed the agency's experts declining to study the potential benefits of safety systems that could have diminished the impact of the Fukushima disaster. The resulting wave of shame-based retirements left the NRA doubly understaffed on matters of technical significance.

The NRA had told the IRRS team that staffing the conformity reviews was difficult. The IRRS, in response, described the NRA's chicken-and-egg dilemma: An understaffed review must rely on the soundness of its policies, but in many instances the conformity reviewers didn't have updated policies to resort to, and the agency was too overburdened to develop them.

At risk was the NRA's ability to establish independence from the expertise provided by the utilities, which NISA's staff had relied on unabashedly. Confronted with a stack of simultaneous restart applications, under-qualified and overworked NRA staff were increasingly accepting—as their predecessors had—the utilities' version of reality. Hokkaido Electric had risked submitting a minuscule sample of volcanic ash in response to active fault concerns at Tomari, hoping the reviewers wouldn't understand that the sample was worthless. It's hard to say how the NRA would have treated this submission if Yoko Ono's antinuclear cousin hadn't intervened. Worse evidence had been accepted at Tsuruga during the NISA era, and Shimazaki's objection to ground motion calculations at Ōi had revealed the extent to which the NRA presented the tentative assessments of outside experts as the agency's own firm conclusions.

For NRA staff whose responsibilities and language skills brought them into regular contact with IAEA activity, the IRRS report resembled recommendations issued to countries where the nuclear industry had been established within the last decade. The IRRS team, nineteen nuclear experts from Western nations, perhaps unfamiliar with the administrative biases of Japanese bureaucracy (which favor the preservation of institutional relationships), had produced a report that failed to obscure their bewilderment at having found Japanese regulators treating the nuclear industry as permissively as they had before Fukushima.

The report reached the apex of its bewilderment in Chapter 7—Inspection, among the following findings:

The NRA's inspection program omitted any authority to evaluate the performance of the utilities and their personnel.

The NRA's inspection program did not include any participation in the periodic safety review of the plant. This review was left to the utilities, and later endorsed by the NRA so long as it had been carried out to the utilities' satisfaction.

The NRA had not developed an integrated baseline inspection program. Inspection tasks were completed and paperwork filed, but the results did not inform future inspections, and were not reviewed to discern patterns of potential concern, nor did inspection results inform relevant tasks in other NRA departments.

Inspectors across departments and sites lacked a mechanism for maintaining routine communication, let alone coordinating and unifying their practices.

NRA inspectors received one week of training. This training was not designed to ensure or test competence.

The NRA's facility inspections were limited to reactor outages. During outages, NRA access to inspection sites was governed by site access agreements the NRA had voluntarily adopted. These agreements ensured that the timing and circumstances of inspections were decided by the utilities, whose formal inspection requests supplied the legal basis of the inspectors' subsequent visits. In the IRRS team's words, this was "highly unusual."

Daily visits to nuclear facilities by NRA personnel were considered assessments, not inspections, and lacked the limited authority available during inspections. If an assessment uncovered a safety issue, the only mechanism for follow-up was an appeal to the voting power of the NRA's five commissioners, or an emergency order from the director general of the NRA's regulatory branch. In the IRRS team's words, this constituted an "unusual practice." This point was immediately repeated, almost verbatim, and followed by the report writer's lament that the same concerns were raised during the previous IRRS mission in 2007. The writer noted that any solution would be further delayed because inspection authority was statutory; only parliament could expand it.

Recommendation R9 in the 2016 report instructed the NRA to redesign their inspection program to pursue "risk-informed regulation of nuclear and radiation safety." A lay reader might ask: What other kind of regulation is there? In the NRA's case, the IRRS report appeared to suggest that "World's Strictest" had meant "The World's Most Equipment and Highest Margins on Paper," without addressing the deficiencies the IAEA and the National Accident Commission considered most important, which were matters of how the agency obtained information, digested it, discussed it with licensees, and followed up.

Chapter 7 also included a stinging eyewitness description of an inspection. "The IRRS team witnessed how inspectors in the control room of the NPP were just checking status of certain indicators on the control panel as it was prescribed in their checklist while paying no attention to numerous alarms and activities by operators that were going on at the same time in the control room." The inclusion of this anecdote became the object of intense chagrin among NRA staff, who felt (although some were glad of it) that the agency had been deliberately emasculated in the eyes of industry stakeholders, the primary audience for IRRS reports.

"In America, inspectors are beat detectives. They work in pairs, with authority and expertise appropriate to their position," said a former NISA official who answered the hotline inspectors would call if they discovered a safety issue. "In Japan, they're monkeys with clipboards. The more expert you are, the more likely you're behind a desk in Tokyo, in a position of authority. When inspectors reported potential issues, I often found myself thinking, *I'm not sure you understand how a nuclear plant works.*" I asked whom he trusted more, his own inspectors or utility personnel. "Utilities," he said. "Not even close."

According to the international standards of the nuclear industry, an excellent safety culture evaluates *all hazards.* For instance, the risk of a natural disaster shouldn't be taken more seriously than the risk of major human error. An excellent safety culture is *integrated.* All inputs enter the same system, where processes have been established to examine them in relation to one another, to discern patterns, and to disseminate up-to-date information across departments and agendas. An excellent safety culture uses a *graded approach.* Inputs are prioritized in a simple, consistent manner according to industry-specific criteria. Outputs also occur according to grade; issues at certain levels of significance trigger grade-specific responses. Finally, an excellent safety culture is actively *tracked and monitored.* The system is not a database; appropriate time frames are established for continuous attention to safety issues on the basis of their importance, and these time frames are followed until the issue is resolved or its priority diminishes.

Examining the administrative processes critical to the NRA's institutional integrity (management, safety review, inspection, and enforcement), the IRRS had found that the agency failed to adopt an integrated, graded, tracked and monitored approach, and failed to evaluate all hazards. In the media, the IRRS report was a one-day story, its findings obscured by complimentary language in the IAEA's accompanying press release. "The team concluded that the [NRA] had demonstrated independence and transparency since it was set up in 2012." For NRA staff, who possessed the necessary context, the report revealed that the NRA had repeated the failures of safety culture that caused the Fukushima disaster. "Parts of the IRRS report were like the National Accident Commission report, with our name in place of NISA," said one regulator who had worked for international safety organizations for several years before returning to Japan. "Which meant we had accomplished almost nothing."

Presumably, the IRRS team had observed NRA inspectors ignoring alarms in the control room at Takahama, the only plant the team was recorded visiting. After the report's publication, some NRA staff were convinced the anecdote was a disguised reference to a similar incident that occurred outside the scope of the IRRS review, during control room exercises at an idled reactor. This rumor, recalled a Takahama safety reviewer, revealed the staff in a state of anomie. "You made a mistake, but you convince yourself it was two mistakes and your parent figure is too embarrassed to scold you twice. Doesn't anybody tell the truth?"

37.

After the IRRS report was submitted to the NRA, I contacted Iriya to tell him the restart process had exited its post-Fukushima period, and I could say what I was likely to include in my book. For the first time since we began meeting, he was not actively embattled by an adversarial dialogue with regulators. He chose a restaurant farther from his office than our usual place, in a neighborhood that predated corporate Tokyo. It was good, he said when we sat down, to rejoin the world, to remember how strange, how contingent and individualized, were the rules and goals that characterized the milieu where the last few years of his life had occurred.

It was May. In July, Genden would rotate its employees. Iriya had reached the age when upwardly mobile engineers are promoted into management. Having guided a small team against long odds during the active fault review, he could expect to receive significant supervisory responsibilities. This was not a scenario he welcomed. The managerial ability he had demonstrated was the result of an intense personal connection to his research, to tangible outputs he could count and own. You can dress like a Gilded-Age dandy, he joked, slick your hair instead of cutting it, decline to press your clothes, and wear your tie like a necklace, and the bosses won't flinch, as long as your achievements belong to you more than they belong to the company.

In one respect, I told him, my book would agree with his bosses: The events surrounding his work had determined the shape of nuclear safety in Japan after Fukushima. A few people on both sides had been faced, before there was time to prepare, with decisions of historic significance.

"Am I the mouse that bit the cat?" he said.

I told him Hoshino thought of himself like that.

Iriya repeated a characterization of his and Hoshino's diverging experiences I had heard from each of them before. "In the worst times, I was forced to work, which causes depression. Hoshino was forced to fight, which causes adrenaline." Adrenaline defers the personal consequences of conflict, Iriya explained, and the consequences are altered by deferment. Two people might experience the same difficulties at the same time, in each other's presence, but the toll on each is distinct.

After trailing his father's job for several itinerant years, Iriya reminded me, his family settled in Tokyo and enrolled him in a suburban school where transfer students, presumed less affluent, were expected by their peers to prove the attached stereotypes—to fight. Iriya was the youngest of three boys, but they were separated by only twenty-six months, none big enough to protect the others. Three decades later, Hoshino protected him. Seeing how it affected Hoshino, he was glad his brothers hadn't.

Iriya asked if I had promised the NRA a sympathetic portrayal of the active fault review. He imagined it would have been difficult to develop sources otherwise. NRA staff, I explained, were rarely acquainted with the science in the Tsuruga findings. Their interpretation was, "We wanted more safety than Genden did," and nothing I had learned would threaten that perspective.

The difference between talking to Genden and talking to the NRA, I explained, was a difference of candor. My Genden sources knew there was less to lose for a small utility already convicted of trying to run a plant on a fault and were confident my inquiries would reveal they had bested the NRA, forcing Tanaka to annul the authority of the active fault review. The NRA, on the other hand, remained a self-conscious beneficiary of public esteem. My NRA sources wanted the book to be accurate, even if the results were unflattering, and some believed I had developed an affection for their mission. But most were unwilling to affix their names to an account of a conflict that demonstrated the agency's fallibility.

The panelists, I clarified, were less likely to request anonymity. If a question touched the substance of their research, they honored their responsibilities as public intellectuals.

"And Shimazaki?" Iriya said.

"More panelist than staff."

"We blamed him. Sometimes, he deserved it. Other times, you wonder. When Tanaka said we couldn't apply for restart, that was Shimazaki's panel, Shimazaki's meeting."

"But Tanaka talking."

"Thinking back, it was probably Tanaka's mistake," Iriya said. "I went years believing Shimazaki planned it." He was cheerfully perplexed, a person who has failed a riddle but enjoyed the answer. "Before this experience, I thought I understood the expression 'an open secret.' My understanding has matured. There are degrees of openness, multiple audiences, and we often obstruct our own view."

Two of Genden's independent experts had told Iriya about their experience of watching news coverage together when the Fukushima disaster was unfolding. The nuclear professionals who appeared on the newscasts gave modest estimations of the worst case, conditioned by decades of using science to vindicate the industry. Genden's experts were crestfallen. Anyone familiar with nuclear plants knew a catastrophe was imminent. Why exacerbate the crisis of credibility that would follow?

Iriya remembered this now, the initial dishonesty, the stubborn self-delusion. "That's always the first instinct. Fukushima really was, for a time that seems long when I remember it, the event capable of replacing that instinct with better notions." His career was built on his belief in the role of reform, he reflected. Shortly after his arrival, Genden had concealed the seismic risks he was hired to mitigate. The regulator's experts balked, and he resumed his work, his sense of mission restored.

"I still believe the past and its consequences are in proportion," he said. "I don't know whether solutions necessarily follow."

We took a second dinner closer to Genden's offices, with Iriya's team. During a recent meeting at the NRA, they had been asked to explain why their investigation at Tsuruga addressed 5 faults of 160 identified. The answer was "per NISA instructions," but preliminary meetings had revealed the NRA staff's aversion to any mention of their former employer, and Iriya couldn't frame his answer otherwise; he had never obtained or evaluated NISA's original rationale. At dinner, Iriya and his team related this setback amiably. If NRA staff weren't familiar with the technical and regulatory underpinnings of the process, at least they were prepared to demand comprehensive, contextualized submissions.

Later in the evening we went to karaoke at a hostess club, where a conservatively dressed hostess, perhaps reserved for designated drivers, talked with me while I waited for Iriya and his colleagues to stop caring whether I stayed. The engineer who would inherit Iriya's job in July shared our corner of the wraparound couch. He grew up on Atomic Broadway, he said. I hoped we'd talk about the mechanical coelacanth at the Japan Atomic Energy Agency's abandoned aquarium, but the ultimate subject was the lack of local employment for nuclear contractors. The youngest member of Iriya's team had recently married a Tsuruga native whose friends and relatives had lost their jobs. I mentioned my host father in Tsuruga, who found work at Takahama but drank himself down a flight of dormitory stairs between shifts. The plant's call to his wife, in retrospect, could have waited until morning. It took several minutes to convince her he wasn't dead or irradiated. I was trying for levity, but Iriya's team could imagine their wives taking the call and received the story as an object of pathos.

Iriya sat apart from the group, increasingly morose and fixated on his forthcoming transfer, cataloging his regrets to a bored hostess whose attire (pink lamé and rhinestones) recalled the era when nuclear company managers were glamorous clients. Every song he picked, he stood up to sing, eventually explaining to his hostess, "I speak at public meetings."

Iriya's expectations for the future of nuclear safety in Japan, and his role in shaping it, had begun to lose their adrenaline-coated clarity six months earlier, after his first meeting at the NRA to discuss the Tsuruga 2 conformity review, when the new year—his transfer year— was approaching, and NRA staff proclaimed their willful ignorance of the active fault research that filled his life after Fukushima. A week later we met at our usual restaurant near Genden, and he provided several quotations that may upset his employer, a risk he acknowledged before speaking. A few are combined here.

"I don't feel it's important whether the K fault is determined to be active. What's important is the NRA's ability to make that determination with scientific and administrative integrity, and to repeat the quality of that determination at other sites. In a case like Tsuruga 2, where the evidence isn't complete and the science isn't clear, extremely robust and well-conceived regulatory measures should be available, and followed with care. If the real problem is a larger threat nearby, like the Urasoko fault, let's approach it directly. If that's undesirable because inherited regulations for large-scale risks are insufficient, or because nuclear companies tend to exploit that insufficiency, then everything should come to a complete stop and we shouldn't operate nuclear plants on risky sites until the rules are good enough, the industry has learned to respect them, and regulators are able to respect the industry.

"I think Tsuruga 2 can run safely. I chose to work at Tsuruga because the site presents serious risks and I believe they can be managed. But I would have been happier than I am today if trenching at Tsuruga had revealed conclusive evidence that K was active. My employer would have been bankrupted, but I would have been spared the difficult experience of watching brilliant scientists cobble together a conscience-driven response to inconclusive evidence. The damage to the institutionalization of the NRA that has been caused by the agency's need to defend questionable findings, and to stretch those findings to cover the enormous gap where a real discussion of the Urasoko fault should have occurred—it's severe damage, and it will affect nuclear safety."

In my notes about Iriya I had once written, "The past is a friend?" because he stayed in his maternal grandfather's house from age thirteen to twenty-four, for the entirety of college and graduate school, for the first four years of his friendship with his wife, a former classmate he approached during a reunion at the school where he had been bullied. After leaving his grandfather's house to work for TEPCO, an unwanted job obtained unwittingly, Iriya found a project at Genden worthy of his desire to build an institution that would outlive him. Like the difficulties of the school where he met his wife, the difficulties of entering his career, once accepted and revisited, had become the basis of the life he wanted. His indifference to the embarrassment of quitting TEPCO to work for Genden had protected him from the possibility that he would have been complicit in corporate negligence at Fukushima, or would have been working at the plant when the tsunami struck.

After explaining his concern for the NRA, Iriya talked about his son, who was young enough to be kissed when the active fault review began and was now eleven, annoyed by his father's physical affection. When Iriya watched videos of NRA meetings, his son asked, "Who is the worst person in this video?" Iriya explained the meetings were not like television: no villains or heroes you can recognize by their costumes or the music that plays when they appear. This answer never satisfied, and Iriya supposed there were two confusions: Why would you watch something like that? And if his father was not fighting bad people, how was it possible he was a good person who was gone so long, so often? Iriya continued repeating his original answer, for the simple, important lesson it reinforced, and because he was pleased to find he believed himself. There were no bad people in those meetings.

Iriya was surprised by his desire to discuss family in this setting, a restaurant where Genden held working dinners. "That's how I know my part has ended," he said. The part that began with his desire to transform the practice of nuclear safety and ended with five years of antagonism among the scientists—the engineers, bureaucrats, nuclear executives, and politicians—who received that opportunity after Fukushima. Indicating with a gesture the restaurant and what it was part of, he said, "Stay too long, and you shrink to the size of this world." A world so small, my host father in Tsuruga had joked, it is hardly the size of an atom.

Afterword: The New Century

In November 2017, several Japanese newspapers exposed Genden's mismanagement of their decommissioning fund for Tsuruga 1 and Tokai 1, the company's oldest reactors, marked for closure after Fukushima. Genden had satisfied the letter of the law by adding a token amount to the fund annually, but the balance was diverted to pre-construction expenses for Tsuruga 3 and 4. When Hoshino told me that he would use whatever rationale was available to pursue the construction of additional reactors, his answer was more fraught than he could say: The costs were sunk.

A week after this deception was revealed, Genden applied to extend Tokai 2's operating license beyond the reactor's design life span. Tokai was the nuclear plant closest to Tokyo and the nuclear plant with the most people (960,000) living in its evacuation zone. If Genden's application was successful, Tokai 2 would become the first boiling water reactor (the same category as the Fukushima reactors) to receive a license extension. To avoid controversy, Genden had previously claimed they were not certain they would attempt to restart Tokai 2.

The timing of the Tokai 2 application, while revelations about Genden's decommissioning fund remained in the headlines, was deliberate. Information about the decommissioning fund had been passed to cooperative journalists, whose colleagues were led to believe it was a genuine leak. TEPCO sources were quoted in several of the resulting articles, lamenting Genden's poor financial health and lack of access to credit. They warned that the company could collapse, exposing the decommissioning of Tsuruga 1 and Tokai 1 to administrative entropy. Tokai 2's license extension was promptly approved, and TEPCO pledged to cover Genden's restart costs, a promise TEPCO could afford because taxpayers had been responsible for the company's bottom line since the Fukushima disaster.

In 2018, the village of Tokai declared that it could not presume to speak for the 960,000 people in Tokai 2's evacuation zone. Genden signed an agreement allowing additional municipalities to join restart consultations, and the six participating governments vowed to issue unanimous decisions. By the end of the year, Tokai's mayor began denying that he would permit other signatories to exercise veto power. The agreement contains no language about split-opinion scenarios, leading industry critics to remark that Genden has gained leverage by diluting local restart authority.

Tsuruga 2's conformity review has progressed slowly. When it began, my NRA sources insisted they would flunk the reactor based on hazards identified during the active fault review. At Ōi, the flexibility of ground motion calculations had been used to permit restart. At Tsuruga, the same flexibility would be turned against Genden. Shimazaki, one member of the NRA's legal staff insisted, would be proud. I'm not certain he would. Science and prediction overlap less than we think, and there is never a shortage of opportunities to confuse them. It's an old story, and in its present telling, nothing happens. To quote from press coverage, "Restart at Tsuruga remains imponderably distant."

As I write this, Hoshino is site superintendent at Tokai—the plant's chief executive. To keep him at Tokai as long as possible, Genden has exempted him from the company's regular rotation of senior managers. As far as I am aware, no other nuclear executive has remained in the critical, decision-making layer of management for an entire decade; they are always promoted into corporate roles. Hoshino's mechanical engineering background is unusual among nuclear executives. He might be the only person in Japan who is qualified to operate a nuclear plant and manage its restart negotiations simultaneously. His colleagues say he's unflagging and enjoys every element of his work: the choreographed subterfuge of the TEPCO funding, the diplomacy required by the multiparty restart consultations, and the technical challenges of bringing an older boiling water reactor online when no other utility has dared it.

I rarely hear from Iriya. Since February 2020, he has been embroiled in a dispute with the NRA that appeared in the media as a "data falsification" scandal.

Shortly before the scandal erupted, Iriya's team audited their Tsuruga 2 research and discovered roughly a thousand mistakes resulting from database management errors. In subsequent consultations with the NRA, Genden was advised to submit a corrected edition of their research and to avoid pointing out their corrections within the new document, which could be compared with previous versions if Genden's changes required scrutiny. Four months later, a senior NRA official—who was present when Genden received these instructions, but felt that Genden had applied them in a self-serving manner— used a public meeting to accuse Genden of falsifying data, and the resulting headlines made it difficult for the NRA to take responsibility for shaping Genden's submission.

NRA commissioner Ishiwatari Akira (Shimazaki's replacement) has been forced to chart a delicate course, neither endorsing these accusations, which he knows to be inaccurate, nor faulting his staff for making them. Sensing an opportunity to restore the culture of reciprocal favors that characterized the relationship between nuclear utilities and regulators before Fukushima, Genden's current leadership (key officers were replaced in 2019) has made a concerted effort to avoid undermining Ishiwatari's position. Iriya must fall on his sword.

Since retiring, Shimazaki increasingly avoids public controversy, but in 2018 he testified in the criminal trial of three former TEPCO executives. Prosecutors had declined to file charges against the executives for delaying, diminishing, and eliminating countermeasures that could have prevented the Fukushima disaster. An inquest committee (similar to a grand jury) returned two consecutive indictments, forcing the district court to prosecute. Shimazaki testified that government panels reviewing tsunami risk in Fukushima had been repeatedly pressured to alter their findings for TEPCO's benefit. All three defendants were acquitted in September 2019. The court determined that the severity of the tsunami was unforeseeable, even though several experts—including Shimazaki—had foreseen it.

While the TEPCO trial dominated coverage of the nuclear industry, Fukui established itself as the hub prefecture for restart. Once Ōi and Takahama were operating, other utilities couldn't match Kanden's rates. On March 23, 2018, Japan's primary stock index declined by 4.5 percent over fears of an impending trade war. A handful of stocks added value anyway: tobacco companies and Kanden.

In autumn 2019, dozens of Kanden executives, including the company's president and the chairman of its board, were caught taking bribes from a small-town influence peddler in Takahama. They had been doing it for decades. The company's stock plunged, its restart agenda was suspended, and every prominent executive with ties to the company's nuclear division was forced to resign. The scandal engulfed Fukui's prefectural government, where 109 officials had also accepted bribes. The corruption came to light because its scale expanded after Fukushima, when construction money poured into Takahama to support the restart process, and because the town's former deputy mayor, who directed the scheme, lost his grip on the regional tax bureau shortly before dying at age ninety.

The scandal prompted Tanaka Shunichi, the NRA's former chairman, to declare that "Japan's nuclear industry has no future." These were the words, one of Tanaka's former advisers explained, "of a man who wanted to rehabilitate his industry, but the industry would not be rehabilitated."

I didn't interview Tanaka until I had written most of this book. My NRA sources described him as brilliant and defensive—someone who would interfere with my research. When I contacted him in February 2020, he was guileless, inviting me to his cabin on the basis of a brief introduction. The coronavirus pandemic made this visit unwise (Tanaka is seventy-six), and we spoke instead by videoconference.

It was immediately clear that Tanaka did not take criticism gracefully. Shimazaki's decision to contradict the NRA's ground motion calculations was "incomprehensible." Any issues Shimazaki wanted to raise, Tanaka insisted, "belong in an academic journal. Talking to the press, testifying in court—that's not science." Judge Higuchi's Takahama injunction was "lacking even in common sense." Tanaka still hadn't read the ruling, an admission he defended by arguing that "the judiciary's independence shouldn't be mistaken for competence."

Yet I left the interview feeling that I understood why Shimazaki and the legal staff had protected Tanaka. I would have made the same choice, and not purely out of a desire to preserve the NRA's reputation. His critics had accused him of allegiance to Japan's nuclear industry, but his statements during our interview never borrowed the industry's language, and his frame of reference was strikingly narrow: He referred to only one event that occurred before the Fukushima disaster, which was the event that inspired him to accept his nomination to the NRA. Tanaka had endeavored to reinvent himself in the mold of the agency's mission; if his flaws survived the process, this seemed like an acceptable compromise.

When I inquired about his verbal decommissioning of Tsuruga 2, Tanaka explained that he attended the site's first review meeting because the Ōi meeting, a month earlier, had been a shambles, dominated by Watanabe's antinuclear rhetoric on the one hand and Okada's hostility toward the NRA's regulatory obligations on the other. He admitted that Shimazaki and the panelists had not known he would attend the Tsuruga meeting, and when he spoke at the end of the meeting, he was "deeply suspicious that the fault in question was active." When I engaged him about possible translations of his closing comments, however, and their effect on Genden's restart prospects, he adhered to the language developed by the NRA's legal staff years earlier: Nobody paid attention to what he said, it was a vague notion he happened to utter, and perhaps a few newspaper outlets thought the plant wouldn't be able to restart, but that was an idea gleaned from the other panelists' comments. He concluded by daring me to state that "Japanese society was bold enough to decide that my opinion was consequential."

Industry partisans claim the pace of conformity review has diminished under Tanaka's successor, Fuketa Toyoshi. Japan will likely fail to meet the LDP's goal of generating 20–22 percent of the nation's electricity at nuclear plants by 2030. Fuketa has demonstrated willingness to heed the industry's critics, but the pace of the restart process is not necessarily a result of his deliberative leadership style. The reactors screened during Tanaka's tenure were strong candidates for restart. Fuketa's agenda, filled with license extensions and antiterrorism upgrades, demands a methodical approach.

And the NRA is increasingly capable of thorough reviews. In 2017, after the IRRS report made the NRA's lack of inspection authority a source of international embarrassment, Japan's parliament overhauled the relevant statutes. The NRA subsequently embarked on a campaign to transform its workforce. New hires and mid-career regulators are now removed from their job rotations for up to two years to complete technical training.

Pending the results of this transformation, the NRA's inconsistent assessment of seismic hazards continues to invite judicial intervention. Shikoku Electric has been ordered to shut down their Ikata plant twice since 2017. The more recent ruling, in January 2020, focused on earthquake risks, including the possible presence of undetected faults linked to Japan's largest fault system, a four-hundred-kilometer network of crush zones that bisects the southern half of the country. Risks like these—remote, unsubstantiated, but never ruled out entirely—are the reason why American electricity companies will no longer operate reactors in seismically active regions after 2025. (The entirety of Japan is seismically active.)

The city of Tsuruga has lost all of its reactors. Tsuruga 1 entered decommissioning in 2017. Tsuruga 2 remains in purgatory. Before he retired, Chairman Tanaka shuttered Monju permanently. The last time I stopped in Fukui to visit Kawase Kazuharu, Tsuruga's mayor emeritus, he declined to discuss the nuclear industry, preferring the triumph of Tsuruga Pines High School at the national baseball championship and the possibility that a new bullet train route would include multiple stops in Reinan.

The bullet train was on his mind when I filmed a subsequent interview with him in Tokyo, where he was exploring a possible run for parliament. "We'll get the bullet train because the reactors are hopeless," he said when I stopped recording. "They have to give us something. Fukui is the best place to be a politician." His next words tumbled out in an excess of enthusiasm. "Look at Takagi."

Takagi Tsuyoshi, Reinan's representative in parliament, had run essentially unopposed for twelve years, even though he had been implicated in corruption scandals and exposed as a serial panty thief. These embarrassments hadn't prevented the LDP from making him minister of nuclear disaster recovery. *Look at Takagi.* Look how far a mediocre politician can go when a single vote in his electoral district equals three votes elsewhere, because Japan's districts were drawn when the nation's population was overwhelmingly rural. Fukui benefits from this imbalance more than any other prefecture. In 2019, a survey of LDP parliamentary candidates showed that only one out of three agreed with the party's pro-nuclear platform. "Which three?" was the canny response of one Tsuruga politician. In the group that matters, the results would be different.

As we parted, Kawase gave me his old business card (*MAYOR of TSURUGA*), which is a tiny pop-up book. He appears in the centerfold, in cartoon form, dressed as a boat captain. *Please come and see Tsuruga in all seasons.* A picture of cherry blossoms for spring. Seafood delicacies for winter. A sandy island in the harbor for summer. And in the bottom left, representing an indeterminate season, *Nuclear power plant.* Kawase's card is the only promotional material from Tsuruga I have seen that treats the plants themselves—not their public relations facilities—as a first-order tourist attraction.

"Before Fukushima, it was heaven," he said.

"You mean the city?"

"I mean for me."

If I wasn't confusing dates, I said, the Monju accident occurred at the beginning of his first term. "Was it heaven after Monju?"

"Monju was different," he said. "I understood what the facility meant to Japan's nuclear industry. I regarded it as a necessary evil."

Outside, the heat of summer was approaching. Where we stood, in the atrium of a conference center, the air was cool and carried the smell of carpet shampoo. Peak electricity use would begin next month. I had seen Kawase sweat in warm rooms, and sweat in the close air of a press conference. Presently his face had the powdered, set-in-place look some male politicians achieve without wearing makeup.

"What should I call you, now that you're not mayor?" I said.

That drew a chuckle; once a mayor always a mayor, especially in Japan's small cities. In Tsuruga, I knew Kawase's high school classmates, who worked on his first political campaign. I knew his closest childhood friend. I knew his favored contacts at Genden, and I knew what these people would call him if they couldn't call him mayor: a son of Reinan, bearing its flaws and virtues. I also knew what his answer would be. I had heard him give it to someone the last time I was with him, in the hallways of Tsuruga Pines High School, where his honorary title ranked him above the principal.

"Think of me as a salesman."

"I will," I said, and wished it weren't so easy.

During my intermittent residency in Wakkanai, I joined the Japan Firefly Society. Mayor Yokota and Monma Katsuhiko (the former Antarctic explorer) were friends of the Wakkanai chapter's chairman, Hiranuma Michihiro, a retired construction contractor suffering from advanced Parkinson's disease. When he was a boy in his father's marine engine workshop, Hiranuma told me, grasshoppers were attracted to the heat of the workshop's fire, and he would pass the hours by catching, observing, and releasing them. His interest in the natural world recurred when his children were in elementary school and a fellow PTA member mentioned that her daughter had never seen fireflies; the population vanished after Wakkanai urbanized. This disappointed Hiranuma more than he expected, stirring regrets about the competitive, solipsistic society his generation had passed to their children. He visited his children's teachers, to propose firefly breeding as a class project. "Kids today would let a goldfish die," one teacher replied. He wondered if this teacher had tried keeping one in the classroom, or only imagined the outcome. The latter possibility, the presumption, offended him.

Hiranuma opened the Wakkanai chapter's meetings by reading from the first book on firefly entomology he acquired, which survived a fire in his house. The borders of the pages were brown and brittle where the flames had touched them. Favorite passages were followed by his own commentary: "If a firefly enters the beam of a passing car's headlights, it never illuminates again." "Fertilizers, pesticides, and the fumes of industrial chemicals disrupt illumination. Disoriented chemically, a male and female firefly could land on opposite sides of a leaf and never notice each other." I didn't look these notions up or verify them with an entomologist. They belonged to my experience of Hiranuma, whose body revolted against his presence in the meetings. His left arm moved involuntarily, and he sat with it pinned between his knees. His left eye wouldn't open as wide as the right, which produced the sensation (he complained) of viewing his surroundings through a transparent place in his eyelid. When he spoke, saliva pooled in his cheeks, making it difficult to enunciate.

A few of his gestures retained the physical vitality of his former profession, and this incongruity made us feel as if we were witnessing the first day of his illness.

Hiranuma had assigned himself the goal of holding the Japan Firefly Society's national conference in Wakkanai, a city with no fireflies and a difficult city to reach. His bid was supported by the southernmost chapter, in Okinawa, which insisted it would host the following year's conference, and Hiranuma would be healthy enough to attend. Wakkanai was selected to host in 2018, and Hiranuma died shortly after the selection process concluded. He was sixty-six.

Among the retirement-age residents of Wakkanai I befriended, many of whom were in their nineties and living in care facilities, Hiranuma was the only one who passed away. I saw him shortly before his death, in Wakkanai Station. He was leaning on a vending machine to keep his balance, watching the local train depart, lifting his cane to salute the conductor. We spoke, but he was disoriented and didn't remember who I was. "I'll tell you how the Ainu describe fireflies in their epic songs," he offered, after I identified myself as a member of the firefly society. "The gods of light that begins and ends."

Hundreds of firefly enthusiasts traveled to Wakkanai for the 2018 conference and were taken to the neighboring town of Toyotomi for firefly viewing. Opening ceremonies occurred at Toyotomi's gleaming new high school, a testament to the stability of the town's dairy industry. The school's band played, and a few students were still present when the ceremonies ended, filling the hallways with youthful laughter, completing the sense that we had traveled to a type of institution that would never exist in Wakkanai again.

When hundreds of people gather to commemorate a mutual friend—when they file into the darkness of a spring evening, when they speak in whispers and stand shoulder to shoulder in a soft rain, observing the illuminations of the evening's first fireflies—a sense of that friend's presence is a ritual certainty. "These fireflies must be related to the larvae Hiranuma hatched," Mayor Yokota remarked, and his tone was not wishful, though his notion was. Only one firefly was observed alighting on a conference attendee, and that attendee was Hiranuma's widow, Asako.

Asako managed the city's welfare program for infants and their mothers. She had planned to retire in March 2019, to care for her husband full-time, and mourned the loss of this difficult experience, wishing she could have become the person it would have made her. During my final period in Wakkanai, I became friends with her son, Masahiro, who lived in Sapporo and worked for a supplier of construction materials. His dated house in a distant suburb received the best light on its block, which filled the tatami room where his daughters (four and one) spent their days. "That's where they're from—that room, this city," he told me. "When they're older, one will remind them of the other. If you drive to Wakkanai, there's a moment when you come into the flat terrain of Soya, and the horizon is suddenly much farther away. I am always overwhelmed by my involuntary recognition of that place."

On the evening before conference attendees departed, the firefly society's Wakkanai chapter invited the Okinawa chapter to sing karaoke. I sat next to Yokota for most of the night, talking about population decline and Japan's aging society. The birthrate had plateaued far below government targets, and domestic migration was drawing women of childbearing age to the cities where starting families was most difficult. "This conversation makes me so depressed," Yokota complained, "that I'm repaying you with a melancholy song." He insisted we perform a duet of American folksinger Hedy West's ode to homesickness, "500 Miles." I told Yokota I didn't know it, but he played it anyway, and I surprised him by following the tune. I couldn't understand why I wept when I listened to the song later in my hotel room, as soon as I realized the original recording had been performed by a woman. My mother died shortly after and I inherited her dulcimer. She had only ever practiced a few simple arrangements of popular songs, and in the book she played from, only one page was dog-eared, which was the tune I remembered when Yokota asked me to sing.

I spent my final night in Wakkanai with Steve Tamaki, whom I'd met when I was working on my first feature story and he was adjusting to life in Wakkanai after forty-four years in America. His mother had died in February 2018, at age ninety-five. Other family members had considered placing her in a care facility, but Steve, who was living with her, honored her desire to die at home. The last few months of her life were exhausting. When she was tired, she would ask the same question hundreds of times, forgetting the answer in a matter of seconds. She stopped when he shouted at her. She would say, "Thank you, I was stuck," but the act of deliberately frightening her left Steve emotionally disoriented. "Then I'm understand," he reflected. "Already I'm one miracle, come back Wakkanai. Even I'm hard time with Mom, make a wrong decision—okay, I'm still a son who did his duty."

After his mother died, Steve filled his time with day labor, joining a friend's crew on landscaping jobs and building demolitions. He was seventy and his working years were over, whether or not he knew it. He was losing his eyesight and his cuticles were bruised from misplaced hammer blows (they barely held his fingernails in place). Three cups of instant coffee, with several spoonfuls of crystals in each of them, didn't stop him from nodding off periodically during our conversation, which started while the sun was up.

We were in his living room. Because Steve behaved like a working-class American, mutual acquaintances of ours in Wakkanai (Japanese people) assumed he lived like a working-class American. His home would have surprised them. His furnishings were sparse but adequate, and the house was broom clean. A bowl on the living room table was filled with fresh fruits. Portraits of his parents adorned the family altar in the neighboring room, which was otherwise empty, in the Japanese style. The house could have belonged to any lifelong resident of Wakkanai.

Steve inherited it from his mother. He didn't expect it to be valuable, but he was surprised to learn it was worthless. It couldn't be sold, couldn't be rented, and the cost of knocking it down would exceed the value of the land, even though the house was located within walking distance of downtown. Still, the property was subject to taxation, and as Wakkanai's population declined, each home-owner's share of the tax burden expanded. Steve, who had no idea whether he was entitled to access Japan's crumbling national pension system, was working to keep a roof over his head, as he always had.

He observed that the pace of decline in Wakkanai was accelerating. The Fukuko Ichiba dining complex—the city's main redevelopment of downtown property—was closing for lack of business. There would be no Sakhalin ferry next summer. Wakkanai Hokusei Gakuen University had warned students that it might cease operations. And,

as Steve discovered when other family members tried to institution-
alize his mother, changes in the national government's system of
subsidies had left Wakkanai's elder-care facilities overwhelmed with
difficult cases. One of the front-runners in Wakkanai's forthcoming
mayoral election, Steve pointed out, openly opposed the construction of a municipal curling rink—and by extension the enormous
subsidies the project would attract—in favor of directing the city's
resources toward an expansion of elder care.

Before leaving, I asked Steve if he would tell me his given name.
He evaded the question by telling stories I hadn't heard, about serving
in the Vietnam War, about securing a transfer out of Vietnam after
seeing another soldier lose an arm in a firefight, and about spending
the rest of the war on an American base performing the lowest duties
his commanding officers could find for him. I was surprised when he
explained that the base was in Japan. Soldiers of Japanese ancestry, he
told me, were forbidden to mix with locals, so his time there never
felt like an interruption of his American period. I imagined him sitting in the basement of the officers' mess, peeling a crate of potatoes,
affected by his American uniform and his disdain for the war far
more than he was affected by the country he was stationed in, or the
fact that he was born there.

I didn't know Steve's name, I didn't know if he had married the
mother or mothers of his children in America, and I didn't know
what he intended by telling me about his military service at this late
moment in our friendship. I didn't feel I needed to. People in Wakkanai found Steve odd, but I'd come to regard him as part of the
scenery. Japan is a society that values the quiet endurance of hardship, especially in depopulating rural areas. Steve cared for his mother
when nobody else was willing. He worked in physically demanding
jobs when other people would have sought family or government
support. He aged the way the city aged: Before his time, and it was
the first thing you noticed about him.

We parted in laughter, because I knelt down to cluck at an
approaching shadow on the street in front of his house, and Steve
knew I had mistaken a northern fox for a house cat. The fox crossed
the street to avoid me, treading on a pile of crushed scallop shells.
The white dust from the shells left a trail of paw prints leading to the
porch of an abandoned home, where the fox disappeared through a

broken panel in the door. "If you build a city, someone can always use it," Steve observed, and joined me on the street to count the number of houses on his block that had their lights on, in the hours of the evening when people were awake after sunset.

Events described in this book appear as they would in a work of literature. My knowledge of these events was gained by research. I have tried to make the distinction between literary elements and research evidence simultaneously subtle and plain. The tension between these qualities is acute.

Be wary of accepting quotations. Many were given in Japanese, then recorded in my notes in English. Sometimes I worked with an interpreter, whose translations influenced or superseded mine. The process of interpreting an interview subject's words, checking your interpretation after the interview concludes, and writing it into prose is a process that erases the spontaneity of speech.

If a person is depicted in a context that is primarily literary, the resulting descriptions may be attributed to my independent recall of their words and behaviors. I did not rely on memory alone to describe events or discussions whose factual qualities determine their significance. Where I suspected my impressions of my sources might differ from their own, I used the first person.

My use of literary license with journalistic material is limited and explicit. I confess one exception. The words I attributed to Mayor Yokota's speech during his inaugural visit to Sakhalin were not spoken during that trip. They were provided seventeen years later, when he gave an eloquent summary.

Anonymous sources appear more frequently than I would like, the result, in part, of Japan's *tokumei shakai* (anonymous society). Japan's libel laws are restrictive, no laws have been enacted to punish frivolous litigation, and highly esteemed journalists have lost cases even when their sourcing was tight. The result is a pool of potential sources in government and the private sector who demand opaque attribution. Wherever possible, my unnamed sources are described in a manner that suggests their agenda (or their employer's agenda). To an extent I wouldn't typically consider, I have afforded Tomo Genshiryoku the fictionalization of his identifying details. Without him, readers couldn't observe the effects of the nuclear restart process on the private lives of industry employees. He deserves as much anonymity as nonfiction permits, and more.

Prosper was an active participant in the development of his story. I can't verify everything he told me, and didn't always try. On the day of the earthquake, as he recalls, he found himself in an unusual discussion with a police officer who brought him to an interrogation room so he wouldn't be subjected to post-disaster paranoia by xenophobic Tokyo residents. This officer, Prosper believes, was at least partly Korean, a heritage he would have kept secret, because Japan's police force is aggressively mono-ethnic. I could learn if this was true. The variations of description that would arise in the officer's account, however, would oblige me to write a version that is less Prosper, less expressive of his experience of the disaster. In this instance, I defer to the material, to the half of it I was given.

Until coronavirus cleared out Tokyo's red-light districts in early 2020, Prosper was working two jobs, a night shift at a hostess club and a morning shift at a restaurant. He sent 10 percent of his earnings to Keiji, who finished high school a year late with the support of a sympathetic teacher and no longer exhibits symptoms of social withdrawal. Keiji works at an upscale café in Sendai, where his English—nearly fluent now—earned him a place in the café's managerial training program.

Prosper sees Keiji every few months. He accepts no credit for Keiji's decision to finish high school, a choice Keiji made shortly after the day they spent together in Odaiba, at the indoor theme park. Prosper admits that the emotions revealed by Keiji's behavior on that day might have been difficult for his Japanese mother and grandparents to understand, "but he show them to me as gift; nothing I have done to earn them." In lieu of disputing this observation, I'll add another: When Keiji speaks English, he retains the tonal lilt of a native Igbo speaker. Considering the limited time he has spent with his father, this is difficult to explain unless it's intentional.

Prosper sees Yumiko on the rare occasions when she visits Tokyo. She has returned to work, as an administrative assistant at a car dealership. Since her mother died in 2019, she cares for her elderly father, whose memory and mobility are declining. At present, Prosper, Keiji, and Yumiko have not spent time together. Somewhere between the possibility of becoming strangers and the possibility of becoming a family, they have become people who know each other. This offers meaningful consolation in a diaspora community where men frequently disclose stories of divorce to establish common experience, and the first question is always "Any children?" followed by "Do you know them?"

Prosper observes that the circumstances leading to the formation of families in Japan's Nigerian community have not recently changed. In 2019, the Abe administration enacted the first legislation in Japanese history that encourages large-scale immigration from developing countries, but the effects so far have been superficial. Foreign workers continue to arrive primarily through trainee programs that predate the recent reforms. Most of the Nigerians Prosper meets have entered Japan on visitor visas and intend to marry Japanese citizens. If they're detained before they normalize their immigration status, they avoid deportation by applying for asylum.

Japan has attempted to deter the arrival of asylum applicants, primarily by eliminating work permits for immigrants who file their asylum applications before the expiration of their entry visas, a change that affects a slim minority of Nigerian asylum seekers. Another change in detention policy has posed more serious consequences: The interminable cycle of incarceration and provisional release was ended by executive memorandum in 2018 and replaced with a system of indefinite detention. When I reported for *The Japan Times,* getting sick was a good way to obtain provisional release. For the past three years—until shortly before this book's publication—it was the only way.

Suicide attempts, self-harm, and hunger strikes soon became the new normal at detention centers. One detainee I spoke with observed his cellmate slashing his own gums and inducing vomiting so he could summon staff and claim he was vomiting blood. On a day when I visited the Immigration Bureau in Shinagawa, I watched a group of young immigrants guzzle water together; each of them planned to urinate in his pants if he was threatened with detention, because it would suggest that he was too emotionally unstable to safely detain. By forcing immigrants to choose between instant deportation and imprisonment without end, Japan's immigration authorities ensured that the line between detainees who needed immediate support and detainees who were desperate enough to exaggerate their difficulties was impossible to distinguish.

In this environment of despair and confusion, detainees died unnecessarily. In June 2019, a detainee starved to death over the course of four weeks. I heard about the death in January 2020, when Prosper called to express his shock that an Igbo detainee had been the first person to die as a result of hunger strikes in Japan's detention centers, and word of the death had not circulated in Tokyo's Igbo community. Prosper had learned about it when he was throwing out a pile of old newspapers in his apartment and noticed an article that included a photograph of a man he met once, many years earlier.

On the day they met, Prosper had a rare weekend off from the factory where he was working and visited a nearby town to view the cherry blossoms. Waiting for his train home, he overheard a conversation occurring in Igbo at the far end of the platform. He moved closer and found two Igbo men in the company of two Japanese women, one of whom was pregnant. The men appeared to be younger than Prosper, and their discussion revealed that one of them was in the early stages of establishing a container business. The pregnant woman's movements struck Prosper as darting and anxious. The longer he watched, the more natural her gestures seemed, and he realized she was maintaining a visual inventory of her surroundings. She was deaf.

Prosper introduced himself to the men and asked about the women. The man with the container business said they were his wife and mother-in-law. Prosper explained that he had lived in Japan for more than a decade and during that time never advised anyone to avoid criminal activity or the professions that invited it, knowing his advice couldn't change a person's circumstances. "But I begged him that he should take different road." While he spoke to the men, Prosper found himself close to tears, affected by an image of the deaf woman enduring the trials of motherhood alone after her husband inevitably shipped stolen goods, then was jailed and deported.

The husband, Prosper recalled, listened patiently and responded with a brotherly smile. "How work?" he said, and Prosper told him about the factory. As Prosper's train departed, he could see the two men succumb to a fit of mirth about what they had been told to do: Work hard, earn little, look forward to nothing. Having encountered

Prosper, they would take whatever risks were necessary to avoid becoming him.

Many years later, the husband starved to death in detention, and Prosper couldn't understand how his picture came to be in the newspaper. The authorities hadn't released his name, and Prosper found it difficult to believe that an Igbo family would provide a portrait of their deceased relative to the Japanese media. The question of the dead man's picture tangled itself in his feelings of helplessness about Uche's detention, which was now indefinite. He tried talking to the activists, journalists, and academics who visited the Ushiku detention center but was unable to communicate his simple wish to understand; they thought he was out for justice, the way they were.

The Immigration Services Agency (ISA)—the institution that resulted from a cosmetic restructuring of the Immigration Bureau's responsibilities in 2019—had issued a report on the man's death, clearing themselves of wrongdoing. I was reluctant to investigate. After I stopped writing for *The Japan Times,* I heard occasional rumors about the politicization of the detention system, which suggested that desperate behavior was proliferating on all sides. "Let him have a name," Prosper implored me. Without one, the deceased existed only as a picture of a "Nigerian" man, a description of his criminal record, and in the most sympathetic articles a few quotations about his kindness to other detainees. No ethnicity, no village of origin, no family history—none of the symbols that make Igbo people human in the eyes of their compatriots.

According to his relatives, his name was Gerald "Sunny" Okafor. His village was Umuhu Okabia, in Imo State. His matrilineal ancestry led—within two generations—to the Nwokeji family, the traditional rulers of Akuma, another town nearby. Sunny's generation of the family had established itself in the Igbo entrepreneurial class of Lagos, and he was the first member of his immediate family to expatriate. He told his friends in Japan that he previously lived in Southeast Asia.

He reached Japan in 2000 and found work in a leather tanning facility, then used his savings to start a container business in Hyogo Prefecture, where he lived with his wife and daughter. Initially he exported electronics, but struggled to earn a profit, and began exporting cars. After living in Japan for roughly a decade, he was arrested on charges the Immigration Services Agency later described in vague terms: "theft," "drugs," "organized crime." These were serious, premeditated offenses, the ISA insisted, and that was why Sunny's requests for provisional release were never granted. But court records and interviews I conducted with Sunny's associates indicate that his crimes were less than cunning. The police became aware of him after someone reported the theft of a generator from his scrapyard. They later investigated him for marijuana possession, and eventually caught him with three stolen vehicles, which he didn't realize were traceable when one of his colleagues purchased them from low-level yakuza contacts.

Sunny's convictions resulted in the loss of his immigration status, and in 2015 he was transferred to a detention center in Osaka, close enough for family members living in the Kansai region—including his brother Peter—to visit him. He was also able to maintain a close relationship with his former wife, Hisako, who divorced him after she was pressured by immigration officials and immediately regretted it. She had been a victim of domestic violence in previous relationships, and felt that the improbable love she shared with Sunny had freed her from thinking of herself as a stereotype of a battered woman.

Sunny's Nigerian relatives, including the two who were living in Japan, urged him to return to Nigeria. His deportation order was unlikely to be lifted, they pointed out, and even if it was, Sunny would be a paroled criminal with no immigration status. His relatives operated a warehouse in Lagos and were prepared to involve him in the business so he could return to Nigeria with reliable income. Hisako had expressed willingness to visit him after he repatriated, and their daughter would soon be old enough to accompany her.

Sunny refused to accept deportation and explained that he was participating in a political struggle over Japan's detention policies. Peter sympathized with the goals of this struggle but resented the involvement of his brother, who would gain nothing from it and whose claims of injustice included several that weren't credible. For instance, Sunny stressed that he couldn't leave his daughter, and Peter observed that Sunny made it sound as if he would never be able to communicate with her again, an exaggeration that fellow detainees had advised him to adopt for the sake of his provisional release applications.

In 2016, Sunny was transferred to Omura Detention Center, in Nagasaki. Friends and family could no longer visit him regularly due to the expense and distance involved, and he could no longer seek emotional support from Hisako, whose hearing difficulties made their phone calls one-sided (she spoke, he listened). Disheartened by his isolation, Sunny called his Nigerian relatives to express doubt about the possibility of remaining in Japan. He asked these relatives to put extra money on the phone card he used for calling Nigeria, and in 2018, after several years of fighting their attempts to repatriate his assets, he allowed these efforts to proceed, lamenting to a friend that there wasn't much left to repatriate and most of it would be squandered by the time he was able to access it. He made no further attempts (none after June 2018) to obtain provisional release. In early 2019, his family members living in Japan tried to renew their legal efforts on his behalf, but he declined their help.

On May 30, 2019, detention center staff were informed that Sunny had begun a hunger strike. When they questioned him, he begged for his freedom, explicitly asking to be deported. The ISA's investigation of his death documents this request but diminishes its significance by referring to Sunny's previous insistence (in January) that he was not ready to cooperate with deportation. Sunny was initially taken to a hospital, where he accepted an IV. After June 5, he refused further medical attention. Reviewing the ISA's investigation, one independent medical expert (who published his own analysis) asked readers to consider a critical omission: "The report offers no comment about this important change. *Why* did he refuse treatment that he was previously willing to accept?"

ISA and/or hospital sources (under conditions insisted upon by these sources, I cannot identify their employer or employers) report that a person or persons working for the detention center, the hospital, or both, engaged in a conversation (or conversations) with Sunny that influenced his subsequent refusal to accept treatment. The precise contents of this interaction (or these interactions) were not recorded, but their nature and effect was a matter of consensus among those who described them: Sunny had interpreted his hospitalization as a sign that he was being stabilized in preparation for a return to Nigeria. Someone (or some people) explained to him that they were not aware of any plans to end his detention.

He died of starvation in solitary confinement nineteen days later. According to the ISA, no one witnessed his final moments.

Media coverage of Sunny's death depended on the handful of details the ISA released and statements provided by fellow detainees, who related Sunny's inaccurate claims about losing contact with his daughter. In the version that prevailed, a Nigerian detainee, estranged from his Japanese family, died of starvation after insisting that he would never agree to be deported. Peter received the news from Hisako in fragments of broken English sent mostly by text message. By the time Peter organized a wake keeping, Sunny's family and friends were desperate to know more about the circumstances of his death.

Before the wake keeping, which was held at Sunny's old scrapyard, Sunny's family was contacted by a Nigerian woman who described herself as a representative of a nongovernmental organization that advocated for detainees. She asked to attend and surprised the event's patrons (senior officials of Nigerian civic associations in western Japan) by arriving with two cameramen. One was an employee of Japan's national broadcaster. The other was filming a documentary about the NGO worker, later released by Yahoo.

This woman's name is Elizabeth. I've known her since 2014. Her life has been difficult, and she previously engaged in selfless work on behalf of vulnerable people. I omit her last name because it will follow her, and I hope her recent conduct will not.

When I met Elizabeth, refugee narratives were occasionally—but not often—in Japan's news cycle. Elizabeth and I shared a sense of how dramatically this was about to change, because Nigerians in Japan were early adopters of the asylum process as an immigration back door. For two years I took her calls, arranging press coverage for the events and organizations she supported. If a Nigerian civic association approached me with questions about detention, I referred them to Elizabeth, who was always familiar with the current state of play.

Elizabeth was an asylum applicant on provisional release, living in open defiance of rules that prevented former detainees from engaging in activism. (A note from an early meeting with Elizabeth reads: "Very poor. Fearless.") When she was detained again in 2016, I visited her at Ushiku. She told me the Immigration Bureau had revoked her provisional release in an attempt to curtail her organizing work.

I believed her, and I was aware that other reporters were planning articles based on her claims.

In 2020, a regular visitor to the detention centers warned me that Elizabeth was "not the person you knew." I had no desire to learn the reasons for this statement, but Elizabeth's role in events that took place following Sunny's death left me without a choice.

Activists close to Elizabeth all told the same story. When an Ushiku-based NGO stopped delivering phone cards to detainees in 2015, Elizabeth began selling them at a modest markup. Soon after, she befriended an immigration attorney and charged detainees for access to his services (the attorney denied knowing, but a few of his clients told me they informed him). For a separate fee, she would arrange false addresses for detainees who needed to meet release requirements. As a result, several detainees became homeless after release, and detention centers were forced to tighten these requirements, which were already onerous.

In one incident recounted by fellow activists, Elizabeth misappropriated funds held by the Immigration Bureau on a detainee's behalf, and the affected parties contacted the authorities. This preceded her re-detention in 2016. Her friends reacted to these events with sadness and relief. If she knew she had been caught, they reasoned, she would stop mixing activism with self-enrichment.

While she was detained, several media outlets published articles describing her as a political prisoner. When she was released, she resumed her opportunistic behavior, promising detainees she could expedite their release if they paid her, and using press coverage—including old clips of mine—to make her claims appear legitimate. Her genuine advocacy on behalf of detainees did not cease, but it was difficult to know whom she helped and whom she harmed. Payments she collected from detainees were often used to provide financial assistance to other detainees, partly to ensure that they would defend her against accusations from people who felt she had cheated them.

In the course of investigating Sunny's death, I met several detainees who wanted Elizabeth to return their money, having realized, after months or years of frustration, that the results she promised were unrealistic. Despite multiple attempts—including one I made in person, in the waiting room at Ushiku—I was unable to convince Elizabeth to discuss these problems. Instead, when she learned that I was in contact with Sunny's family, she began to approach other journalists at the detention centers, and warned them not to trust me.

After Sunny's death in 2019, Elizabeth told friends that she traveled to his wake keeping on the budget of the filmmaker who was following her for Yahoo. This filmmaker had helped her obtain a new apartment, she said, and he was pitching her story to major television networks. (The filmmaker confirmed he was shopping Elizabeth's story, but denied covering her expenses or helping with her apartment; other media professionals who requested interviews with Elizabeth during the same period reported that she would not participate unless they paid her.)

I watched a recording of the wake keeping. To an audience desperate for information about the death of their relative and friend, Elizabeth immediately made the baseless accusation that Sunny's cremation by the Japanese government had been conducted to prevent an autopsy, which might have revealed a cause of death more nefarious than starvation. This inaccurate statement had enough staying power that none of Sunny's friends and family had seen his autopsy report until April 2020, when I hand delivered it.

One of the wake keeping attendees was Sunny's closest friend, Obinna, whom Prosper remembers as the other Nigerian on the train platform the day he met Sunny. Alarmed by Elizabeth's pronouncement, Obinna pleaded with her to help Sunny's family pursue justice. Elizabeth insisted that the possibility of legal action had been permanently foreclosed by the cremation of Sunny's body. "I want to use this opportunity to advise our brothers here," Obinna replied, "if this type of thing happens again, let us not destroy that evidence, that corpse, that body." He quoted an Igbo proverb: *The fowl that is caught by the hawk is not crying out to be spared, but for people to know that he will not be.*

Igbo funerary tradition abhors cremation. Hisako, after seeing Sunny's body, had felt cremation was appropriate. I've seen pictures, and the evidence of starvation is so dramatic in Sunny's exhausted features, his face a mere coincidence around the bulges of his skull, that it is easy to imagine the body causing more grief than burial rites could relieve. Nothing about Sunny's cremation, however, reduced his family's legal options. Litigation resulting from the death of a diabetic detainee at Ushiku had recently affirmed the right of deceased detainees' families to seek judicial intervention, even if they did not reside in Japan. That litigation had focused on the detention center's failure to monitor a detainee's health over the course of three days. Sunny's health had deteriorated over the course of four weeks. Had his family chosen to file a wrongful death suit, human rights lawyers say it could have been the strongest lawsuit ever filed against Japan's detention system.

And Sunny's family might have filed a lawsuit, even after Elizabeth said it wasn't possible, if that was all she had said.

At the wake keeping, when she discovered that Peter knew little about the circumstances of Sunny's death and couldn't provide her with the authority to involve her lawyer, Elizabeth snapped. She had believed (as she later explained to Sunny's friends) that the death of a fellow Nigerian would enable her to launch a high-profile attack on the Immigration Services Agency. Instead, she encountered a grieving, bewildered family who could not understand why activists like Elizabeth thought people like Sunny should remain in Japan. When she warned attendees that any of them could die in detention, a voice off camera immediately replied, "I won't stay there one day. I go back to my country."

"You neglect your brother," she shouted at Peter. She blamed him for failing to prevent Sunny's cremation and making it impossible to hold immigration authorities accountable. She criticized him for neglecting to visit Sunny after he was transferred to Nagasaki, implying—and later stating—that she had previously visited the facility (footage of her visit after Sunny's death demonstrates that she had never been there before). She insisted that a more attentive family could have secured Sunny's freedom, naming several detainees whose releases were allegedly the results of valiant behavior by their loved ones, apparently unaware that Sunny's family had taken the steps she subsequently accused them of omitting. "She presented herself as a hero," one wake keeping attendee recalled. "When she blamed Peter, you could see she was someone who had been traumatized."

Her accusations caused several loud disagreements to erupt among attendees, until Obinna rose from his seat to accost Peter. "You left my brother to die," Obinna shouted, menacing Peter physically. Other attendees dragged Obinna away. "You killed a free-born man who grew from the soil of my mother's village," he called out, in Igbo, choking back tears. Peter spent the rest of the wake keeping defending his actions during Sunny's detention, with the help of another family member in attendance.

Elizabeth's criticisms destroyed Peter's reputation in western Japan's Igbo community, in part because the scene I've just described was included in the Yahoo documentary. Before Sunny's death, Peter had sold Sunny's business and remitted the proceeds to Nigeria. After the wake keeping and the release of the documentary, Sunny's friends accused Peter of theft, arguing that Sunny would have wanted the funds to go to his daughter. The wake keeping itself had raised so much money for Sunny's daughter that Igbo Union Kansai was cash-strapped for several months, and questions began to swirl about whether Peter had exaggerated the cost of holding the event, in order to keep an inappropriate share of the donations.

With Hisako's permission and support from his church, Sunny's other relative in Japan explored the possibility of filing a lawsuit, but Peter—the senior, decision-making member of the family after Sunny's death—returned to Nigeria, and the family there, after learning they had been accused of misappropriating Sunny's assets, stopped taking calls from Japan. According to the chairman of Igbo Union Kansai, who emphasized that he did not think Peter exaggerated the cost of the wake keeping, "If Peter now say he make case, people will say: He don't visit brother in Nagasaki. It is only after Sunny die that he want to eat money from court case." When I first reached Peter by phone, he was so bruised by Elizabeth's accusations that several minutes of explaining were necessary before he realized I was not calling to reiterate them.

The Immigration Services Agency published its investigation of Sunny's death in October 2019. The medical examiner quoted in the report took the unusual step of admitting that medical intervention could have saved Sunny's life, but this information was included primarily to refute its relevance: Even if the intervention had occurred, the report stated, its chances of success would have been low. The ISA ultimately concluded that detention center staff had acted appropriately in the weeks leading to Sunny's death.

Following the deaths of other detainees in Japan, accountability has been pursued through civil litigation, filed on behalf of detainees' families by human rights attorneys who coordinate their work through the Japan Federation of Bar Associations (JFBA). Because immigration authorities have consistently refused to submit to third-party review, judicial review provides the sole opportunity to prove wrongdoing. When I asked one of the JFBA's human rights attorneys what his colleagues' conversations with Sunny's family had entailed, the attorney said, "I heard he has a brother in Osaka." In other words, what conversations?

The JFBA's human rights committee was not aware of Elizabeth's inaccurate statements about Sunny's cremation and autopsy, but members were aware that she attended a memorial service held by Sunny's family. An opinion about potential litigation was sought from the attorney who was closest to Elizabeth. This attorney is a lifelong advocate for the rights of immigrants; she is also a trusting soul who shed tears when I told her that my mother spent her legal career representing the disabled children of poor people. After Sunny's death, she was unable to reach his family through the lawyer who had helped him apply for provisional release, and trusted the information Elizabeth circulated about his family's wishes: They did not want the public attention that would accompany a lawsuit, specifically because Sunny's criminal history would be scrutinized.

Both of Sunny's relatives living in Japan denied telling Elizabeth that they wanted to avoid public attention. They pointed out that Sunny's criminal convictions had already been described in major media outlets, in reports accompanied by his picture (which the Yahoo filmmaker provided to a leading newspaper). Peter said he did not push for a lawsuit because Elizabeth's comments at the wake keeping convinced him that litigation was impossible, and her attack on his character made it difficult to continue representing his family's interests in Japan. Sunny's other relative, who had considered litigation, declined to pursue it because his Japanese spouse was vehemently opposed. "If I really believed I could get justice," he said, "I would have told my wife she could leave me." Hisako, he observed, was not intimidated by the prospect of a long trial, and she alone among the Japanese women who had married into his family should decide how Sunny would be remembered.

In March 2020, I arranged a meeting with one of the JFBA's human rights lawyers to ask what the organization knew about the ISA's failure to act on Sunny's request for deportation. The lawyer mentioned a passage in the ISA report that claimed an attempt to deport Sunny had failed because the Nigerian authorities were slow to approve it, and said the claim was accurate. When I asked for the lawyer's reasoning, the lawyer said the relevant information was sensitive and colleagues would need to be consulted before it could be disclosed. I made it clear that I was willing to give the JFBA the right to retract the information—as though I had never heard it—if the JFBA's human rights committee did not feel that the way I intended to use the information would further the interests of justice. The lawyer promised I would receive a decision shortly. I followed up multiple times and never heard back.

This made me curious. A man had starved to death after begging for deportation, during an era of detention policy when Japan's immigration commissioner had declared that anyone who agreed to deportation would be released "as soon as tomorrow." The claim mentioned by the JFBA lawyer was the only language in the ISA report that justified the agency's failure to return Sunny to Nigeria. It amounted to accusing the Nigerian government of manslaughter by bureaucratic ineptitude.

This accusation was deployed in a barrage of ambiguous language. A close translation would read, "Japan negotiated with Nigeria about repatriation. People who refuse to be repatriated are accompanied by Japanese immigration officials, who hand these people to officials in their home country. Without [Sunny's] consent, it was difficult to send him back to his country." Notable omissions include: the date when negotiations occurred, the names of the people or institutions that represented the Nigerian government, and whether Sunny's case in particular was discussed.

In Japan, forced deportations of Nigerians require the assistance of the Nigerian embassy. Immigration officials accompanying the deportee need entry visas, or at least the assurance of visas on arrival, and in many cases (Sunny's would have been one of them) the deportee's Nigerian passport has expired, requiring the issuance of an Emergency Travel Certificate (ETC) so the airline can carry out the boarding process. The embassy arranges entry visas as a courtesy; only the passport details of the immigration officials are needed. The ETC requires an application, which immigration officers are permitted to complete if the deportee refuses. The completed application must contain the detainee's flight itinerary.

For as long as I have been reporting on the Nigerian embassy, ETC applications have been handled by the same Japanese employee. This employee checks the application and attachments, prepares the ETC, prints it, then passes it to a consular officer for final approval. ETCs are usually sought two weeks in advance of the deportation date. Whenever necessary, they are issued on the next business day after the application is received. Under urgent circumstances, they have been issued in a matter of hours. ETC requests submitted by Japan's immigration authorities (as opposed to requests submitted by individual Nigerians) are processed at the highest level of priority.

If the ISA is determined to deport a Nigerian citizen, there are no negotiations and no reason for them to occur. There is only the routine process of applying for the deportee's ETC, by which time the immigration authorities have scheduled the deportation, selected the personnel who will accompany the deportee, and purchased flight reservations. According to every current and former Nigerian consular official I contacted (all of whom worked for the embassy in Tokyo), completed ETC applications for deportees in Japanese custody have never been rejected or strategically delayed, and the embassy has never refused to provide entry permission for immigration personnel accompanying deportees. Nor has the embassy ever attempted to control the way deportations are executed; there would be no legal basis for interfering with the enforcement of Japan's immigration policies.

In the words of the Nigerian diplomat who approves the ETCs issued in Japan, "The ISA does not need Nigeria's permission to deport." Any discussion of related procedures, this diplomat insisted, would not prevent the ISA from carrying out deportations. Esther Ogundipe, the consular officer who was responsible for communicating with the ISA about Sunny's death, was equally direct: "Those are pure lies by the Japanese government." She pointed out that the ISA often fails to report back to the embassy about the status of deportees, revealing how insignificant the involvement of the Nigerian authorities becomes once travel documents have been issued, but before the deportee arrives on Nigerian soil.

I challenged my ISA sources to describe an instance when the Nigerian embassy failed to provide travel documents in time for a planned deportation. They couldn't. The report's language about negotiations with Nigeria, they confirmed, was deliberately vague so it could be attributed to a variety of interactions with Nigerian officials, taking place at any time during Sunny's detention. For example, a significant percentage of Nigerian detainees claim they can't afford their own repatriation flights, and the ISA regularly communicates with the Nigerian embassy about reducing this percentage. After a decade of incarceration, Sunny was part of this group. Other routine discussions—about verifying the nationality of Nigerian deportees, or planning deportations to Nigeria via connecting flights—could also be invoked. And Japan, like Germany, recently pressured the Nigerian government to permit forced deportations to occur without embassy-issued travel documents. These discussions are not specific to Sunny's case, my ISA sources acknowledged, and they did not pose a meaningful obstacle to deporting him.

Strikingly, the ISA report also declines to specify whether Sunny attempted to use an asylum application to avoid deportation, a step that Hisako says he often considered taking. If an asylum request was involved, the ISA might have asked the embassy to convince Sunny to withdraw it, allowing immigration officials to bypass the complicated process of issuing an expedited decision—or requesting an emergency court judgment—in order to deport Sunny immediately.

Newspapers ultimately paraphrased the ISA's claims in the way the agency hoped for, saying the ISA was "unable" to deport Sunny before negotiating with the Nigerian government about "procedures." Staff at the Nigerian embassy wanted to contradict these statements but the ambassador ordered them to avoid speaking with the press. In communications that I was able to review, he insisted that he would not risk his cordial relationship with the Japanese government—nor the diplomatic and trade agenda that depended on it—by criticizing the ISA for mistreating a convicted criminal who resisted a lawful deportation order.

The JFBA's human rights lawyers were briefed about Sunny's death (by immigration officials) on a confidential basis. According to attorneys with knowledge of the briefing process, JFBA representatives declined to challenge the ISA's claims about negotiating with

Nigerian authorities "because it could have damaged our relationship with the Ministry of Justice and prevented us from receiving similar information in the future." It also would have been inconsistent with the JFBA's opposition to Japan's detention practices, one JFBA representative explained, to demand to know why a vulnerable detainee had not been deported swiftly and forcefully during his most vulnerable moments.

If delays imposed by the Nigerian embassy had not prevented Sunny's deportation, what had? Twenty-five days elapsed after Sunny's request to be deported; nineteen elapsed after he stopped accepting medical treatment, when the seriousness of his situation was clear. There had been time to act.

According to Japanese diplomats who gained knowledge of the ISA's internal investigation of Sunny's death, Sunny's request was never taken seriously. Forced deportations are expensive and complicated, requiring several immigration personnel to board the outgoing flight. The ISA does not schedule them lightly, and Sunny's request had at least three shortcomings: First, it was an utterance of desperation, one among many in the detention center; second, he had previously refused deportation, which the ISA regards as a behavior deserving of punishment; and third, he was Nigerian. "Nigerians are the most opportunistic," alleged one detention center employee questioned during the ISA's investigation. The most likely to misrepresent their intentions, and the most likely to exaggerate their need for medical attention.

According to ISA sources (and several detainees who knew him), Sunny was particularly mistrusted by detention center staff. They appreciated his gentle, cooperative temperament, but based their opinion of his credibility on the notion that he had refused to take responsibility for his crimes, filing an appeal against his theft conviction despite overwhelming evidence of his guilt. They had also heard—and chose to believe—that a police officer was injured during the theft investigation, and Sunny, who witnessed the injury, responded by making an offering to a small idol he carried, then proclaiming that his native gods had punished the officer for interfering with his livelihood. I could only substantiate one element of this story: A police officer had injured himself while attempting to open the door to a storage unit at Sunny's apartment building. Court records don't indicate that Sunny mistreated investigators, and according to his defense attorney, "If that ridiculous story were true, the police would have documented it and the prosecution would have used it."

In addition, detention center staff were aware that Sunny had started a hunger strike in Osaka, but didn't stick to it. They assumed that he was better at talking about hunger strikes than he was at going hungry. According to Sunny's family, Hisako had convinced him to stop. When he was in Omura and Hisako couldn't hear his side of their phone calls, it was nearly impossible for her to learn that he intended to start another hunger strike, let alone intervene.

When I spoke to senior immigration bureaucrats about the tendency of detention center staff to disbelieve Sunny, they downplayed the significance of these perceptions. "Even if [Sunny] had been the most credible detainee in the facility, it's unlikely that anyone working there would have made a vigorous effort to fulfill his deportation request," said one bureaucrat. Sunny was almost certainly the first detainee to go on hunger strike in an attempt to be deported, this bureaucrat pointed out, instead of a single-minded attempt to obtain provisional release. Staff weren't accustomed to this scenario. "They treated him like he was making the request everyone else makes, which was an easy request to ignore."

Human rights attorneys and independent medical experts believe nationality played the critical role in determining how Sunny was treated, from the moment immigration officials decided that he should be separated—legally and geographically—from his Japanese spouse, to the final weeks of his life when the severity of his weight loss was overlooked. "It's racism," said physician Yamamura Jumpei, a frequent visitor to the detention centers. "During his last days, Sunny was in no shape to resist deportation, and there would have been a significant window before he died when he couldn't resist medical intervention." Only prejudice could explain why ISA officials didn't treat Sunny's deteriorating condition as clear evidence that his deportation should be scheduled immediately. Immigration attorney Komai Chie noted that the ISA report's focus on "forced deportation" was more distraction than substance; the associated security measures might not have been necessary during the final month of Sunny's life, after he asked to be sent home. "The number of detainees who change their minds about deportation when the day arrives is not zero, but it is low," she said. "I've never heard of it happening after the detainee begged to be deported."

Sunny died, the ISA's critics contended, because the ISA chose to gamble his life by waiting for him to end his hunger strike and pay for his own flight home, instead of gambling their resources on deporting him.

In the months after his death, Sunny's family received phone calls from detainees who had known him in Nagasaki, offering condolences. "I never knew Sunny was doing hunger strike," his brother Peter said. "How it's possible they can access my number, but I never find out?" Sunny was a prisoner twice over, Peter realized: In the eyes of his adopted home, he could only be a criminal; among his fellow detainees, he could only be the victim of an unjust system—and finally a martyr. "Not Sunny or anybody called me to pay his ticket," Peter said. "I would have saved his life."

Sunny's friends, his relatives, and the Igbo civic associations in Kansai told me they would have done the same. "But sickness made him forget us," remarked Obinna, his closest friend, who met him when they were both new to Japan and working in leather tanneries. "And cruelty is not medicine. It was cruelty that made him sick."

A government panel of immigration experts was convened in response to Sunny's death, and issued its final policy recommendations in June 2020. Although the panel resisted calls to end the practice of indefinite detention, its proposals are likely to result in shorter detention periods for many asylum seekers detained at Japan's airports (in principle, they will be held for no longer than six months). Some immigrants who cooperate with deportation orders will now be less likely to find themselves permanently unable to reenter Japan, and immigrants from war-torn countries will receive special consideration in an attempt to avoid trapping them in an endless cycle of deportation orders and asylum applications.

For Sunny's loved ones, the proposed reforms are an object of ambivalence. "If other immigrants are treated like humans because of Sunny, that's wonderful," Hisako told me. But she also pointed out that none of the reforms would have applied to Sunny. He was a long-term resident of Japan, with a criminal record. Under the new rules, he still would have been stripped of his immigration status and detained indefinitely. As one of Sunny's cousins wondered: "What does this panel have to do with Sunny's death, except to pretend they did something about it?"

In September 2020, Japanese newspapers reported that the ISA planned to adopt several of the panel's recommendations. This information was leaked to the press as part of an ISA public relations effort, in anticipation of the Tokyo Olympics. "By treating the most sympathetic cases more gently, the ISA intends to convince the public that everyone else deserves what they get," explained one immigration official I spoke with. According to this official, pressure from activist groups has already declined. "There is less will to antagonize the government while reform is under way."

The JFBA's human rights attorneys will soon be allowed to seek "certification" from the ISA, which will empower them to "monitor" their own clients for compliance with the new rules. "That's an exceptionally shrewd gesture," noted a former government solicitor who has represented Japan's immigration authorities. "It rewards the JFBA for declining to attack the weaknesses in the ISA's investigation of [Sunny's] death, and simultaneously neutralizes many human rights attorneys, who won't want to handle the ISA roughly if their clients depend on them to maintain a collegial relationship with the agency."

In the opinion of several of my sources in Japan's Ministry of Foreign Affairs, a court could reasonably conclude that ISA staff acted negligently after Sunny asked to be deported. These sources have been advised that the ISA considers this a serious liability. "From the ISA's perspective, reform will be less painful than justice," the former solicitor said. "Their statement is: 'We promise to treat immigrants better. As for the Nigerian who died, there was nothing we could have done.' The second half of that statement is a lie, but the first half makes it easier to believe."

When I shared my findings with Prosper, his interpretation was succinct. "Immigration kill him. Immigration lie how they kill him, and those people who are fighting immigration—whether activist, lawyer, or reporter—help to spread lie, even they don't know they are doing it." His next words were less solemn, verging on laughter. "Only regular Nigerian person have behaved well. Embassy staff never make mistake on Sunny case, and Igbo Union send many donation for Sunny daughter." We were eating in a railway station in Chiba, watching the flow of morning commuters pass through the ticket gates. A few passengers had disembarked from an outbound train and struggled through the crowd toward the exit. "This world now upside down," Prosper concluded.

It seemed appropriate to mention that Obinna had continued working in factories after he met Prosper; the people who enticed Sunny to do otherwise were Japanese as often as they were Nigerian, according to Sunny's family and friends.

"Not everything upside down," Prosper allowed. "Crime is human. Either greed, despair, or any other thing."

From the station, we took a taxi to the recycling facility where Prosper and Uche worked in the 1990s. His former boss was baling cardboard when we arrived, operating a machine that he promised "will settle for a finger, but would prefer your arm." He had a thick Tohoku accent. "Aomori?" I ventured. "In 2010, I'd belt you for that," he said. "Today I guess I'll take it." He was from Fukushima.

The facility was a handful of buildings with corrugated-steel walls. The largest had a second story with a few windows, showing the silhouettes of office furniture pinned to the glass by toppled piles of boxes. A smaller office, including a dust-coated cash register, occupied an outbuilding adjacent to the street. The shelves were lined with fragile items and semi-valuable curiosities recycled mistakenly and discovered during the sorting process. The neighborhood was the type of low-density industrial zone found at the extremes of Greater Tokyo, intermittently silent and eerily peaceful. A block away from the factory, the manicured gardens of a small temple were visible above its stone walls. The location of the temple seemed intentional; the featureless facades of the surrounding industrial buildings would please an uncluttered mind.

Prosper's former boss fit the Japanese stereotype of someone who had lowered himself to manual labor. He was stout and gray-haired, age was prematurely visible in his face, and he emanated a physical vitality that would have felt threatening in a white-collar workplace. Shreds of cardboard fiber from the baler were on his clothes and in the air. While he chatted with Prosper, he inhaled them, periodically sucking phlegm into his throat and spitting. "I have to finish this," he said, indicating the machine. "But go ahead. It's all the same as before." He advised me that I was associating with the world's worst businessman. "Every few weeks he comes up here and looks at the same stuff I'm selling."

Prosper walked me through. Some rooms were empty. Some contained industrial debris, neatly piled in the corners. A few were filled with recycled goods: sorted and baled secondhand clothing, offcuts of PVC piping. None had windows. Light entered when Prosper opened the doors. In one building, the dust was thick enough we left shoe prints. The repetition of form in the rooms that contained recycled goods made the experience becalming and aesthetic. In a room filled with baled clothes on one side and baled shoes on the other, Prosper inspected the bales, leaning to examine the geometry of twisted garments, but never touched them. "If you bring this to Nigeria, you can do many things," he said.

Prosper's former boss bought us canned drinks from a vending machine opposite the temple, where a few meters of vacant land had been ringed with benches so workers from the factories could socialize during their breaks. He asked about Uche. Prosper said Uche had recently been detained for the third time and was still unwilling to accept deportation until he was able to visit his children. "Dreams go with you to prison," Prosper's boss observed. Prosper asked if his boss had been in prison. "No," his boss said. "I came to Tokyo. I wanted to be one of these." He lifted the back of his hand, the first gesture in the greeting exchanged by Igbo titleholders.

"Anyone who has money can buy a chieftaincy title," Prosper said.

"That's what I meant." Prosper's boss dropped the push tab from his drink into the can and rattled it like a beggar's cup. "Nothing has changed."

They sat awhile without talking. As I often did, I imagined the years that had passed between people in the Nigerian community and their Japanese acquaintances before I met them. For an Igbo immigrant recently arrived in a foreign country, the magic of a stable job emanated from the promise of future windfalls, from the village myths surrounding the chiefs who parked white Daimlers in their compounds before anyone else in the village had been close enough to a car to touch it. In these myths, the chiefs were self-taught entrepreneurs, people who had learned their business through proximity. Workers.

In Japan, Prosper and his compatriots discovered they could not transform automobile components or secondhand clothing into money by spending a certain number of hours in the presence of these items. They discovered that the men who owned the factories where they worked were not entrepreneurs either. In Igbo terms, they were petty traders. The goods they sold provided a modest livelihood. They could not offer a private diagram of prosperity. Only wages.

"Do you remember the Gwari boy at the construction site?" Prosper asked me. I remembered, and Prosper told his boss the story, while the blade of a tree pruner bobbed above the temple wall, clipping the limbs of a zelkova. The rhythm of the branches rustling as they fell and Prosper's uninterrupted delivery of his long anecdote led me to feel that the purpose of today's visit had settled on revisiting this memory, something that happened several years ago in Nigeria.

On the day Prosper was describing, three of us—Prosper, myself, and a real estate developer who had lived in Japan—visited a mutual friend of ours, another Japan returnee who was building a hospital annex on the outskirts of Abuja. To get the contract, our friend had promised the hospital's board that he could build the annex without taking a down payment. He paid his foreman out of pocket and told the laborers he hired that pay was coming but never said when. Most laborers cut their losses and left, and on the day we visited, the foreman asked our friend to hire more.

Our friend visited the hospital director's office to ask for an advance against costs, but the director was busy consoling a young man whose one-year-old daughter had been born with a brachial plexus injury. Most injuries of that type eventually healed themselves; in the instances when they didn't, the child often lost use of the affected arm. The young man in the director's office, whose face was smeared with tears and mucus from crying, hadn't known about the injury until today, because he lived and worked abroad. We left without talking to the director, and I was certain we were all thinking of a person we knew in Japan, whose son had been born in Lagos with the same injury, and he wasn't told about it until he visited home the following year. His son's arm never healed.

We drove to a nearby squatter settlement, inhabited by Abuja's internally displaced Gwari people (Gbari, in their preferred pronunciation), whose land was expropriated when the capital was built.

At the vegetable stand where we parked, a teenage boy ignored our friend's inquiries until he flashed a few American bills, more money than a vegetable stand could earn in a month. The boy entered a nearby dwelling and when he emerged he was wearing a tool belt—barely tight enough to stay on his hips. Our friend asked the boy if he stole it, and the boy said it belonged to his late father.

"I wanted you should find a man," our friend told the boy, but he gestured for the boy to enter the car and we drove to the construction site.

Our friend took the boy on a tour of the building's half-finished skeleton, to determine whether the boy could learn to do the work. The sun finished setting while they were gone and the hospital's generators sent power to the floodlights in the parking lot, where the rest of us waited. In the vicinity of the hospital, the lot was the only illuminated place, overwhelming and deepening the darkness around it, attracting insects from the surrounding farmland until the movement around the bulbs of the floodlights was dense and abstract. When our friend and the boy returned, the boy slung his tools over his shoulder, to make for the road and a ride home, but our friend stopped him. "You will see it," he said, and gestured for the boy to watch the floodlights.

The flight of the first bat through the insects was momentary and indistinct. Dozens more appeared, their wings beating silhouettes of empty space around them as they hovered and fed. "These bats are native here," our friend said to the boy. "Same like you." The boy made an effort to show interest, but his expression was impatient. He stood up and attempted to address our friend in Igbo, the words coming out in a halting stutter. Immediately, all three men—native Igbo speakers—were overcome by laughter. I didn't hear what the boy tried to say, but the laughter was too long, too rich with relief and a private understanding, to come from the clumsiness of the boy's speech.

The boy stormed off toward the borrow pits that were excavated when the hospital was under construction, now filled with fetid, water-logged garbage. The three men watched in surprise and amusement as the boy disappeared into one. The boy meant to make them follow him, but only Prosper showed any willingness, removing his dress shirt so he could use it to cover his nose. I asked what the boy had tried to say. "He wants to know if we are taking him home," Prosper said. "But the way he says it, it's like, *Can anyone go home?*"

A flashlight was brought from our friend's car and Prosper went looking for the boy, to offer an apology, but no answer.

Acknowledgments

Measured in hours of effort, anything praiseworthy between these covers belongs to my research assistants, interpreters, and translators more than it belongs to me. In some ways, I was the wrong author for this book, but I was surrounded by people who believed in the value of an outside perspective, if enough layers of ignorance were shed in the process of forming it.

The best advice I received about shedding ignorance came from the late David Wise, whose coverage of intelligence agencies in Washington, DC, made it possible for subsequent generations of writers to turn over rocks that ostensibly didn't exist. At the end of his life, he remained a gracious teacher, and his wife, Joan Wise, a loyal friend.

The critical phase of drafting this book was completed at Twin Oaks, the intended lifetime residence of Robert Penn Warren. I was allowed to use the house by its owner, who did not buy the property to share it with literary seekers but made a generous exception for a determined stranger. Equally generous accommodations were made at Headwaters Farm (now RowanLark). On the Otago Peninsula, hospitality was provided by Neil and Shukuru Munro.

To name a few writers and academics who assisted this project in a variety of ways: Victoria Wyatt, Sigrid Nunez, Ha Jin, Ted Fowler, Ryan Boudinot, Matsumoto Hisashi, Elizabeth Onogwu, Crocker Snow, Julian Dibbell, James MacKinnon, P-J Ezeh, Yuki Miyamoto, James Ogbonna, Sy Hersh, Geraldine Harcourt, Chioma Akaeze, Chinedu Ezebube, Christopher Agbedo, Nicole Fong, Robert Goree, Chris Norwood, Martin Zinggl, Yamada Rie, Nat Philbrick, and the late Robert Bausch. I never had occasion to collaborate with the three anthropologists of my generation who conducted research in Japan's Nigerian community, but their presence was felt and appreciated.

I am grateful to Matthew Mizenko, who helped me sustain the notion that the literary tradition in Japan—and the tradition of foreign writers reading Japanese literature—retains its significance as a way for people and cultures to encounter one another. Newcomb Greenleaf, expert in Japanese temple geometry, taught me to view my environment as an ecology of cultures, nothing isolated, nothing monolithic (Japan is neither of these things). Norma Field wrote the book I wanted to follow and allowed me to believe—however foolishly—that I could follow it.

My writing life has included three great mentors, whose names alone I consider deeply meaningful: Robert Earle, Lynley Hood, and Hideo Levy.

When I was a magazine and newspaper writer in Tokyo, I worked for editors who wanted the best for me: Steve Trautlein, Jeff Richards, Ben Stubbings, and Kitazume Takashi. Working for the Community desk of *The Japan Times* under Ben's supervision was an irreplaceable experience. A lectureship at Boston University, in a program administrated by Sarah Frederick, was equally formative. The influence of my fellow editors at *Kyoto Journal*—John Einarsen, Ken Rodgers, Lucinda Cowing, Susan Pavlovska, and David Cozy—is everywhere in these pages.

The University of Otago has supported my research and writing without fail. Greg Rawlings, head of the Social Anthropology program, is an asset to all who collaborate with him. Over the past four years, economist Nathan Berg often challenged me to reconcile opposing viewpoints without overvaluing their differences, and the late Cyril Schäfer encouraged me to maintain an awareness of the risks involved in every decision an ethnographer makes.

Many members of the Corstorphine Community Hub, as well as its entire leadership team, were enthusiastic supporters throughout the drafting process. As were the Friday Poets of Dunedin, especially Philip Temple and Diane Brown, John Jillett, and Neville Peat.

In Nigeria, I chose the city of Aba as my place of work, and I am indebted to my host, His Royal Highness Eze S.I. Akataobi, for ensuring that I never lacked opportunities to improve my understanding of Igbo culture and society. In Japan, I avoided personal relationships with the leaders of Nigerian civic unions but inevitably came to rely on some more than others. Okeke Kevin Christian, Kennedy Fintan Nnaji (Nahfkenn), Obi Ernest, and Tony Ikeotuonye were especially supportive.

I owe the completion of this book to my family: to my father's patient counsel, my brother's friendship, my father-in-law's humorous analysis of complicated ideas, and my mother-in-law's reliable surplus of insight. Two uncles were valuable sounding boards, especially Michael McCloskey. My mother's contributions were too numerous to describe. My partner, Anastasia, endured the writing of this book with characteristic poise, applying cruelty whenever necessary and love whenever possible.

Close friends were equally candid and helpful. Special thanks are reserved for the novelist Jowhor Ile and the filmmaker Ian Thomas Ash. Sue Wild was my instructor in all matters of kindness, whether in prose or in life, ably assisted by her partner, Joel.

My editors at Pantheon and Knopf Doubleday have been superb: Dan Frank, Vanessa Haughton, and Altie Karper. As far as I know, Ingrid Sterner is the only copyeditor whose suggestions are more compelling than most books. Emily Mahon provided an excellent jacket design in the tradition of great jackets that reference Japan. I have done nothing to deserve my talented publicists, Rose Cronin-Jackman and Suzanne Williams, but somehow they're in the picture. My agent, Kelly Falconer, maintains a list that demonstrates an abiding love of challenging titles, a comfort to any writer. If this book achieves anything through boldness, it is because I had the support of talented attorneys—Claire Leonard, Dan Novack, and Michael Grygiel.

Finally, I want to acknowledge the Argentine intellectual Juan José Sebreli. Translating his work during moments stolen from this project has been a necessary stimulant, ensuring that I was never frustrated by the ineffable qualities of events and ideas I was attempting to describe. As he explained at our first meeting, "There is no reason to be alarmed. Whatever you leave behind is sustenance for shadows."

A Note About the Author

Dreux Richard is an American writer, journalist, and literary translator. From 2011 to 2016, he covered Japan's African community for *The Japan Times,* where he also led a yearlong investigation of the "world's safest" nuclear plant, published as one of the longest print articles in the newspaper's 124-year history. His work has appeared in *CounterPunch, The New York Times,* and *Estadão,* among others. He is a doctorate-by-research candidate at the University of Otago in New Zealand and a visiting researcher at the University of Nigeria, Nsukka, where he studies smuggling and political patronage in the former Biafra region.

A Note on the Type

This book was set in Minion, a typeface produced by the Adobe Corporation specifically for the Macintosh personal computer and released in 1990. Designed by Robert Slimbach, Minion combines the classic characteristics of old-style faces with the full complement of weights required for modern typesetting.

Typeset by Scribe,
Philadelphia, Pennsylvania

Printed and bound by Friesens,
Altona, Manitoba

Designed by Michael Collica

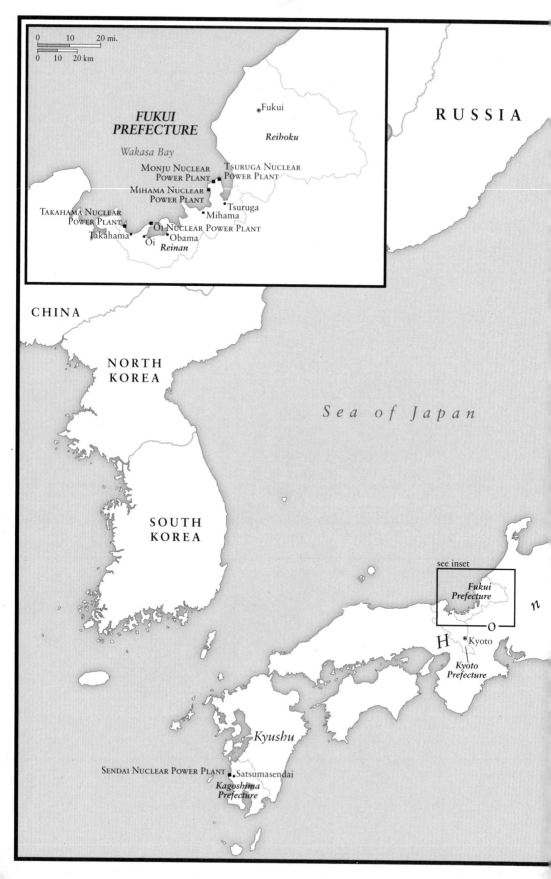

FUKUI PREFECTURE

0   10   20 mi.
0   10   20 km

Wakasa Bay

Fukui

Reihoku

MONJU NUCLEAR
POWER PLANT

TSURUGA NUCLEAR
POWER PLANT

MIHAMA NUCLEAR
POWER PLANT

Tsuruga

Mihama

TAKAHAMA NUCLEAR
POWER PLANT

OI NUCLEAR POWER PLANT

Takahama

Oi

Obama

Reinan

RUSSIA

CHINA

NORTH
KOREA

SOUTH
KOREA

Sea of Japan

see inset

*Fukui
Prefecture*

n

O

H

Kyoto

*Kyoto
Prefecture*

Kyushu

SENDAI NUCLEAR POWER PLANT

Satsumasendai

*Kagoshima
Prefecture*